NEUROMETHODS

Series Editor
Wolfgang Walz
University of Saskatchewan
Saskatoon, SK, Canada

For further volumes:
http://www.springer.com/series/7657

Neuromethods publishes cutting-edge methods and protocols in all areas of neuroscience as well as translational neurological and mental research. Each volume in the series offers tested laboratory protocols, step-by-step methods for reproducible lab experiments and addresses methodological controversies and pitfalls in order to aid neuroscientists in experimentation. *Neuromethods* focuses on traditional and emerging topics with wide-ranging implications to brain function, such as electrophysiology, neuroimaging, behavioral analysis, genomics, neurodegeneration, translational research and clinical trials. *Neuromethods* provides investigators and trainees with highly useful compendiums of key strategies and approaches for successful research in animal and human brain function including translational "bench to bedside" approaches to mental and neurological diseases.

Awake Behaving Mesoscopic Brain Imaging

Edited by

James E. Niemeyer, Hongtao Ma, and Theodore H. Schwartz

Department of Neurological Surgery, Weill Cornell Medicine, New York, NY, USA

Humana Press

Editors
James E. Niemeyer
Department of Neurological Surgery
Weill Cornell Medicine
New York, NY, USA

Hongtao Ma
Department of Neurological Surgery
Weill Cornell Medicine
New York, NY, USA

Theodore H. Schwartz
Department of Neurological Surgery
Weill Cornell Medicine
New York, NY, USA

ISSN 0893-2336 ISSN 1940-6045 (electronic)
Neuromethods
ISBN 978-1-0716-4122-4 ISBN 978-1-0716-4120-0 (eBook)
https://doi.org/10.1007/978-1-0716-4120-0

This Humana imprint is published by the registered company Springer Science+Business Media, LLC, part of Springer Nature.
The registered company address is: 1 New York Plaza, New York, NY 10004, U.S.A.

Preface to the Series

Experimental life sciences have two basic foundations: concepts and tools. The *Neuromethods* series focuses on the tools and techniques unique to the investigation of the nervous system and excitable cells. It will not, however, shortchange the concept side of things as care has been taken to integrate these tools within the context of the concepts and questions under investigation. In this way, the series is unique in that it not only collects protocols but also includes theoretical background information and critiques which led to the methods and their development. Thus it gives the reader a better understanding of the origin of the techniques and their potential future development. The *Neuromethods* publishing program strikes a balance between recent and exciting developments like those concerning new animal models of disease, imaging, *in vivo* methods, and more established techniques, including, for example, immunocytochemistry and electrophysiological technologies. New trainees in neurosciences still need a sound footing in these older methods in order to apply a critical approach to their results.

Under the guidance of its founders, Alan Boulton and Glen Baker, the *Neuromethods* series has been a success since its first volume published through Humana Press in 1985. The series continues to flourish through many changes over the years. It is now published under the umbrella of Springer Protocols. While methods involving brain research have changed a lot since the series started, the publishing environment and technology have changed even more radically. *Neuromethods* has the distinct layout and style of the Springer Protocols program, designed specifically for readability and ease of reference in a laboratory setting.

The careful application of methods is potentially the most important step in the process of scientific inquiry. In the past, new methodologies led the way in developing new disciplines in the biological and medical sciences. For example, Physiology emerged out of Anatomy in the nineteenth century by harnessing new methods based on the newly discovered phenomenon of electricity. Nowadays, the relationships between disciplines and methods are more complex. Methods are now widely shared between disciplines and research areas. New developments in electronic publishing make it possible for scientists that encounter new methods to quickly find sources of information electronically. The design of individual volumes and chapters in this series takes this new access technology into account. Springer Protocols makes it possible to download single protocols separately. In addition, Springer makes its print-on-demand technology available globally. A print copy can therefore be acquired quickly and for a competitive price anywhere in the world.

Saskatoon, SK, Canada *Wolfgang Walz*

Preface

The human brain is composed of nearly one hundred billion neurons of various types. Understanding how these cells interact with one another is critical if we want to untangle the mystery of how the brain functions in both health and disease. While early detectives like Holmes and Watson used their powers of observation aided by magnifying glasses and two-lens microscopes, modern neuroscientists have more sophisticated tools at their disposal. At the finest level, neurophysiologists measure electrical current and voltage changes across the membrane of individual cells. This technique provides granular understanding of the activity of single cells but yields limited information about the brain as a whole. At the other extreme, functional magnetic resonance imaging can measure blood flow throughout the entire brain on a macroscopic level. This method provides some understanding of which brain regions are more or less active under different circumstances, but only at a very large scale: on the order of tens to hundreds of thousands of cells per voxel of data, and typically at a temporal rate on the order of seconds. In between these two extremes is a class of neural recording techniques that function at a level called the "mesoscale." Mesoscale measurements fill the gap between the micro- and macroscale—examining neural activity over portions of the brain at high spatial and temporal resolution. These techniques can rely on arrays of electrodes but for the most part they depend on various forms of neuroimaging.

Since the 1980s, several different imaging modalities, such as intrinsic optical signal imaging, flavoprotein autofluorescence imaging, as well as voltage- and calcium-sensitive dye imaging have been used in neuroscience to measure mesoscale neural activity. More recently, advances in genetics and molecular biology have produced transgenic animals, particularly rodents, that express activity-dependent fluorescent markers. These laboratory animals can be readily acquired, bred, and employed in studies that track neural activity using different imaging techniques. Meanwhile, advances in virus development are producing increasingly diverse options for fluorescence imaging. Currently, fluorescence markers can be restricted to specific cell types, cortical layers, and brain regions and can be made dependent on secondary genetic factors, allowing for conditional expression. More recently, these markers can also vary in fluorescent color and, instead of simply marking neural activity, can provide a readout of specific neurotransmitter release. Accompanying these advances, recent engineering developments in imaging and computer technology have coalesced to make mesoscale imaging an increasingly accessible and important method in neuroscience.

A large body of neuroscientific work has employed these methods in animal models that are head-fixed and anesthetized. However, all anesthetics carry the confound of altering both natural brain activity and vascular reactivity, better known as neurovascular coupling. To better understand neural function in more realistic, ethologically valid, settings, it is critical to apply the above techniques to animals in awake, non-anesthetized, conditions. This is now readily achievable by using the intracranial imaging tools and methods described in this volume.

In Part I of this book, the authors highlight novel imaging modalities, such as neurotransmitter and single-cell resolution imaging and analyses of neuroimaging data in awake behaving rodents. In Part II, technical innovations are emphasized such as novel head-mounted camera systems, multi-site photometry, and concurrent imaging-electrographic probes. Part III presents methods of mesoscale imaging during normal behavior, with a focus on motor systems that are easily accessed in rodents. Finally, Part IV highlights the mesoscale imaging methods used to study neurological diseases and disorders, which are especially critical to observe in the non-anesthetized condition. Altogether, the chapters of this volume will provide an understanding of how mesoscale imaging can be applied in various areas of neuroscientific research. We thank all the contributing authors for their time and efforts in making this exciting and innovative topic available to the reader through this textbook.

New York, NY, USA

James E. Niemeyer
Hongtao Ma
Theodore H. Schwartz

Contents

Contributors

ANNA LETIZIA ALLEGRA MASCARO • *LENS: European Laboratory for Non-Linear Spectroscopy, Sesto Fiorentino, Italy; Neuroscience Institute, National Research Council, Pisa, Italy*

RUSSELL E. CARTER • *Department of Neuroscience, University of Minnesota, Minneapolis, MN, USA*

QIONGHAI DAI • *Department of Automation, Tsinghua University, Beijing, China*

FEI DENG • *State Key Laboratory of Membrane Biology, School of Life Sciences, Peking University, Beijing, China; PKU-IDG/McGovern Institute for Brain Research, Beijing, China*

ANDREW K. DUNN • *Department of Biomedical Engineering, The University of Texas at Austin, Austin, TX, USA*

TIMOTHY J. EBNER • *Department of Neuroscience, University of Minnesota, Minneapolis, MN, USA*

JIESI FENG • *State Key Laboratory of Membrane Biology, School of Life Sciences, Peking University, Beijing, China; PKU-IDG/McGovern Institute for Brain Research, Beijing, China*

FRITJOF HELMCHEN • *Brain Research Institute, University of Zurich, Zurich, Switzerland; Neuroscience Center Zurich, Zurich, Switzerland*

EUNSONG KO • *Department of Mechanical Engineering, University of Minnesota Twin Cities, Minneapolis, MN, USA*

SUHASA KODANDARAMAIAH • *Department of Mechanical Engineering, University of Minnesota Twin Cities, Minneapolis, MN, USA; Department of Biomedical Engineering, University of Minnesota Twin Cities, Minneapolis, MN, USA; Department of Neuroscience, University of Minnesota Twin Cities, Minneapolis, MN, USA*

TAKAKI KOMIYAMA • *Department of Neurobiology, Center for Neural Circuits and Behavior, Department of Neurosciences, University of California San Diego, La Jolla, CA, USA; Halıcıoğlu Data Science Institute, University of California San Diego, La Jolla, CA, USA*

DUYGU KUZUM • *Halıcıoğlu Data Science Institute, University of California San Diego, La Jolla, CA, USA; Department of Electrical and Computer Engineering, University of California San Diego, La Jolla, CA, USA*

MALACHI LEHMANN • *Department of Biomedical Engineering, University of Minnesota Twin Cities, Minneapolis, MN, USA*

YULONG LI • *State Key Laboratory of Membrane Biology, School of Life Sciences, Peking University, Beijing, China; PKU-IDG/McGovern Institute for Brain Research, Beijing, China; Peking-Tsinghua Center for Life Sciences, New Cornerstone Science Laboratory, Academy for Advanced Interdisciplinary Studies, Peking University, Beijing, China*

XIN LIU • *Halıcıoğlu Data Science Institute, University of California San Diego, La Jolla, CA, USA; Department of Electrical and Computer Engineering, University of California San Diego, La Jolla, CA, USA*

JESSICA LUCCHESI • *LENS: European Laboratory for Non-Linear Spectroscopy, Sesto Fiorentino, Italy; Department of Physics and Astronomy, University of Florence, Sesto Fiorentino, Italy*

Part I

Imaging Modalities and Analyses

Chapter 1

Mesoscopic Imaging of Neurotransmitters and Neuromodulators with Genetically Encoded Sensors

Fei Deng, Jiesi Feng, Hao Xie, and Yulong Li

Abstract

Neurotransmitters (NTs) and neuromodulators (NMs) play vital roles in various physiological and pathological processes. To better understand the transmission mechanisms of NTs or NMs (NTs/NMs), it is critical to monitor the dynamics of NTs/NMs with high sensitivity, specificity, and spatiotemporal resolution. With recent advances in genetically encoded NT/NM indicators (GENIs) and mesoscopic imaging, it is now possible to visualize the spatiotemporal dynamics of NTs/NMs in vivo on a large scale, even cover the entire mouse neocortex, enabling simultaneous recording of multiple brain regions to further investigate the relationship among different cortex regions. Here, we present an overview of the principle, development, and application of GENIs, focusing on their joint utilization with mesoscopic imaging.

Key words Neurotransmitter, Neuromodulator, GENI, Sensor, PBP, GPCR, GRAB, Mesoscopic imaging

1 Introduction

As one of the most complex organs in our body, the brain plays essential but diversified functions via interconnected neural networks consisting of billions of neurons and other cells, such as glia. To decipher neural networks, it is critical to monitor neural activities. The recording of neural activities has been revolutionized by advances in voltage and calcium sensors, as discussed in previous chapters, which deepens our understanding of neuromodulation. However, mysteries are still embedded in the nervous system because recording neural activities alone does not capture communications between different cells directly, particularly the molecular dynamics during signal transmission in neural networks.

As we know, neurons communicate with each other via synapses, including electrical synapses and chemical synapses. Chemical synapses, as the predominant synapse, can release and receive neurotransmitters and neuromodulators (NTs/NMs) for

James E. Niemeyer et al. (eds.), *Awake Behaving Mesoscopic Brain Imaging*, Neuromethods, vol. 214,
https://doi.org/10.1007/978-1-0716-4120-0_1,

signal transduction [1]. Typically, NTs/NMs are released from the presynaptic terminal and bind to receptors on the postsynaptic cells, regulating the excitability and gene expression of targeted cells. There are over 100 different NTs/NMs, which can be classified into acetylcholine (ACh), amino acids, monoamines, nucleotides, neuropeptides, and neurolipids [2, 3]. The neurotransmission mediated by NTs/NMs involves many critical physiological processes, such as arousal, attention, appetite, sleep–wake cycles, perception, emotion, reward, learning, and memory [4, 5]. In contrast, the malfunction of neurotransmission is associated with many brain disorders, including depression, addiction, anxiety, epilepsy, schizophrenia, and even Alzheimer's and Parkinson's disease [6–12]. So, there is an urgent need to monitor NTs/NMs with high sensitivity, specificity, and spatiotemporal resolution, which will enhance our comprehension of their functions and may also provide valuable insights into the treatment of related diseases.

Various methodologies have been developed to detect NT/NM dynamics, including traditional methods and newly developed genetically encoded NT/NM indicators (GENIs), which have been extensively reviewed elsewhere [2, 3, 13, 14]. Traditional methods show different limitations, in terms of sensitivity, molecular and cell-type specificity, temporal and spatial resolution, and invasiveness, particularly for in vivo detection. For example, microdialysis has a relatively slow sampling rate, usually about 5–10 min; while fast-scan cyclic voltammetry (FSCV) struggles to distinguish chemicals with similar structures; and electrophysiology requires the application of receptor antagonists to verify molecular specificity. In addition, these three methods are invasive and low-throughput. Although some traditional genetically encoded methods have already been developed, such as SNAP-tag based indicator proteins with a fluorescent intramolecular tether (Snifits [15]), cell-based neurotransmitter fluorescent engineered reporters (CNiFERs [16]), and the TANGO assay [17], they still cannot fulfill the in vivo detection requirements. In contrast, newly developed GENIs offer numerous advantages over traditional methods, including high sensitivity, specificity, and spatiotemporal resolution as well as less invasiveness (Fig. 1). These advantages are especially significant when combined with mesoscopic imaging, which provides a large field of view (FOV) spanning millimeters at video frame rate [18]. According to the Allen brain atlas, there are 43 isocortical areas with distinct functions [19], such as visual, somatosensory, and motor cortex. By simultaneously imaging NT/NM dynamics in different regions under different conditions, we can gain a better understanding of their release patterns and relationships. Moreover, multiple NTs/NMs can be simultaneously monitored with multicolor GENIs in the same cortical region or even same-color GENIs in different regions, benefited from the large FOV of mesoscopic imaging. Overall, GENIs in combination with

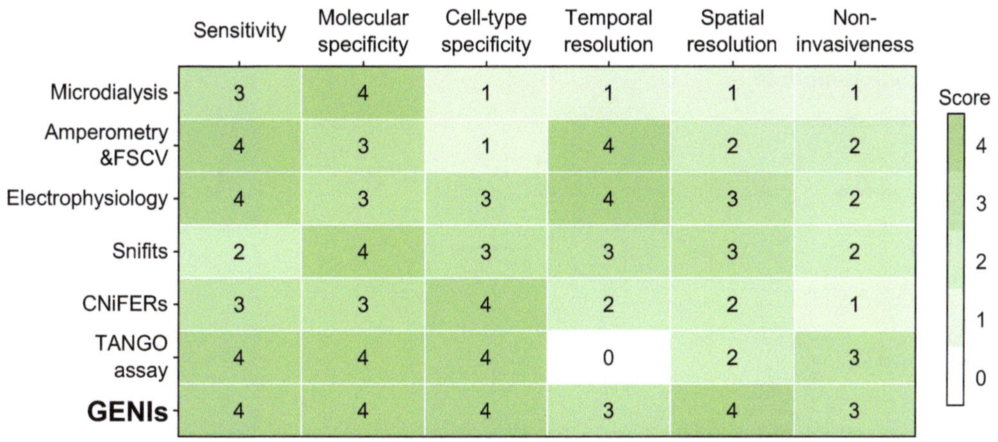

Fig. 1 Current methods for detecting neurotransmitters and neuromodulators. Several current methods are evaluated in different aspects with corresponding scores. The higher score indicates better performance and GENIs exhibit an overall superior performance. FSCV fast-scan cyclic voltammetry, Snifits SNAP-tag based indicator proteins with a fluorescent intramolecular tether, CNiFERs cell-based neurotransmitter fluorescent engineered reporters, GENIs genetically encoded neurotransmitters or neuromodulators indicators

mesoscopic imaging are a powerful technique for investigating the intricate dynamics of NTs/NMs in the brain.

In this chapter, we will introduce GENIs and their fundamental concepts. Additionally, we will discuss how GENIs can be applied in mesoscopic imaging, and provide an example to illustrate the primary workflow.

2 Genetically Encoded Neurotransmitter and Neuromodulator Sensors

Here, we will firstly provide a comprehensive overview of GENIs, including their principles and properties. Next, we will summarize general strategies for developing and optimizing GENIs to meet the requirements for in vivo applications. Lastly, we will discuss different methods for delivering GENIs to animals for whole cortex expression and considerations when using GENIs for mesoscopic imaging.

2.1 Principles and Properties of GENIs

A GENI usually consists of a recognition module and a reporter module, typically connected by linkers (Fig. 2). The recognition module binds the NT/NM molecules and induces conformational changes, which can be transduced to the reporter module to report signal changes by various readouts, such as fluorescence intensity.

So far, two main types of scaffolds have been employed as the recognition module in GENIs, including periplasmic binding proteins (PBPs) and G-protein-coupled receptors (GPCRs). PBPs of gram-negative bacteria belong to a widespread protein superfamily, capable of binding a variety of ligands [20]. Upon ligand binding,

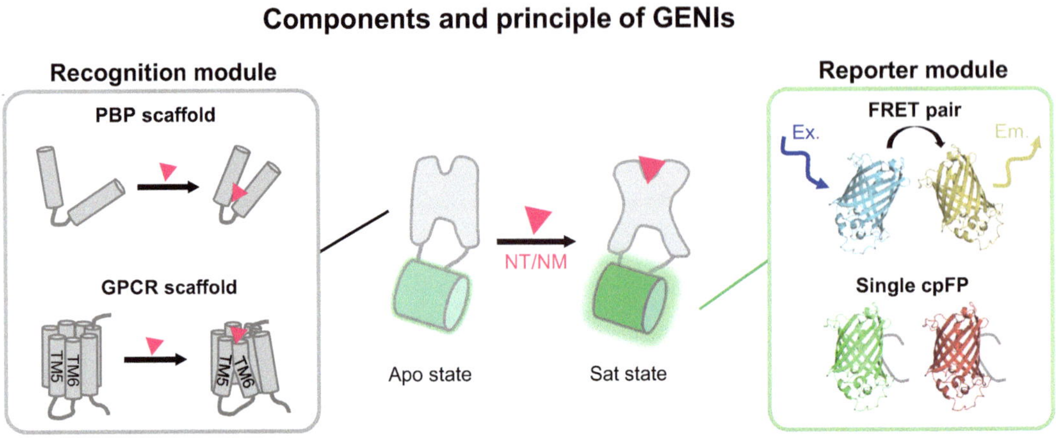

Fig. 2 Components and principle of genetically encoded NT/NM indicators (GENIs). A GENI consists of a recognition module and a reporter module. Upon binding with NT/NM, it can change from the Apo (ligand-free) to Sat (ligand-bound) state, producing changeable fluorescence signals. PBP periplasmic binding protein, GPCR G protein-coupled receptor, FRET fluorescence resonance energy transfer, cpFP circularly permuted fluorescent protein

PBPs change from a ligand-free open state to a ligand-bound closed state, which can transduce large conformational changes to the reporter module. In the past decades, a series of sensors have been developed based on PBPs, such as glutamate [21–24], GABA [25], ATP [26], ACh [27], and serotonin (5-HT) [28] sensors. Another class of recognition modules are GPCRs, which are the most prominent family of membrane receptors and can detect most extracellular signaling molecules, including NTs/NMs [29]. GPCRs share a highly conserved structure, composed of seven transmembrane helices (TM1–TM7), three extracellular loops (ECLs) and three intracellular loops (ICLs), an N-terminal, and a C-terminal. Upon binding its corresponding ligand, the GPCR undergoes a conformational change from an inactive state to an active state, leading to significant structural changes, particularly between TM5 and TM6 [30]. After years of dedicated efforts, many GPCR-based sensors have also been developed, such as ACh [31, 32], dopamine (DA) [33–36], norepinephrine (NE) [37, 38], 5-HT [39–41], adenosine (Ado) [42, 43], adenosine 5′-triphosphate (ATP) [44], and histamine (HA) [45] sensors.

Both PBPs and GPCRs have their own advantages and limitations, including the following:

1. PBP-based sensors can be expressed not only in the cell membrane with some membrane-tethered motifs for detecting extracellular NTs/NMs but also in the cytosol or some organelles for intracellular detections [28]. While most GPCR-based sensors can only localize to cell membrane, which limits their application in intracellular compartments.

2. GPCR-based sensors inherit the pharmacological properties of the endogenous GPCR, which may be a double-edged sword. On the one hand, agonists or antagonists can be used to verify signals of the sensor, serving as an internal control; on the other hand, it is important to avoid using drugs that can directly affect sensor performance when monitoring NTs/NMs in vivo. However, this property can also be leveraged to screen for drugs that can bind to endogenous GPCRs. For example, using a 5-HTR2A-based sensor, PsychLight, Dong et al. successfully identified a non-hallucinogenic psychedelic analog [40]. In contrast, PBP-based sensors are less likely to be affected by some drugs targeting GPCRs since PBPs are derived from bacteria.
3. GPCRs typically have affinities in accord with physiological concentrations, making them more suitable for detecting NTs/NMs in most physiological conditions. Moreover, for some of NTs/NMs, there may be multiple subtypes of GPCRs with varying affinities, which can be developed to sensors with a wide range of affinities. For example, there are 12 subtypes of GPCRs for 5-HT with half-maximal effective concentration (EC_{50}) ranging from nanomolar to micromolar [46], which are appropriate for detecting 5-HT levels both in physiological and pathological conditions [47–49]. Thus, a variety of GPCR-based sensors with different affinities have been developed, with EC_{50} ranging from nanomolar to micromolar [39–41, 50]. And in general, sensors with lower affinities have faster off kinetics, which are more suitable for detecting transients of NTs/NMs. In contrast to GPCR-based sensors, most of PBP-based sensors have low affinities and fast off kinetics.
4. There are limited PBPs that can bind NTs/NMs, which restricts the expansion of PBP-based sensors for most NTs/NMs. It is possible to engineer the binding pocket of an existing PBP to tune its selectivity for other NTs/NMs, as demonstrated in one of the state-of-the-art works that a 5-HT sensor, iSeroSnFR, was redesigned from an ACh sensor, iAChSnFR [28]. However, its sensitivity is still not satisfied for in vivo detection of 5-HT dynamics. Moreover, it is quite difficult and laborious to expand this strategy to other NT/NMs for developing highly selective sensors. In contrast, for most NTs/NMs, there is at least one endogenous GPCR that can bind to NTs/NMs, making it more practical to develop GPCR-based sensors for most NTs/NMs.

The reporter module can be used to transduce conformational changes of the recognition module into various signals, and fluorescence is one of the most commonly used signals. This can be achieved by either a fluorescence resonance energy transfer (FRET) pair or a single fluorescent protein (FP). The reporter module of a FRET-based sensor consists of a FRET pair with both a donor and an acceptor fluorescent protein, such as CFP and YFP [51]. The conformational changes in the recognition module can alter the distance and orientation between the FRET pair, thereby modulating the FRET efficiency and leading to changes in the fluorescent intensity ratio between the donor and acceptor. FRET-based sensors can be used as ratiometric sensors since they rely on two spectral-distinct FPs, making them resistant to artifacts (e.g., movement), and suitable for analyte quantification. However, FRET-based sensors have limited dynamic range due to the steric hindrance of the bulky FP moiety, which prevents the FRET pair from getting close enough to each other. For example, a FRET-based 5-HT sensor, named 5HT-CC, exhibited good membrane localization and high selectivity to 5-HT, but only a ~4% increase in FRET ratio in response to saturated 5-HT [52]. Nonetheless, no in vivo application of 5HT-CC has been reported yet, possibly due to its insufficient sensitivity. Single FP-based sensors typically contain a circularly permuted FP (cpFP) or split FP, and their fluorescence intensities change with the conformational changes of the recognition module. And a circularly permutated enhanced green fluorescent protein (cpEGFP) is widely used in most single FP-based sensors. Unlike FRET sensors, their dynamic ranges are not limited by the FRET efficiency and can show relatively high response, making them suitable for in vivo applications. For example, a green fluorescent 5-HT sensor, named $GRAB_{5\text{-}HT1.0}$, can report endogenous 5-HT release in flies and mice with over 50% fluorescence increases evoked by odor and MDMA application, respectively [39]. Most single FP-based sensors are intensiometric, with one excitation or emission peak. Thus, these intensiometric sensors are also suitable for commonly used and low-cost imaging systems, such as the fiber photometry recording system.

Considering that single FP-based sensors show good performance and have been used in vivo, properties for most GENIs with single chromophore are summarized in Table 1.

2.2 Development and Optimization of GENIs

As mentioned above, GENIs consist of a recognition module, a reporter module, and linkers. To develop a GENI, the first step is to screen a well-performed recognition module and an appropriate reporter module. Then, the best insertion site for the reporter module should be scanned and some critical residues of the sensor can be further optimized. The detailed workflow has been summarized in other reviews using DA sensors as examples [2, 3, 53]. Using native GPCR as the recognition module, a series of GENIs have been developed recently.

Here, we take the $GRAB_{5\text{-}HT}$ sensor [39] (GRAB stands for GPCR activation-based) as an example to illustrate the typical

Table 1
Summary of single FP-based GENIs

NTs/NMs types	Sensor name	Recognition module	Reporter module	Ex/Em (nm)	Responses (max. $\Delta F/F_0$)	Affinity (EC50)	In vivo application	Refs.
Amino acids	*Glutamate sensors*							
	iGluSnFR3.v82	*E. coli* GltI	cpmVenus	512/530	13.1^a	$31.7\ \mu M^a$	Mouse	[90]
	iGluSnFR3.v857	*E. coli* GltI	cpmVenus	502/522	54.0^a	$195.9\ \mu M^a$	Mouse	[90]
	SF-iGluSnFR. A184S	*E. coli* GltI	cpsfGFP	490/510	$3.1^a/0.69^b$	$7.5\ \mu M^a$	Mouse, ferret	[23]
	$iGlu_u$	*E. coli* GltI	cpEGFP	NA	$2.8^a/1.7?^b$	$600\ \mu M^a/53\ \mu M^b$	NA	[91]
	R-iGluSnFR1	*E. coli* GltI	cpmApple	562/588	$-4.9^a/-0.35^b$	$11\ \mu M^a$	NA	[24]
	GABA sensors							
	iGABASnFR	Pf622	cpsfGFP	485/510	$2.5^a/\sim 0.7^b$	$9\ \mu M^a/30\ \mu M^b$	Mouse	[25]
	iGABASnFR. F102G	Pf622	cpsfGFP	NA	$4.5^a/\sim 1.7^b$	$50\ \mu M^a/42\ \mu M^b$	Mouse	[25]
	iGABASnFR. F102Y.Y137L	Pf622	cpsfGFP	NA	$3.5^a/\sim 0.45^b$	$70\ \mu M^a/106\ \mu M^b$	Mouse, zebrafish	[25]
Acetylcholine	*Acetylcholine sensors*							
	$GRAB_{ACh3.0}$	M_3R	cpEGFP	492/510	$2.8^b/\sim 2.7^c$	$2.2\ \mu M^c$	Mouse, fly	[32]
	iAChSnFR	X513-OpuBC	cpsfGFP	485/510	$12^a/10^b/\sim 4.5^c$	$1.3\ \mu M^a/2.9\ \mu M^b/0.4\ \mu M^c$	Mouse, zebrafish, fly, worm	[27]
Monoamines	*Dopamine sensors*							
	$GRAB_{DA2h}$	D_2R	cpEGFP	500/520	$\sim 2.8^b/\sim 2.4^c$	$7\ nM^b/28\ nM^c$	Mouse	[34]
	$GRAB_{DA2m}$	D_2R	cpEGFP	NA	$\sim 3.4^b/\sim 3.1^c$	$90\ nM^b/104\ nM^c$	Mouse, fly	[34]
	$rGRAB_{DA1h}$	D_2R	cpmApple	565/595	$\sim 1.0^{b,c}$	$4\ nM^b/18\ nM^c$	Mouse	[34]
	$rGRAB_{DA1m}$	D_2R	cpmApple	565/595	$\sim 1.5^b/\sim 1.7^c$	$95\ nM^b/ 93\ nM^c$	Mouse, fly	[34]
	dLight1.1	D_1R	cpEGFP	490/516	2.3^b	$330\ nM^b$	Mouse	[35]
	dLight1.2	D_1R	cpEGFP	490/516	3.4^b	$770\ nM^b$	Mouse	[35]
	dLight1.3b	D_1R	cpEGFP	NA	9.3^b	$1680\ nM^b$	Mouse, rat	[14, 35]
	dLight1.4	D_4R	cpEGFP	NA	1.7^b	$4.1\ nM^b$	NA	[35]
	dLight1.5	D_2R	cpEGFP	NA	1.8^b	$110\ nM^b$	NA	[35]
	YdLight1	D_1R	cpEGFP	NA/525	3.06^b	$1.63\ \mu M^b$	NA	[36]

(continued)

Table 1 (continued)

NTs/NMs types	Sensor name	Recognition module	Reporter module	Ex/Em (nm)	Responses (max. $\Delta F/F_0$)	Affinity (EC50)	In vivo application	Refs.
	RdLight1	D_1R	cpmApple	562/588	2.48^{b}/~3^{c}	859 nM^{b}/ 229 nM^{c}	Mouse, rat	[36]
	R-GenGAR-DA1.2	D_1R	cpmApple	566/595	-0.43^{b}/−0.52^{c}	920 nM^{b}/250 nM^{c}	NA	[92]
	Noradrenaline sensors							
	$GRAB_{NE1h}$	$\alpha_{2A}R$	cpEGFP	NA	1.3^{b}/1.5^{c}	83 nM^{b}/ 229 nM^{c}	Mouse, zebrafish	[37]
	$GRAB_{NE1m}$	$\alpha_{2A}R$	cpEGFP	NA	2.3^{b,c}	930 nM^{b}/ 229 nM^{c}	Mouse, zebrafish	[37]
	nLight1.3	$\beta_{2A}R$	cpEGFP	NA/516	~2.1^{b}/1.3^{c}	764 nM^{b}/ 919 nM^{c}	Mouse	[38]
	Serotonin sensors							
	iSeroSnFR	X513-OpuBC	cpsfGFP	NA	8^{a}/17^{b,c}	310 μM^{a}/390 μM^{b}	Mouse	[28]
	$GRAB_{5\text{-}HT1.0}$	$5\text{-}HT_{2C}R$	cpEGFP	NA	2.5^{b}/2.8^{c}	14 nM^{b}/ 22 nM^{c}	Mouse, fly	[39]
	PsychLight2	$5\text{-}HT_{2A}R$	cpEGFP	NA	0.8^{c}	26 nM^{c}	Mouse	[40]
	$GRAB_{HTR6\_1.0}$	$5\text{-}HT_6R$	cpEGFP	NA	1.5^{b}	84 nM^{b}	NA	[93]
	sDarken	$5\text{-}HT_{1A}R$	cpEGFP	NA	-0.71^{b}	127 nM^{b}	Mouse	[41]
	$gGRAB_{5\text{-}HT3.0}$	$5\text{-}HT_4R$	cpEGFP	505/520	13^{b}/27^{c}	150 nM^{b}/ 80 nM^{c}	Mouse	[50]
	$rGRAB_{5\text{-}HT1.0}$	$5\text{-}HT_4R$	cpmApple	560/595	3.3^{b}/4^{c}	790 nM^{b}/ 600 nM^{c}	Mouse	[50]
	Histamine sensors							
	$GRAB_{HA1h}$	hH_4R	cpEGFP	505/520	3.7^{b}/1.8^{c}	17 nM^{b}/ 19 nM^{c}	Mouse	[45]
	$GRAB_{HA1m}$	wbH_1R	cpEGFP	505/520	5.9^{b}/3.2^{c}	380 nM^{b}/ 400 nM^{c}	Mouse	[45]
Nucleotides	*Adenosine sensors*							
	$GRAB_{Ado1.0}$	$A_{2A}R$	cpEGFP	NA	1.2^{b}/2^{c}	60 nM^{c}	Mouse	[42, 43]
	$GRAB_{Ado1m}$	$A_{2A}R$	cpEGFP	NA	3.5^{c}	3.6 μM^{c}	Mouse	[42, 43]
	ATP sensors							
	iATPSnFR1.0	F_0F_1-ATP synthase ε subunit	cpsfGFP	490/512	~2.4^{a}/1.0^{b}/1.1^{c}	~120 μM^{a}/350 μM^{b}/400 μM^{c}	NA	[26]
	ATPOS	F_0F_1-ATP synthase ε subunit	Cy3	556/565	~1.8^{a}/1.0^{b}/1.1^{c}	~150 nM^{a}/350 μM^{b}/400 μM^{c}	Mouse	[94]
	$GRAB_{ATP1.0}$	$hP2Y_1$	cpEGFP	500/520	5^{b}/7.8^{c}	~6.7 μM^{b}/77 nM^{c}	Mouse, zebrafish	[44]
	$GRAB_{ATP1.0\text{-}L}$	$hP2Y_1$	cpEGFP	NA	~10^{c}	97 μM^{b}/32 μM^{c}	Zebrafish	[44]

Neuropeptides	*Opioid sensors*							
	kLight1.2a	KOR	cpEGFP	NA	0.6^b	NA	Mouse	[35, 95]
	GrpLight1.3ER	GRPR	cpEGFP	NA	$\sim 8^c$	$198\ nM^c$	Mouse	[96]
	M-SPOTIT	MOR	cpGFP	NA	$12.5^b/4.6^c$	$15\ nM^c$	NA	[97]
	Orexin sensor							
	OxLight1	OX_2R	cpEGFP	490/512	9.06^b (orexin-A), 8.59^b (orexin-B)	$75\ nM^b/127\ nM^c$ (orexin-A), $47\ nM^b/110\ nM^c$ (orexin-B)	Mouse	[98]
	Oxytocin sensors							
	$MTRIA_{OT}$	meOTR	cpGFP	NA	7.35^b	$\sim 20\ nM^{b,c}$	Mouse	[99]
	$GRAB_{OT1.0}$	bOTR	cpEGFP	505/520	$\sim 4^b/4.5^c$	$1.6\ nM^b/3\ nM^c$	Mouse	[100]
	Corticotropin-releasing factor sensor							
	$GRAB_{CRF1.0}$	CRF1R	cpEGFP	500/520	$15.6^b/13.3^c$	$33\ nM^b/18\ nM^c$	Mouse	[101]
	Somatostatin sensor							
	$GRAB_{SST1.0}$	SSTR5	cpEGFP	500/520	$4.2^b/4.5^c$	$70\ nM^b/130\ nM^c$ (SST-14), $13\ nM^b$ (SST-28)	Mouse	[101]
	Cholecystokinin sensor							
	$GRAB_{CCK1.0}$	CCKBR	cpEGFP	505/520	$7.3^b/9^c$	$4\ nM^b/5.5\ nM^c$	Mouse	[101]
	Neurotensin sensor							
	$GRAB_{NTS1.0}$	NTSR1	cpEGFP	NA	3.8^b	$6\ nM^b$	NA	[101]
	Neuropeptide Y sensor							
	$GRAB_{NPY1.0}$	NPY1R	cpEGFP	500/520	3^b	$40\ nM^b/0.7\ nM^c$	NA	[101]
	Substance P sensor							
	$GRAB_{SP1.0}$	TACR1	cpEGFP	NA	0.7^b	NA	NA	[101]
	Urocortin sensor							
	$GRAB_{UCN1.0}$	CRF2R	cpEGFP	NA	11.9^b	NA	NA	[101]
	Vasoactive intestinal peptide sensor							
	$GRAB_{VIP1.0}$	VIPR2	cpEGFP	500/520	4.9^b	$120\ nM^b/19\ nM^c$	NA	[101]
	Parathyroid hormone-related peptide sensor							
	$GRAB_{PTH1.0}$	PTH1R	cpEGFP	NA	3.8^b	NA	NA	[101]
	Glucagon-like peptide-1 sensor							
	GLPLight1	GLP1R	cpEGFP	500/512	$5.3^b/4.6^c$	$28\ nM^b/9.3\ nM^c$	NA	[102]
Neurolipids	*Endocannabinoid sensor*							
	$GRAB_{eCB2.0}$	CB1R	cpEGFP	500/520	$3.3^b//9.5^c$ (2-AG), $2.8^b/5^c$ (AEA)	$3.1\ \mu M^b//9.0\ \mu M^c$ (to 2-AG), $0.2\ \mu M^b/0.8\ \mu M^c$ (to AEA)	Mouse	[67]

NA data not available
[a]Measured in purified proteins
[b]Measured in cell lines
[c]Measured in cultured neurons

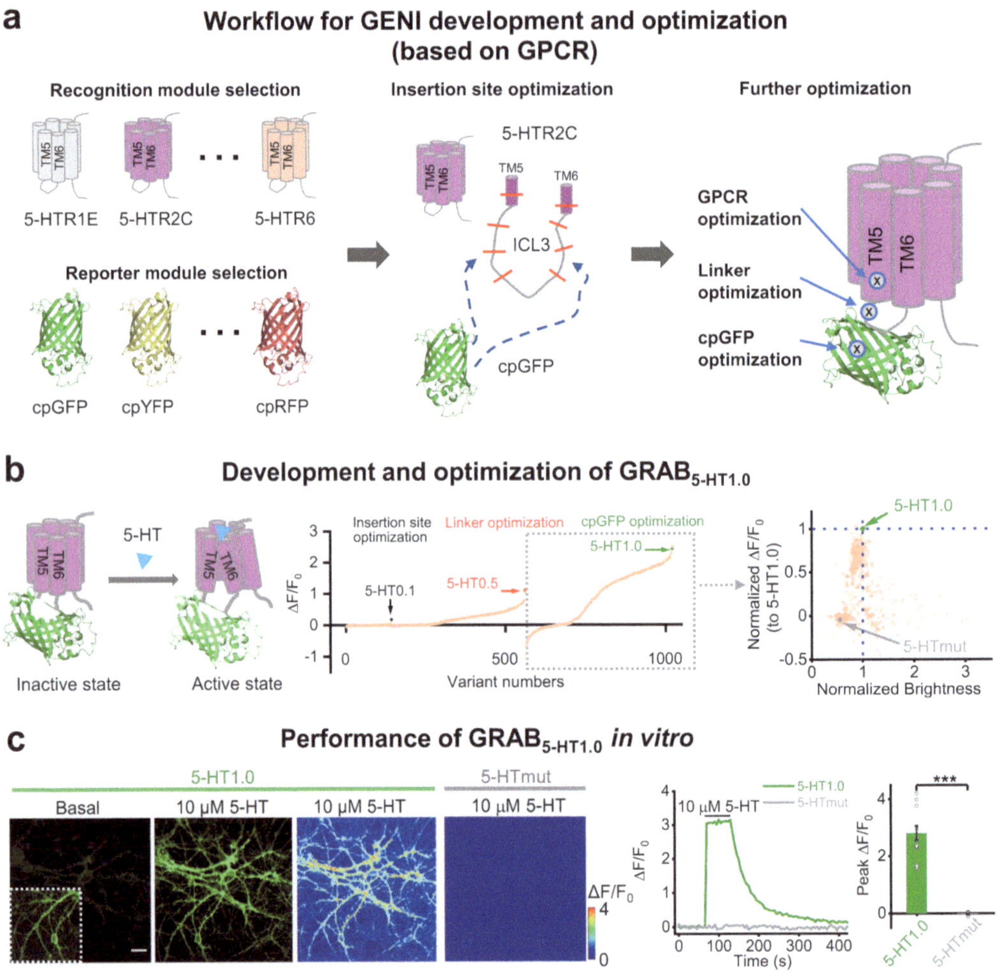

Fig. 3 Development and optimization of genetically encoded NT/NM indicators (GENIs). (**a**) Schematic diagram depicting the general workflow for GENI development and optimization based on GPCR scaffolds, taking the $GRAB_{5\text{-}HT1.0}$ sensor as an example. (**b**) Development and optimization of $GRAB_{5\text{-}HT1.0}$ (adapted from Wan et al. [39]). (**c**) Performance of $GRAB_{5\text{-}HT1.0}$ in cultured rat cortical neurons (adapted from Wan et al. [39]). TM5/TM6 transmembrane domains 5/6, ICL3 the third intracellular loop, cpGFP/cpYFP/cpRFP circularly permuted green/yellow/red fluorescent protein

workflow for developing and optimizing GENIs based on GPCR scaffolds in details (Fig. 3a, b). To choose a suitable GPCR scaffold as the recognition module (*see* **Note 1**), several 5-HT receptor subtypes have been fused with a cpEGFP, examining based on the membrane trafficking and affinities. The 5-HTR2C-based sensors showed the best membrane trafficking and was selected as the scaffold. To better transduce the GPCR conformational changes into the cpEGFP, insertion sites for cpEGFP and linkers were systematically screened. Since the N- and C-terminal linkers between GPCR and cpEGFP are critical for the efficiency of conformational transduction, linkers have been optimized by site-directed random mutagenesis (*see* **Note 2**). In addition, some sites in cpEGFP, which are contributed to fast GFP folding and high brightness [54, 55], have been optimized as well. After the above rounds of iterative optimization, the finalized best candidate was named the $GRAB_{5\text{-}HT1.0}$ sensor, which showed a ~280% increase in fluorescence ($\Delta F/F_0$) with saturated concentration of 5-HT application. Additionally, a 5-HT-insensitive sensor was generated by introducing the $D134^{3.32}Q$ substitution, which serves as a negative control (Fig. 3b). The $GRAB_{5\text{-}HT1.0}$ sensor also showed good membrane localization and large response in cultured neurons (Fig. 3c).

2.3 Expression of GENIs In Vivo

To deliver GENIs in the mammalian system for in vivo application, two approaches are commonly used, including viral-mediated gene delivery and GENIs knock-in animals.

Viral-mediated gene delivery is a convenient, fast, and versatile approach to achieve efficient expression of GENIs. One of the most popular viral tools is the recombinant adeno-associated virus (rAAV), which is modified from the wild-type AAV, a member of the nonpathogenic parvovirus [56–58]. GENIs can be expressed in specific cell types with different serotypes or promoters, i.e., using the AAV2/9 pseudotype and CaM kinase II alpha (CaMKIIα) promoter to drive GENIs expression in excitatory neurons. Local injections of AAVs are commonly used to achieve regional expression of GENIs in the brain, but global expression in the animal's brain for mesoscopic imaging requires multiple injections due to the limited diffusing distance [58]. Although multiple-site injections in the brain can drive robust expression of GENIs, it is laborious, time-consuming, and hard to make a relatively uniform expression of sensors (*see* **Note 3**). By engineering AAV capsids, some AAV pseudotypes, can efficiently cross the blood-brain barrier and infect the central nervous systems, such as PHP.eB, which can infect more than half of cortical and striatal neurons by intravenous administration [59]. PHP.eB-mediated expression of a calcium sensor, GCaMP6s, works well in mesoscopic imaging [60], but no PHP.eB-mediated expression of GENIs has been reported yet, possibly due to the low expression level and signal-to-noise ratio (SNR). Given that the AAV9 has a capacity to cross the blood-brain barrier (BBB) of mice, especially in neonate [61], a novel method called neonatal sinus injection (short for n-SIM) has been reported recently, which employs transverse sinus injections of AAV9 at P0 pups to efficiently deliver gene across the central nervous system, leading to the whole-brain expression [62]. The n-SIM is suitable for expressing GENIs in animals for mesoscopic imaging, as demonstrated in a study using an acetylcholine (ACh) sensor, $GRAB_{ACh3.0}$, to monitor ACh dynamics in mice across the neocortex [63].

Another approach to driving the expression of GENIs is generating GENI knock-in animals. Compared with viral-mediated expression, transgenic methods expand the ability to monitor and manipulate neuronal activities with increased cell-type specificity and express gene cassettes with larger sizes. Transgenic animals also enable a uniform expression of GENIs in the entire dorsal neocortex, making them ideal for mesoscopic imaging. By genetic manipulation, some GENIs knock-in animals have been developed and are usually in a recombinase-dependent manner, such as the iGluSnFR knock-in mice [64], a fluorescent glutamate sensor line, which has also been successfully used in mesoscopic imaging [65]. With the development of gene editing tools [66] and advancements in the development of a series of GENIs [31–37, 39, 44, 67], it is possible to generate transgenic animals for expression of various GENIs. By crossing with different driver lines, we are also able to express GENIs in specific cells for monitoring NTs/NMs. For example, Feng et al. measured cell type-specific NE signals in response to tactile stimuli by expressing the $GRAB_{NE2m}$ sensor in excitatory neurons or astrocytes via crossing the $GRAB_{NE2m}$ knock-in mice with CaMKIIα-Cre or GFAP-Cre mice, respectively [68]. Although GENI knock-in animals exhibit many advantages, compared with viral-mediated expression, generation of these animals is time-consuming and relatively expensive.

2.4 Considerations for Using GENIs

With the advances in GENIs, a variety of NTs/NMs can be imaged in vitro or in vivo. These sensors have their own advantages but also some limitations. Thus, for the applications of GENIs, especially in vivo, it is important to carefully choose the most appropriate sensor for the specific scenario since there is almost no "one-size-fits-all" sensor [3, 14, 69].

The first thing that needs to be considered is the sensitivity, which is determined by multiple factors including the response, affinity, and brightness, and can be evaluated by signal-to-noise ratio (SNR) [3]. For example, compared with dLight1.3b, although $GRAB_{DA2m}$ has a smaller maximum response ($GRAB_{DA2m}$: ~220%, dLight1.3b: ~450%), it shows a higher SNR (~2-fold to dLight1.3b) owing to higher basal and maximum brightness in HEK293T cells [34]. Importantly, for in vivo application, effective sensitivity is also affected by the resting ligand concentration [2]. So, selecting a sensor with an appropriate affinity for detecting NT/NM levels is crucial to maximize its sensitivity, especially when the sensor's EC_{50} matches the basal ligand concentration.

Another important consideration is the specificity. Although most of GENIs show high specificity to corresponding ligands, some of them also respond to other molecules with similar structures. For example, the excellent sensor engineering leads to an ACh sensor, iAChSnFR, whose specificity is shifted from choline toward ACh with ~35-fold selectivity but still responds to choline when the concentration of choline is higher than 1 μM [27]. Since ACh released in synapses can be rapidly hydrolyzed to choline by acetylcholinesterase (AChE) [70], which may disturb the interpretation of signals from iAChSnFR. In contrast, the GPCR-based sensor, $GRAB_{ACh}$, does not respond to choline in high concentration even at 100 μM, showing a much more specific response to ACh [31, 32]. Another case is the DA sensor: although GPCR-based DA sensors, including $GRAB_{DA}$ and dLight, have a higher selectivity for DA over NE (~15-fold for $GRAB_{DA2m}$ and ~60-fold for dLight1), they still respond to NE when the NE concentration is higher than 0.1 μM for $GRAB_{DA2m}$ and 1 μM for dLight1, respectively [33–36]. So, it is important to be careful when choosing GENIs and interpreting signals of GENIs that may respond to other molecules.

Third, when selecting GENIs for a specific application scenario, it is important to consider the kinetics of the sensor. For capturing rapid transients of NTs/NMs, GENIs with fast kinetics are needed. However, there is always a tradeoff between a sensor's affinity and off kinetics. That is to say, sensors with fast off kinetics may have a low affinity, which may limit their sensitivities. If fast kinetics are not necessary, GENIs with slow off kinetics but high affinity could also be beneficial, since they accumulate more photons and contribute to higher SNR [39].

Fourth, different pharmacological properties of GENIs should be considered, especially for GPCR-based sensors since most of them inherit the pharmacological properties of parent GPCRs. On the one hand, agonists or antagonists can be used to manipulate sensor signals for validation. On the other hand, it is important to avoid using compounds that interact with sensors when recording the signals of endogenous NTs/NMs. Fortunately, the problem can be circumvented by developing different subtype-based sensors. For example, DA sensor $GRAB_{DA}$ and dLight are based on D_2R and D_1R, respectively.

Last but not least, the expression of GENIs should not affect cell physiology, especially for long-term expression. For GPCR-based sensors, one potential concern is that they may couple with downstream pathways, including G protein and β-arrestin pathways. Fortunately, most GPCR-based sensors have shown negligible downstream signaling, possibly due to the steric hindrance of the bulky cpFP that replaces the ICL3 of the GPCR, which precludes the interaction between the GPCR and G protein or β-arrestin [44]. However, a few GPCR-based sensors still, to some extent, coupled with downstream signaling, such as the $GRAB_{ACh2.0}$ sensor [31]. But this G_q protein coupling was almost eliminated in an optimized version, the $GRAB_{ACh3.0}$ sensor [32]. In addition, overexpression of GENIs may cause potential buffering effects, which may compete with endogenous GPCRs and affect cell physiology. It is worth noting that current results indicate no obvious buffering effects both in vitro and in vivo [34, 39]. Nevertheless, it is advisable to use more sensitive sensors (higher SNR) with lower affinity and reduce expression levels to minimize the potential buffering effects.

3 Mesoscopic Imaging with GENIs

NTs/NMs are widely distributed throughout the brain and play important roles, such as reward, emotion, learning, and memory. For example, acetylcholine (ACh) and monoamines are very vital NTs/NMs that involve almost all the brain functions and are produced by different neurons. These neurons project to and innervate the entire cortex, which consists of dozens of brain regions (Fig. 4a), according to the Allen CCFv3 atlas [71]. GENIs provide a powerful toolbox for high spatiotemporal imaging of a variety of NTs/NMs. Traditional microscopy, such as

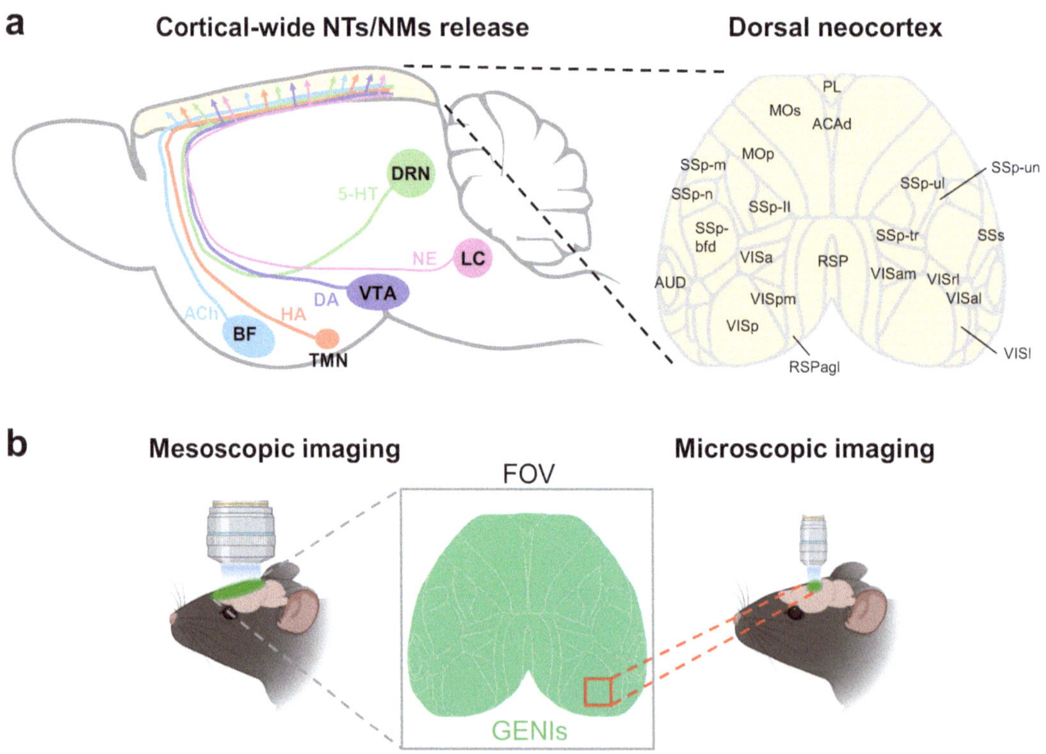

Fig. 4 Combination of mesoscopic imaging and GENIs. (**a**) Schematic diagram depicting the projection of different monoaminergic neurons in the cortex (left) and parcellation of cortical areas (right). (**b**) Comparison of the FOV (field of view) between mesoscopic imaging and microscopic imaging. The black square indicates FOV for mesoscopic microscope; the red square indicates FOV for traditional microscope. BF basal forebrain, DRN dorsal raphe nucleus, LC locus coeruleus, VTA ventral tegmental area, TMN tuberomammillary nucleus

confocal and two-photon microscopes, can provide cellular and subcellular resolution but often with a limited FOV of several hundred microns [18], in which it is hard to monitor the global spatial dynamics of NTs/NMs. Simultaneously imaging NTs/NMs across multiple brain regions can enhance the understanding of NT/NM dynamics and functions in the brain. Mesoscopic imaging, which spans a FOV larger than 1 cm × 1 cm and covers the entire dorsal neocortex of the mouse, offers a powerful tool for large-scale imaging (Fig. 4b).

The combination of GENIs and mesoscopic imaging has greatly facilitated the understanding of NT/NM dynamics, which can lead to new biological discoveries and ideas. Here, we will show some examples of mesoscopic imaging with GENIs.

The earliest mesoscopic imaging with GENIs may be traced back to the mesoscopic imaging of iGluSnFR [65]. Xie et al. expressed iGluSnFR in mouse cortex by AAV injection or transgenic mice and visualized glutamate dynamics with mesoscopic imaging. They detected glutamate release in specific brain regions evoked by multiple sensory stimuli in anesthetized and behaving mice. Moreover, the kinetics of glutamate release reported by iGluSnFR are comparable to signals detected by voltage-sensitive dye (VSD) and much faster than calcium signals reported by GCaMP3 and GCaMP6s. Moreover, using a miniaturized head-mounted microscope, Rynes et al. monitored glutamate dynamics in freely moving mice during the sleep–wake cycle [72].

In addition, it is also important and feasible to detect two different neurochemicals simultaneously with dual-color mesoscopic imaging (*see* **Note 4**). Lohani et al. co-expressed the green fluorescent ACh sensor $GRAB_{ACh3.0}$ and the red fluorescent calcium sensor jRCaMP1b in the mouse dorsal neocortex to simultaneously monitor cholinergic signaling and cortical activity [63]. They found that the ACh signal in the neocortex is highly dynamic and spatially heterogeneous, which can link behavioral fluctuations to the functional reorganization of cortical networks.

Although the combination of mesoscopic imaging with GENIs is a powerful tool for understanding NT/NM dynamics, there are still some limitations and challenges. Firstly, the preparation for animals expressing GENIs uniformly (*see* **Note 3**) may be a rate-limiting step, since n-SIM and crossing of transgenic mice cost more than 6 weeks from the AAV injection or mice crossing, which is much slower than traditional local AAV injection that takes about 3 weeks for expression. Secondly, there are potential signal contaminations from hemodynamic changes [73, 74] and light scattering. For hemodynamic artifacts, if it is not negligible, hemodynamic correction is needed (*see* **Note 5**). In general, there are two correction strategies [18] based on a reference channel. One is using the isosbestic point of the GENI to correct the standard wavelength excited signal. Another is using a reflectance of green (~530 nm) and red (~630 nm) illumination to measure the hemodynamic changes for correction [18, 73]. Thirdly, most mesoscopic imaging systems are custom-built, which limits the availability of mesoscopic imaging with GENIs for more laboratories. Fortunately, with the generous sharing of imaging system building experience [75] and GENIs, it will be easier for more laboratories to adopt mesoscopic imaging with GENIs for more biological discoveries. Fourthly, signals acquired by mesoscopic imaging are limited in the surface of the sample, which indicates that it is mainly used for imaging the dorsal cortex instead of other deep brain regions. In addition, even in the dorsal cortex, signals from the superficial layer are dominant, which may lead to some bias across different layers.

Despite these challenges, mesoscopic imaging with GENIs can monitor the dynamics of NTs/NMs on a large scale and may provide new insights into their important roles in the nervous system.

4 Case Study: Mesoscopic Imaging of Cortical 5-HT Dynamics in Mouse

Here, an optimized green fluorescent 5-HT sensor, named $gGRAB_{5\text{-}HT3.0}$ (referred to hereafter as simply g5-HT3.0 [50]), is used as an example to demonstrate the monitoring of NT/NM dynamics by mesoscopic imaging.

5-HT is a crucial monoamine that regulates lots of essential biological processes, such as appetite, sleep–wake cycles, emotion, reward, learning and memory [76–78]. Meanwhile, the malfunction of the serotonergic system is associated with many psychiatric disorders, including anxiety, addiction, and depression [6–8]. Moreover, some psychotropic drugs targeting the serotonergic system are widely used to treat brain disorders, e.g., some selective serotonin reuptake inhibitors (SSRIs) are used as antidepressants by inhibiting the reuptake of 5-HT and elevating the extracellular 5-HT level. So, monitoring the dynamics of 5-HT is critical for understanding the serotonergic system. Combining the 5-HT sensor with mesoscopic imaging, there is a great opportunity to visualize the spatiotemporal dynamics of 5-HT across multiple brain regions [79, 80] in real-time.

Since the dorsal raphe nucleus (DRN) is the primary source of serotonergic neurons [81], a Cre-dependent red-shifted optogenetic actuator, ChrimsonR [82], was expressed in the DRN of *Sert-Cre* mice [83] to optically activate serotonergic neurons for 5-HT release to validate the sensor's reliability. Besides artificially evoked 5-HT release in the cortex, considering the important role of 5-HT in regulating sleep–wake cycles [39, 84], the physiological 5-HT dynamics were monitored during sleep–wake cycles using mesoscopic imaging, while simultaneous electroencephalography (EEG)

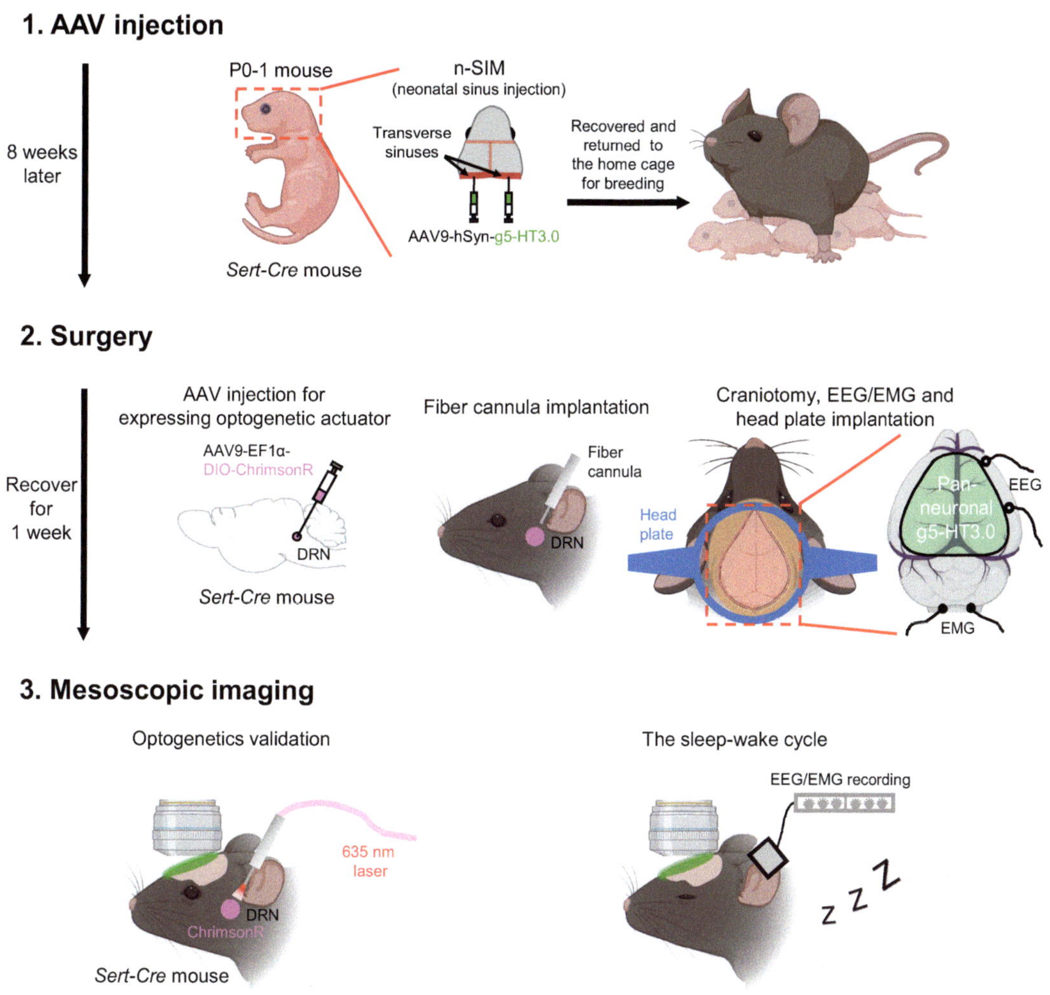

Fig. 5 Workflow of mesoscopic imaging with g5-HT3.0 sensor during optical stimulation of DRN and sleep–wake cycles. Typical workflow for preparing mice for mesoscopic imaging with GENIs, using g5-HT3.0 sensor as an example

and electromyography (EMG) recordings were performed to identify the sleep–wake state of mice.

The example workflow for mesoscopic imaging with a g5-HT3.0 sensor includes three main steps (Fig. 5). The first step involves AAV injection for expressing GENIs. To achieve uniform expression of the sensor, AAVs were injected using the n-SIM method [62] as described above. About 8 weeks later, the second step was performed—the surgery is applied to mice injected with AAVs expressing the g5-HT3.0 sensor. In order to optically activate serotonergic neurons in DRN, AAV (AAV9-EF1a-DIO-ChrimsonR) was injected into the DRN of *Sert-Cre* mice, and an optical fiber cannula was inserted for the delivery of light. Then a large craniotomy was made above the dorsal cortex to create an optical

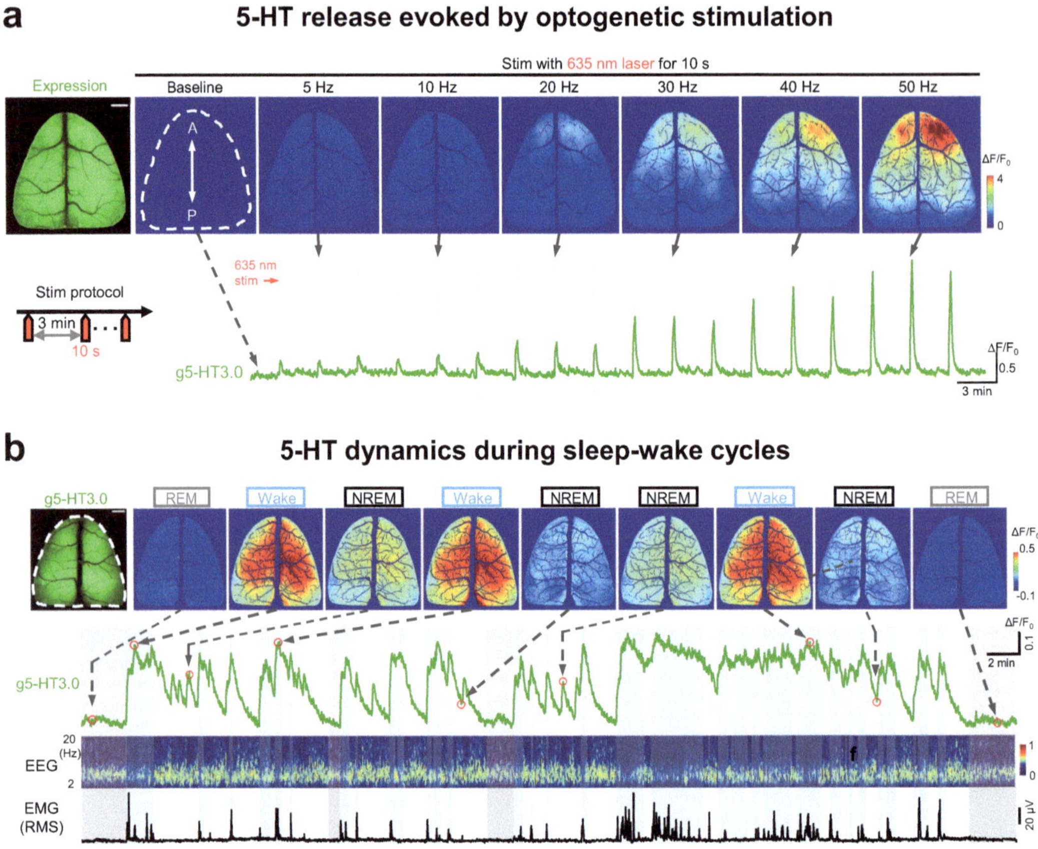

Fig. 6 Mesoscopic imaging of spatiotemporal 5-HT release in mouse dorsal cortex. (**a**) Example images showing the g5-HT3.0 expression and response (top); representative traces of averaged g5-HT3.0 fluorescence changes in the dorsal cortex (bottom). The dashed white outline indicates the ROI for traces analysis. Scale bar, 1 mm. (**b**) Representative fluorescence image of g5-HT3.0 and pseudocolor images of fluorescence changes during sleep–wake cycles (top); representative traces of averaged g5-HT3.0 signals, EEG and EMG signals. (Adapted from Deng et al. [50])

window for imaging most of the dorsal cortex. Finally, EEG/EMG electrodes were implanted for EEG/EMG recording and a metal chamber was affixed to the skull surface. After a week of recovery and habituation to head-fixed conditions, the third step experiments were implemented, including mesoscopic imaging of 5-HT release evoked by optogenetic stimulation and during sleep–wake cycles.

Activation of DRN serotonergic neurons elicited robust g5-HT3.0 signal across the dorsal neocortex and it exhibited a gradient pattern with decreased signals along the anterior-to-posterior axis (Fig. 6a). Meanwhile during the sleep–wake cycle, according to g5-HT3.0 signals, in the mouse dorsal cortex, the 5-HT level was highest during wakefulness and lowest during REM states; and it exhibited obvious oscillations during the NREM sleep state. In addition, signals in different brain regions were relatively homogenous (Fig. 6b) (*see* **Note 3**). Taken together, these results suggest that the 5-HT release has potential to be spatially heterogenous in the dorsal cortex, which is consistent with the heterogeneity of serotonergic projection density in the cortex [71]. But its release may be globally regulated during the sleep–wake cycle, which leads to relatively homogenous 5-HT release.

5 Notes

1. *Selection of a suitable GPCR scaffold* (In Subheading 2.2)

   It is crucial to choose an appropriate GPCR scaffold for the development of a sensor. If membrane trafficking is not good, sensors tend to remain in the cytosol, which hinders their activation by extracellular ligands and consequently leads to a relatively low response ($\Delta F/F_0$). To address this issue, sensor developers can explore various GPCR scaffolds derived from different subtypes or species. In addition, sensors based on different scaffolds may exhibit distinct responses ($\Delta F/F_0$) and affinities, which can provide the optimal sensor prototype.

2. *Site-directed random mutagenesis* (In Subheading 2.2)

   The site-directed random mutagenesis can be achieved by constructing a library using degenerate primers (e.g. base sequence "NNB" for the mutation site). Considering the existence of 20 different amino acids, it is advisable to screen approximately three times the number of possibilities, resulting in around 60 candidates, which accounts for the random mutations and ensures comprehensive coverage of most possibilities. Alternatively, the library can also be constructed by introducing mutations of 20 different amino acids one by one following PCR transfection for testing in cultured cells [85].

3. *Spatial uniformity of mesoscopic imaging* (In Subheadings 2.3, 3 and 4)

   To better interpret the sensor signals from the entire cortex, it is important to optimize the mesoscopic imaging system to achieve uniform imaging. On the one hand, the sensor should be expressed uniformly. The n-SIM method [62] and the utilization of transgenic mice have shown promise in achieving relatively uniform sensor expression. On the other hand, uniform illumination plays a vital role. The illumination uniformity can be assessed by employing uniform samples such as fluorescent plastic plates or white papers [75].

4. *Spectral unmixing in dual-color mesoscopic imaging* (In Subheading 3)

   For accurate dual-color imaging, it is essential to address the potential bleed-through in spectra. The dual-color imaging system can employ two independent channels and two sensors with different colors can be sequentially illuminated. Alternatively, if simultaneous imaging of both channels is a priority, two sensors can be excited simultaneously, followed by spectral unmixing in the data processing, such as by linear unmixing [50, 86].

5. *Hemodynamic correction* (In Subheading 3)

   The detected fluorescent signals are not only from ligand concentration-dependent sensor signals but also from other artifacts, particularly those arising from hemodynamics. To evaluate the performance of hemodynamic correction, a negative control can be used, such as membrane-tethered EGFP or a nonbinding mutant sensor, which should show no obvious fluctuations after the application of hemodynamic correction.

6 Summary and Outlook

In this chapter, we have overviewed the principle, development, and properties of GENIs, followed by the application of GENIs with mesoscopic imaging for monitoring NT/NM dynamics, specifically in a case study imaging 5-HT dynamics by the g5-HT3.0 sensor.

Progresses in sensor development will no doubt accelerate the mesoscopic imaging of NTs/NMs with a variety of GENIs. In the future, considering the requirements for mesoscopic imaging, GENIs may be developed and optimized in the following aspects (Fig. 7):

1. Developing more sensors for detecting more NTs/NMs. Based on the understanding of the GENI's principle and the generalizability of sensor developing strategies, a large repertoire of sensors can be developed for detecting more NTs/NMs to expand the toolbox of GENIs.
2. Improving sensitivities of GENIs. For mesoscopic imaging, sensitivity is quite important, especially for detecting NTs/NMs with minor changes. Actually, sensitivities of some GENIs for detecting vital NTs/NMs have been largely improved after iterative optimization, such as GENIs for glutamate [22, 23], DA [33–36], ACh [31, 32], and 5-HT [39, 50].
3. Expanding spectral profiles of GENIs. Most current GENIs are green fluorescent sensors, which precludes multiplexed imaging of multiple neurochemicals simultaneously due to spectral overlap. Some pioneering works have developed red-shifted GENIs, such as R-iGluSnFR1 for glutamate [24], $rGRAB_{DA1m}$, $rGRAB_{DA1h}$, and RdLight1 for DA [34, 36] and

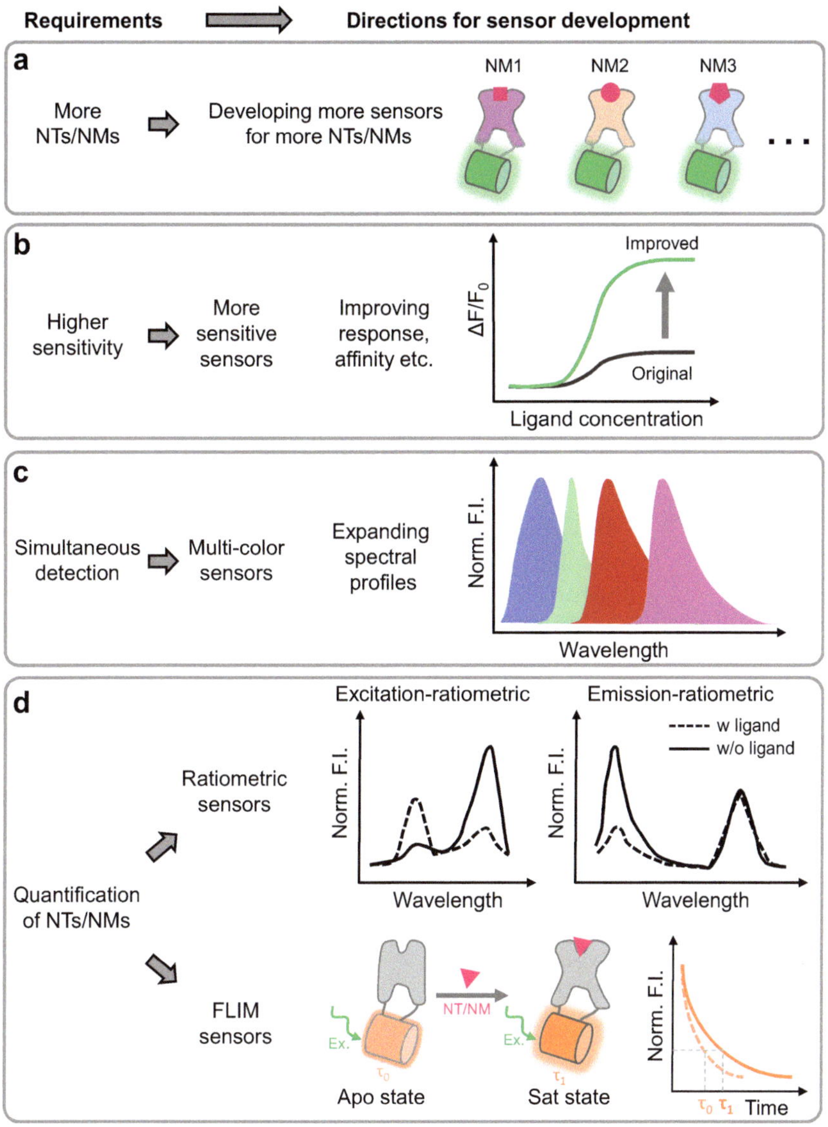

Fig. 7 Future directions for the development of GENIs. Schematic illustrating main directions for further development and optimization of GENIs for four different requirements of mesoscopic imaging. FLIM fluorescence lifetime imaging microscopy

rGRAB$_{5\text{-HT1.0}}$ for 5-HT [50]. Similar strategies can be used to develop red-shifted or even far-red/near-infrared GENIs.

4. Developing GENIs for quantitative imaging. Quantitative imaging is essential for measuring the concentration of NTs/NMs accurately, particularly with changes of tonic level. However, most GENIs are single FP-based and intensiometric. Although they can sensitively report the changes of NT/NM levels, they cannot perform quantitative measurements. At present, there are two kinds of sensors that are capable for

quantitative imaging, including ratiometric sensors and fluorescence lifetime imaging (FLIM)-based sensors. Ratiometric sensors, including excitation-ratiometric and emission-ratiometric sensors, can be excited or emit at distinct wavelengths with or without the ligand binding. Some genetically encoded ratiometric sensors have already been developed, such as some calcium sensors, GEX-GECO1 and GEM-GECO1 [87]. FLIM sensors can report NT/NM levels by fluorescence lifetime instead of fluorescence intensity, which may be ideal sensors for quantification [88], but there are only limited FLIM sensors, largely due to the challenges in sensor development and optimization. Nevertheless, it may be possible to quantitatively measure the dynamics of NTs/NMs in the entire dorsal neocortex by using ratiometric or FLIM GENIs in mesoscopic imaging.

Besides progresses in GENIs, advances in mesoscopic imaging techniques, such as increasing the resolution [89], developing 2-photon/3-photon, or wireless head-mounted mesoscopic imaging systems, will also facilitate the application of GENIs and better understanding of neurotransmission.

References

1. Pereda AE (2014) Electrical synapses and their functional interactions with chemical synapses. Nat Rev Neurosci 15(4):250–263. https://doi.org/10.1038/nrn3708
2. Dong C et al (2022) Fluorescence imaging of neural activity, neurochemical dynamics, and drug-specific receptor conformation with genetically encoded sensors. Annu Rev Neurosci 45(1). https://doi.org/10.1146/annurev-neuro-110520-031137
3. Wu Z, Lin D, Li Y (2022) Pushing the frontiers: tools for monitoring neurotransmitters and neuromodulators. Nat Rev Neurosci 23(5):257–274. https://doi.org/10.1038/s41583-022-00577-6
4. Lovinger DM (2010) Neurotransmitter roles in synaptic modulation, plasticity and learning in the dorsal striatum. Neuropharmacology 58(7):951–961. https://doi.org/10.1016/j.neuropharm.2010.01.008
5. Ma S et al (2018) Dual-transmitter systems regulating arousal, attention, learning and memory. Neurosci Biobehav Rev 85:21–33. https://doi.org/10.1016/j.neubiorev.2017.07.009
6. Lesch KP et al (1996) Association of anxiety-related traits with a polymorphism in the serotonin transporter gene regulatory region. Science 274(5292):1527–1531. https://doi.org/10.1126/science.274.5292.1527
7. Theodore WH, Juhasz C, Savic V, Drevets W (2005) Serotonin, depression, and epilepsy. Epilepsia 46:3–3
8. Li Y et al (2021) Synaptic mechanism underlying serotonin modulation of transition to cocaine addiction. Science 373(6560):1252–1256. https://doi.org/10.1126/science.abi9086
9. Sarter M, Bruno JP, Parikh V (2007) Abnormal neurotransmitter release underlying behavioral and cognitive disorders: toward concepts of dynamic and function-specific dysregulation. Neuropsychopharmacology 32(7):1452–1461. https://doi.org/10.1038/sj.npp.1301285
10. Higley MJ, Picciotto MR (2014) Neuromodulation by acetylcholine: examples from schizophrenia and depression. Curr Opin Neurobiol 29:88–95. https://doi.org/10.1016/j.conb.2014.06.004
11. Nitsch RM (1996) From acetylcholine to amyloid: neurotransmitters and the pathology of Alzheimer's disease. Neurodegeneration 5(4):477–482. https://doi.org/10.1006/neur.1996.0066
12. Lotharius J, Brundin P (2002) Pathogenesis of parkinson's disease: dopamine, vesicles and

α-synuclein. Nat Rev Neurosci 3(12): 932–942. https://doi.org/10.1038/nrn983

13. Wang H, Jing M, Li Y (2018) Lighting up the brain: genetically encoded fluorescent sensors for imaging neurotransmitters and neuromodulators. Curr Opin Neurobiol 50:171–178. https://doi.org/10.1016/j.conb.2018.03.010
14. Sabatini BL, Tian L (2020) Imaging neurotransmitter and neuromodulator dynamics in vivo with genetically encoded indicators. Neuron 108(1):17–32. https://doi.org/10.1016/j.neuron.2020.09.036
15. Brun MA et al (2011) Semisynthesis of fluorescent metabolite sensors on cell surfaces. J Am Chem Soc 133(40):16235–16242. https://doi.org/10.1021/ja206915m
16. Nguyen Q-T et al (2010) An in vivo biosensor for neurotransmitter release and in situ receptor activity. Nat Neurosci 13(1):127–132. https://doi.org/10.1038/nn.2469
17. Barnea G et al (2008) The genetic design of signaling cascades to record receptor activation. Proc Natl Acad Sci 105(1):64–69. https://doi.org/10.1073/pnas.0710487105
18. Cardin JA, Crair MC, Higley MJ (2020) Mesoscopic imaging: shining a wide light on large-scale neural dynamics. Neuron 108(1): 33–43. https://doi.org/10.1016/j.neuron.2020.09.031
19. Wang QX et al (2020) The Allen mouse brain common coordinate framework: a 3D reference atlas. Cell 181(4):936–953.e920. https://doi.org/10.1016/j.cell.2020.04.007
20. Dwyer MA, Hellinga HW (2004) Periplasmic binding proteins: a versatile superfamily for protein engineering. Curr Opin Struct Biol 14(4):495–504. https://doi.org/10.1016/j.sbi.2004.07.004
21. Okumoto S et al (2005) Detection of glutamate release from neurons by genetically encoded surface-displayed FRET nanosensors. Proc Natl Acad Sci 102(24): 8740–8745. https://doi.org/10.1073/pnas.0503274102
22. Marvin JS et al (2013) An optimized fluorescent probe for visualizing glutamate neurotransmission. Nat Methods 10(2):162–170. https://doi.org/10.1038/nmeth.2333
23. Marvin JS et al (2018) Stability, affinity, and chromatic variants of the glutamate sensor iGluSnFR. Nat Methods 15(11):936–939. https://doi.org/10.1038/s41592-018-0171-3
24. Wu JH et al (2018) Genetically encoded glutamate indicators with altered color and topology. ACS Chem Biol 13(7): 1832–1837. https://doi.org/10.1021/acschembio.7b01085
25. Marvin JS et al (2019) A genetically encoded fluorescent sensor for in vivo imaging of GABA. Nat Methods 16(8):763–770. https://doi.org/10.1038/s41592-019-0471-2
26. Lobas MA et al (2019) A genetically encoded single-wavelength sensor for imaging cytosolic and cell surface ATP. Nat Commun 10(1):711. https://doi.org/10.1038/s41467-019-08441-5
27. Borden PM et al (2020) A fast genetically encoded fluorescent sensor for faithful in vivo acetylcholine detection in mice, fish, worms and flies. bioRxiv:2020.2002.2007.939504. https://doi.org/10.1101/2020.02.07.939504
28. Unger EK et al (2020) Directed evolution of a selective and sensitive serotonin sensor via machine learning. Cell 183(7):1986–2002.e1926. https://doi.org/10.1016/j.cell.2020.11.040
29. Rosenbaum DM, Rasmussen SGF, Kobilka BK (2009) The structure and function of G-protein-coupled receptors. Nature 459(7245):356–363. https://doi.org/10.1038/nature08144
30. Latorraca NR, Venkatakrishnan AJ, Dror RO (2017) GPCR dynamics: structures in motion. Chem Rev 117(1):139–155. https://doi.org/10.1021/acs.chemrev.6b00177
31. Jing M et al (2018) A genetically encoded fluorescent acetylcholine indicator for in vitro and in vivo studies. Nat Biotechnol 36(8):726–737. https://doi.org/10.1038/nbt.4184
32. Jing M et al (2020) An optimized acetylcholine sensor for monitoring in vivo cholinergic activity. Nat Methods 17(11):1139–1146. https://doi.org/10.1038/s41592-020-0953-2
33. Sun F et al (2018) A genetically encoded fluorescent sensor enables rapid and specific detection of dopamine in flies, fish, and mice. Cell 174(2):481–496.e419. https://doi.org/10.1016/j.cell.2018.06.042
34. Sun F et al (2020) Next-generation GRAB sensors for monitoring dopaminergic activity in vivo. Nat Methods 17(11):1156–1166. https://doi.org/10.1038/s41592-020-00981-9

35. Patriarchi T et al (2018) Ultrafast neuronal imaging of dopamine dynamics with designed genetically encoded sensors. Science 360(6396):10.1126/science.aat4422
36. Patriarchi T et al (2020) An expanded palette of dopamine sensors for multiplex imaging in vivo. Nat Methods 17(11):1147–1155. https://doi.org/10.1038/s41592-020-0936-3
37. Feng J et al (2019) A genetically encoded fluorescent sensor for rapid and specific in vivo detection of norepinephrine. Neuron 102(4):745–761.e748. https://doi.org/10.1016/j.neuron.2019.02.037
38. Oe Y et al (2020) Distinct temporal integration of noradrenaline signaling by astrocytic second messengers during vigilance. Nat Commun 11(1). https://doi.org/10.1038/s41467-020-14378-x
39. Wan J et al (2021) A genetically encoded sensor for measuring serotonin dynamics. Nat Neurosci 24(5):746–752. https://doi.org/10.1038/s41593-021-00823-7
40. Dong C et al (2021) Psychedelic-inspired drug discovery using an engineered biosensor. Cell 184(10):2779–2792.e2718. https://doi.org/10.1016/j.cell.2021.03.043
41. Kubitschke M et al (2022) Next generation genetically encoded fluorescent sensors for serotonin. Nat Commun 13(1):7525. https://doi.org/10.1038/s41467-022-35200-w
42. Peng WL et al (2020) Regulation of sleep homeostasis mediator adenosine by basal forebrain glutamatergic neurons. Science 369(6508):1208. https://doi.org/10.1126/science.abb0556
43. Wu Z et al (2023) Neuronal activity-induced, equilibrative nucleoside transporter-dependent, somatodendritic adenosine release revealed by a GRAB sensor. Proc Natl Acad Sci 120(14). https://doi.org/10.1073/pnas.2212387120
44. Wu Z et al (2021) A sensitive GRAB sensor for detecting extracellular ATP in vitro and in vivo. Neuron 110:770–782.e775. https://doi.org/10.1016/j.neuron.2021.11.027
45. Dong H et al (2023) Genetically encoded sensors for measuring histamine release both in vitro and in vivo. Neuron 111(10):1564–1576.e1566. https://doi.org/10.1016/j.neuron.2023.02.024
46. Hannon J, Hoyer D (2008) Molecular biology of 5-HT receptors. Behav Brain Res 195(1):198–213. https://doi.org/10.1016/j.bbr.2008.03.020
47. Bunin MA, Wightman RM (1998) Quantitative evaluation of 5-hydroxytryptamine (serotonin) neuronal release and uptake: an investigation of extrasynaptic transmission. J Neurosci 18(13):4854–4860. https://doi.org/10.1523/jneurosci.18-13-04854.1998
48. Thorré K et al (1997) Differential effects of restraint stress on hippocampal 5-HT metabolism and extracellular levels of 5-HT in streptozotocin-diabetic rats. Brain Res 772(1):209–216. https://doi.org/10.1016/S0006-8993(97)00841-X
49. Hashemi P et al (2009) Voltammetric detection of 5-hydroxytryptamine release in the rat brain. Anal Chem 81(22):9462–9471. https://doi.org/10.1021/ac9018846
50. Deng F et al (2024) Improved green and red GRAB sensors for monitoring spatiotemporal serotonin release in vivo. Nat Methods 21(4):692–702. https://doi.org/10.1038/s41592-024-02188-8
51. Kremers G-J, Goedhart J, van Munster EB, Gadella TWJ (2006) Cyan and yellow super fluorescent proteins with improved brightness, protein folding, and FRET Förster radius. Biochemistry 45(21):6570–6580. https://doi.org/10.1021/bi0516273
52. Candelario J, Chachisvilis M (2012) Mechanical stress stimulates conformational changes in 5-hydroxytryptamine receptor 1B in bone cells. Cell Mol Bioeng 5(3):277–286. https://doi.org/10.1007/s12195-012-0232-0
53. Patriarchi T et al (2019) Imaging neuromodulators with high spatiotemporal resolution using genetically encoded indicators. Nat Protoc 14(12):3471–3505. https://doi.org/10.1038/s41596-019-0239-2
54. Bajar BT et al (2016) Improving brightness and photostability of green and red fluorescent proteins for live cell imaging and FRET reporting. Sci Rep-Uk 6(1):20889. https://doi.org/10.1038/srep20889
55. Pédelacq J-D et al (2006) Engineering and characterization of a superfolder green fluorescent protein. Nat Biotechnol 24(1):79–88. https://doi.org/10.1038/nbt1172
56. Kaplitt MG et al (1994) Long-term gene expression and phenotypic correction using adeno-associated virus vectors in the mammalian brain. Nat Genet 8(2):148–154. https://doi.org/10.1038/ng1094-148
57. Wang D, Tai PWL, Gao G (2019) Adeno-associated virus vector as a platform for gene therapy delivery. Nat Rev Drug Discov 18(5):358–378. https://doi.org/10.1038/s41573-019-0012-9

58. Nectow AR, Nestler EJ (2020) Viral tools for neuroscience. Nat Rev Neurosci 21(12): 669–681. https://doi.org/10.1038/s41583-020-00382-z
59. Chan KY et al (2017) Engineered AAVs for efficient noninvasive gene delivery to the central and peripheral nervous systems. Nat Neurosci 20(8):1172–1179. https://doi.org/10.1038/nn.4593
60. Michelson NJ, Vanni MP, Murphy TH (2019) Comparison between transgenic and AAV-PHP.eB-mediated expression of GCaMP6s using in vivo wide-field functional imaging of brain activity. Neurophotonics 6(2):ARTN 025014. https://doi.org/10.1117/1.NPh.6.2.025014
61. Foust KD et al (2009) Intravascular AAV9 preferentially targets neonatal neurons and adult astrocytes. Nat Biotechnol 27(1): 59–65. https://doi.org/10.1038/nbt.1515
62. Hamodi AS et al (2020) Transverse sinus injections drive robust whole-brain expression of transgenes. Elife 9:ARTN e53639. https://doi.org/10.7554/eLife.53639
63. Lohani S et al (2022) Spatiotemporally heterogeneous coordination of cholinergic and neocortical activity. Nat Neurosci 25(12): 1706–1713. https://doi.org/10.1038/s41593-022-01202-6
64. Madisen L et al (2015) Transgenic mice for intersectional targeting of neural sensors and effectors with high specificity and performance. Neuron 85(5):942–958. https://doi.org/10.1016/j.neuron.2015.02.022
65. Xie YC et al (2016) Resolution of high-frequency mesoscale intracortical maps using the genetically encoded glutamate sensor iGluSnFR. J Neurosci 36(4):1261–1272. https://doi.org/10.1523/Jneurosci.2744-15.2016
66. El Marjou F, Jouhanneau C, Krndija D (2021) Targeted transgenic mice using CRISPR/Cas9 technology. In: Ancelin K, Borensztein M (eds) Epigenetic reprogramming during mouse embryogenesis: methods and protocols. Springer US, New York, pp 125–141
67. Dong A et al (2021) A fluorescent sensor for spatiotemporally resolved imaging of endocannabinoid dynamics in vivo. Nat Biotechnol:787–798. https://doi.org/10.1038/s41587-021-01074-4
68. Feng J et al (2024) Monitoring norepinephrine release in vivo using next-generation $GRAB_{NE}$ sensors. Neuron 112:1930–1942.e6. https://doi.org/10.1016/j.neuron.2024.03.001
69. Labouesse MA, Patriarchi T (2021) A versatile GPCR toolkit to track in vivo neuromodulation: not a one-size-fits-all sensor. Neuropsychopharmacology 46(12): 2043–2047. https://doi.org/10.1038/s41386-021-00982-y
70. Soreq H, Seidman S (2001) Acetylcholinesterase – new roles for an old actor. Nat Rev Neurosci 2(4):294–302. https://doi.org/10.1038/35067589
71. Oh SW et al (2014) A mesoscale connectome of the mouse brain. Nature 508(7495): 207–214. https://doi.org/10.1038/nature13186
72. Rynes ML et al (2021) Miniaturized head-mounted microscope for whole-cortex mesoscale imaging in freely behaving mice. Nat Methods 18(4):417–425. https://doi.org/10.1038/s41592-021-01104-8
73. Ma Y et al (2016) Wide-field optical mapping of neural activity and brain haemodynamics: considerations and novel approaches. Philos Trans R Soc B: Biol Sci 371(1705): 20150360. https://doi.org/10.1098/rstb.2015.0360
74. Valley MT et al (2020) Separation of hemodynamic signals from GCaMP fluorescence measured with wide-field imaging. J Neurophysiol 123(1):356–366. https://doi.org/10.1152/jn.00304.2019
75. Couto J et al (2021) Chronic, cortex-wide imaging of specific cell populations during behavior. Nat Protoc 16:3241–3263. https://doi.org/10.1038/s41596-021-00527-z
76. Berger M, Gray JA, Roth BL (2009) The expanded biology of serotonin. Annu Rev Med 60:355–366. https://doi.org/10.1146/annurev.med.60.042307.110802
77. Li Y et al (2016) Serotonin neurons in the dorsal raphe nucleus encode reward signals. Nat Commun 7:10503. https://doi.org/10.1038/ncomms10503
78. Portas CM et al (1998) On-line detection of extracellular levels of serotonin in dorsal raphe nucleus and frontal cortex over the sleep/wake cycle in the freely moving rat. Neuroscience 83(3):807–814. https://doi.org/10.1016/S0306-4522(97)00438-7
79. Ferezou I et al (2007) Spatiotemporal dynamics of cortical sensorimotor integration in behaving mice. Neuron 56(5):907–923. https://doi.org/10.1016/j.neuron.2007.10.007
80. Vanni MP, Murphy TH (2014) Mesoscale transcranial spontaneous activity mapping in GCaMP3 transgenic mice reveals extensive

reciprocal connections between areas of Somatomotor cortex. J Neurosci 34(48): 15931–15946. https://doi.org/10.1523/jneurosci.1818-14.2014
81. Ishimura K et al (1988) Quantitative analysis of the distribution of serotonin-immunoreactive cell bodies in the mouse brain. Neurosci Lett 91(3):265–270. https://doi.org/10.1016/0304-3940(88)90691-X
82. Klapoetke NC et al (2014) Independent optical excitation of distinct neural populations. Nat Methods 11(3):338–346. https://doi.org/10.1038/nmeth.2836
83. Zhuang X et al (2005) Targeted gene expression in dopamine and serotonin neurons of the mouse brain. J Neurosci Methods 143(1):27–32. https://doi.org/10.1016/j.jneumeth.2004.09.020
84. Xu M et al (2015) Basal forebrain circuit for sleep-wake control. Nat Neurosci 18(11): 1641–1647. https://doi.org/10.1038/nn.4143
85. Villette V et al (2019) Ultrafast two-photon imaging of a high-gain voltage indicator in awake behaving mice. Cell 179(7): 1590–1608.e1523. https://doi.org/10.1016/j.cell.2019.11.004
86. Zimmermann T (2005) Spectral imaging and linear unmixing in light microscopy. Adv Biochem Eng Biotechnol 95:245–265. https://doi.org/10.1007/b102216
87. Zhao Y et al (2011) An expanded palette of genetically encoded Ca2+ indicators. Science 333(6051):1888–1891. https://doi.org/10.1126/science.1208592
88. Suhling K et al (2015) Fluorescence lifetime imaging (FLIM): basic concepts and some recent developments. Med Photonics 27:3–40. https://doi.org/10.1016/j.medpho.2014.12.001
89. Fan J et al (2019) Video-rate imaging of biological dynamics at centimetre scale and micrometre resolution. Nat Photonics 13(11):809–816. https://doi.org/10.1038/s41566-019-0474-7
90. Aggarwal A et al (2023) Glutamate indicators with improved activation kinetics and localization for imaging synaptic transmission. Nat Methods 20(6):925–934. https://doi.org/10.1038/s41592-023-01863-6
91. Helassa N et al (2018) Ultrafast glutamate sensors resolve high-frequency release at Schaffer collateral synapses. Proc Natl Acad Sci 115(21):5594–5599. https://doi.org/10.1073/pnas.1720648115
92. Nakamoto C et al (2021) A novel red fluorescence dopamine biosensor selectively detects dopamine in the presence of norepinephrine in vitro. Mol Brain 14(1):173. https://doi.org/10.1186/s13041-021-00882-8
93. Sheu S-H et al (2022) A serotonergic axon-cilium synapse drives nuclear signaling to alter chromatin accessibility. Cell 185(18): 3390–3407.e3318. https://doi.org/10.1016/j.cell.2022.07.026
94. Kitajima N et al (2020) Real-time in vivo imaging of extracellular ATP in the brain with a hybrid-type fluorescent sensor. elife 9. https://doi.org/10.7554/elife.57544
95. Abraham AD et al (2021) Release of endogenous dynorphin opioids in the prefrontal cortex disrupts cognition. Neuropsychopharmacology 46(13): 2330–2339. https://doi.org/10.1038/s41386-021-01168-2
96. Melzer S et al (2021) Bombesin-like peptide recruits disinhibitory cortical circuits and enhances fear memories. Cell 184(22): 5622–5634.e5625. https://doi.org/10.1016/j.cell.2021.09.013
97. Kroning KE, Wang W (2021) Designing a single protein-chain reporter for opioid detection at cellular resolution. Angew Chem Int Ed 60(24):13358–13365. https://doi.org/10.1002/anie.202101262
98. Duffet L et al (2022) A genetically encoded sensor for in vivo imaging of orexin neuropeptides. Nat Methods 19(2):231–241. https://doi.org/10.1038/s41592-021-01390-2
99. Ino D, Tanaka Y, Hibino H, Nishiyama M (2022) A fluorescent sensor for real-time measurement of extracellular oxytocin dynamics in the brain. Nat Methods 19(10): 1286–1294. https://doi.org/10.1038/s41592-022-01597-x
100. Qian T et al (2023) A genetically encoded sensor measures temporal oxytocin release from different neuronal compartments. Nat Biotechnol. https://doi.org/10.1038/s41587-022-01561-2
101. Wang H et al (2023) A tool kit of highly selective and sensitive genetically encoded neuropeptide sensors. Science 382(6672): eabq8173. https://doi.org/10.1126/science.abq8173
102. Duffet L et al (2023) Optical tools for visualizing and controlling human GLP-1 receptor activation with high spatiotemporal resolution. elife 12:RP86628. https://doi.org/10.7554/eLife.86628

Chapter 2

Analysis of Mesoscope Imaging Data

Yongxu Zhang and Shreya Saxena

Abstract

Mesoscope imaging enables the recording of neural activity projections in the dorsal cortex of behaving subjects through photon excitation and fluorescent indicators that measure intracellular calcium movements. Here, we introduce comprehensive methods for analyzing mesoscope imaging data.

This chapter begins with the essential preprocessing steps, including normalization, denoising, and hemodynamic correction. Various decomposition techniques to characterize the spatial and temporal information of mesoscope imaging data are covered, such as region of interest analysis, independent component analysis, singular vector decomposition, nonnegative matrix factorization, and localized semi-nonnegative matrix factorization. Additionally, this chapter explores spatiotemporal flow analysis methods to further understand the dynamics within mesoscope imaging data. The functional connectivity derived from mesoscope imaging data reveals the interaction between different brain regions. This chapter introduces key measurements of functional connectivity, including correlation, centrality, and Granger causality. Artificial intelligence models trained on neural activity and behavior can uncover the hidden connections between neural activity and behavior, while offering profound insights into cognitive and motor-related tasks. This chapter highlights effective behavior decoding models such as linear regression, support vector machines, and recurrent neural networks. Moreover, we discuss neural encoding studies that reveal how sensory information is translated into neural activity patterns in the brain. We end with open questions in the analysis of mesoscopic imaging data.

Key words Mesoscopic imaging, Decomposition, Dimensionality reduction, Spatiotemporal analysis, Modeling, Behavior

1 Introduction

Mesoscope imaging enables recording a projection of the activity in the dorsal neocortex in behaving subjects such as rodents. Photon excitation and fluorescent indicators are used to measure the movement of intracellular ions, e.g., calcium (Ca^{2+}). In neurons, calcium plays a dual role as an intracellular messenger, and its movement indirectly represents neural activity because of the storage and release of calcium during neural spiking. Measuring intracellular calcium concentration has long been a powerful method to monitor neural activity by taking advantage of optics which provides a

James E. Niemeyer et al. (eds.), *Awake Behaving Mesoscopic Brain Imaging*, Neuromethods, vol. 214,
https://doi.org/10.1007/978-1-0716-4120-0_2,

fast readout of the calcium concentration change. The neural activity is reflected through the change of intracellular calcium concentration. In this chapter, we introduce methods to analyze mesoscope imaging data with fluorescent-based calcium imaging, i.e., wide-field calcium imaging (WFCI), as an example. We start from preprocessing of the WFCI including normalization, denoising, and hemodynamic correction. Following preprocessing, WFCI is typically reduced to a lower dimensionality to enable more efficient downstream operations. Next, decomposition approaches are typically used to split the WFCI into spatial and temporal components. Here, we present region of interest (ROI), independent component analysis (ICA), singular vector decomposition (SVD), nonnegative matrix factorization (NMF), and localized seminonnegative matrix factorization (LocaNMF). Furthermore, methods of physical flow analysis can also be used to analyze the dynamics of WFCI.

The exploration of functional connectivity in WFCI helps illuminate how different regions of the brain communicate and interact with each other and uncovers the underlying network dynamics and information flow on fundamental brain processes such as perception, cognition, and behavior. Once the data is decomposed into spatial and temporal components, the functional connectivity between different brain regions can be generated by the relationship between Ca^{2+} activity in different regions. Here, we mainly introduce common measurements of functional connectivity, such as correlation, centrality, and causality, as well as discuss open challenges in this field.

With the advent of the era of artificial intelligence (AI), many hidden connections between neural activity and behavior in settings including but not limited to cognition and motor-related tasks have been revealed by training AI models using recorded neural activity. These models can also be trained to predict the behavior prior to its onset using sequential neural activity. Models that have been shown to be efficient in behavior decoding from neural activity include linear regression, support vector machines, and recurrent neural networks, which are trained to predict behaviors using sequential multiregional neural activity or its low-dimensional transformation as input. Furthermore, neural encoding studies provide a window into the process by which sensory information is translated into patterns of neural activity in the brain.

2 Preprocessing

Raw WFCI data consists of fluorescence values *F* from each pixel. Noise and error are common issues in recording imaging data, due to the background noise, optical aberrations, hemodynamic effect, and so on. In order to improve the quality of the recorded data, preprocessing is necessary to reduce the negative effects of the recording approaches on the data. We summarize the preprocessing as normalization, denoising, and hemodynamic correction.

2.1 Normalization

With raw WFCI data F, preprocessing includes normalization, denoising, and correction. Firstly, normalization of the original data is a fundamental preprocessing step, which is based on a "baseline" F_0, and the normalized data is obtained by $\Delta F/F_0$, where $\Delta F = F - F_0$ [1]. F_0 can be different representations in different experiments, although the mean-adjusted fluorescence value is usually taken. Indeed, different baselines may not result in a huge difference in results in practice; [1] reported no statistical differences in results between using mean pixel baseline and filter baseline. Altogether, the normalization process is to avoid the data having a large difference in scale and, therefore, to measure the change from a baseline.

2.2 Denoising

Denoising is a crucial preprocessing method which can be applied either before or after normalization. Noise usually comes from hemodynamics of neural activity and optical equipment [2]. Principal component analysis (PCA) is a widely used denoising method for imaging data [3]. For WFCI data, PCA is also efficient, e.g., as in [1, 4]. It identifies a linear transformation of principal components, also termed basis vectors, which are ordered by their corresponding eigenvalue from largest to smallest. The components that correspond to the smallest variances/eigenvalues often do not significantly impede an approximate reconstruction of the data. Furthermore, with the assumption that noise has a smaller variance than the signal, removing components with the smallest eigenvalues from the original data can help eliminate substantial noise. Likewise, many papers have combined the denoising process with other necessary steps, such as decomposition [5]. For example, in [6], the authors proposed a novel algorithm, called constrained nonnegative matrix factorization (CNMF), to simultaneously denoise, deconvolve, and demix WFCI; in [7] and [8], low-pass filters are used to denoise the WFCI, with the assumption that noise lies in the high-frequency part of the spectrum. Additionally, deep learning methods have been shown as powerful in denoising calcium imaging data. In [9], convolutional neural networks (CNNs) are trained to reconstruct high signal-to-noise ratio (SNR) fluorescent images from low SNR images. In other words, CNNs are trained to reconstruct denoised fluorescent images from original fluorescent images with noise; as a result, the trained CNN can be used for denoising.

2.3 Hemodynamic Correction

A slow fluorescent change in WFCI may be caused because of bleaching of fluorescence, and at the same time, photons are absorbed by hemoglobin broadly, with variations depending on its oxygenation state [10]. As a result, the excitation and emission of photons may be disrupted by fluctuations in total blood flow and blood oxygenation, which can result in distortion of the signal and lead to possible confounds [7]. Therefore, it is necessary to correct the WFCI to eliminate the effects described above. One capable method is a regression-based method described explicitly in [2] and [7], the process consists of applying a low-pass filter to a hemodynamic channel, e.g., 405 nm illumination channel in [2] and 488 nm standard blue channel in [7], following by subtracting this signal from the channel of GCaMP which is a genetically encoded fluorescent calcium indicator, and then the hemodynamic dependent signal is removed from the WFCI. Alternatively, [7] introduced a comprehensive technique for hemodynamic correction, wherein green and red photons with distinct absorption coefficients in oxygenated and deoxygenated hemoglobin are utilized for reflectance measurements, allowing for direct quantification of hemodynamic absorption. The proposed technique employs alternating frames of blue, green, and red exposures to capture emitted and reflected signals, which correspond to uncorrected calcium-dependent fluorescence, total hemoglobin, and reduced hemoglobin, respectively. The signals can then be utilized to create a hemodynamic absorption correction factor that is more precise. Nevertheless, hemodynamic correction is not ubiquitous in WFCI research, and [1] reported the unavailability of hemodynamic correction as the experiment only used single wavelength excitation.

3 Spatiotemporal Analysis

Typically, WFCI data has a high dimensionality. Following preprocessing, WFCI is reduced to low dimensionality, whereby pixels are consolidated into cohesive regions [7]. Here, we define the preprocessed WFCI as $X \in R^{N \times T}$, where N is the total number of WFCI pixels and T is the total WFCI time points. Researchers decomposed the WFCI into the multiplication of spatial components and temporal components.

$$X = AC \tag{1}$$

$$A = \{a_k(n)\},\ C = \{c_k(t)\} \tag{2}$$

where $A \in R^{N \times K}$ is the set of each spatial component $a_k(n)$, and $C \in R^{K \times T}$ is the set of each temporal component $c_k(t)$. The temporal components here are the low-dimensional sequential neural activity which can be used to decode behaviors. Many research groups have been actively engaged in developing methods for parcellation. Here in this section, the most popular methods are discussed.

3.1 Regions of Interest

Although many distinct parcellation techniques have shown their capability in capturing spatio–temporal information, the most common method is using an anatomical reference atlas, e.g., the Allen Institute CCFv3 [11]. The individual brain is aligned to the reference atlas, and then the cortex is subsequently partitioned into discrete subregions utilizing a combination of projection patterns and molecular markers. The alignment between the atlas and individual WFCI is by means of landmark alignment, for example, in [2] and [12], four anatomical landmarks were used: the left, center, and right points where the anterior cortex meets the olfactory bulbs and the medial point at the base of the retrosplenial cortex. Afterward, pixels are assigned to a particular subregion based on where they are aligned. The regions and subregions identified by the atlas are called regions of interest (ROIs). Then, low-dimensional temporal signals can be extracted by averaging across the corresponding temporal neural activity of each pixel within ROIs [8, 13]. For example, in one ROI k with N pixels, for each $n \in N$, we have $c_n(t)$ as temporal activity. Then, the temporal component of ROI k is

$$c_k(t) = \frac{1}{N} \sum_{n=1}^{N} c_n(t) \tag{3}$$

However, the accuracy of ROI for individual animals can be varying in different cases, and aligning individual brains precisely to a reference can be difficult, as well as the different resolutions between WFCI and the atlas may cause difficulty. Therefore, researchers developed other functional approaches that are discussed below (Fig. 1).

3.2 Independent Component Analysis

Independent Component Analysis (ICA) is a statistical technique widely used to extract underlying independent signals from a set of observed signals. The underlying principle of ICA is to separate a mixture of signals into their independent components $s = (s_1, s_2, \ldots, s_n)^T$ by assuming that the observed signals $x = (x_1, x_2, \ldots, x_m)^T$ are linear combinations of the independent sources

$$x = Ws \tag{4}$$

The minimization of mutual information and the maximization of non-Gaussianity are the two widest definitions of independence for ICA. For decomposition, either spatial components or temporal components can be independent; hence, two kinds of ICA are

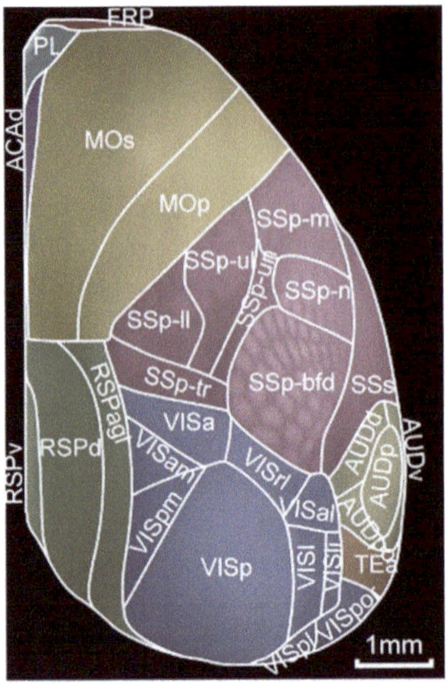

Fig. 1 CCFv3. FRP: Frontal pole, cerebral cortex; PL: Prelimbic area; ACAd: Anterior cingulate area, dorsal part; MOs: Secondary motor area; MOp: Primary motor area; SSp-m: Primary somatosensory area, mouth; SSp-ul: Primary somatosensory area, upper limb; SSp-n: Primary somatosensory area, nose; SSp-ll: Primary somatosensory area, lower limb; SSp-un: Primary somatosensory area, unassigned; SSs: Supplemental somatosensory area; SSp-tr: Primary somatosensory area, trunk; SSp-bfd: Primary somatosensory area, barrel field; AUDv: Ven tral auditory area; VISa: Anterior visual area; RSPd: Retrosplenial area, dorsal part; VISam: Anter omedial visual area; VISrl: Rostrolateral visual area; AUDd: Dorsal auditory area; AUDp: Primary auditory area; RSPagl: Retrosplenial area, lateral agranular part; VISpm: Posteromedial visual area; VISp: Primary visual area; VISal: Anterolateral visual area; AUDpo: Posterior auditory area; VISl: Lateral visual area; VISli: Laterointermediate area; TEa: Temporal association areas; VISpor: Postrhinal area; VISpl: Posterolateral visual area. Figure source: [11]

needed—spatial ICA that returns independent spatial components and temporal ICA that returns independent temporal components. Conventionally, the use of ICA relies on the spatial and temporal resolution of the data. When the data has lower amount of temporal than spatial information, in other words, $T \ll L \times W \times H$, spatial ICA (sICA) is the method that dominates the application of ICA and vice versa. However, in the context of experimenting with ICA, the primary consideration should be whether the signals are heavily dependent on either time or space. If the dependency of the signals is unclear, it is recommended to either not apply ICA or interpret the results carefully [14]. Figure 2 shows an example of spatial ICA on WFCI, each of the spatial components is independent to each other, which means no overlap. The corresponding temporal

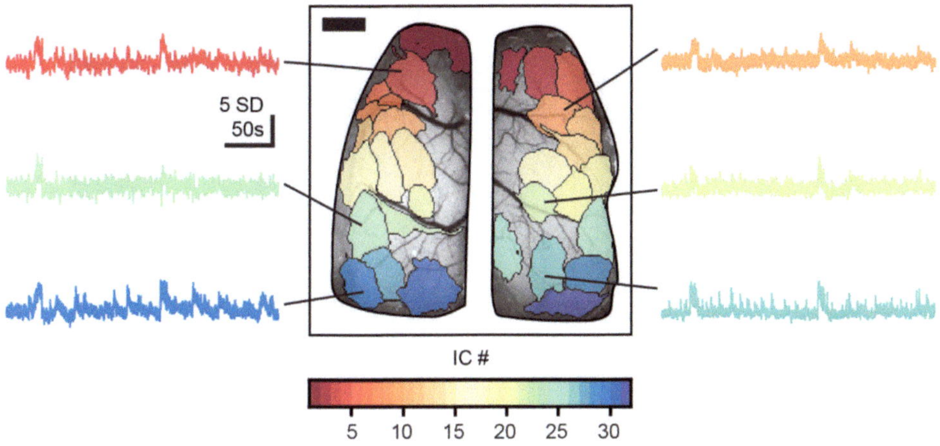

Fig. 2 Spatial independent components (ICs; colored regions) and time courses (colored traces) after independent component analysis of wide-field Ca^{2+} imaging data using the JADE algorithm superimposed on the cerebral cortical surface. Figure source: [5]

components measure the temporal patterns of neural activity associated with their independent spatial area.

3.3 Singular Value Decomposition

In order to decompose the data into temporal components and spatial components, PCA can be utilized by computing the SVD of the WFCI video.

$$X = USV^T \tag{5}$$

where U and V^T are two unitary matrices, and S is a rectangular diagonal matrix of singular values. The rows of V^T are the eigenvectors of the covariance matrix of X. Thereafter, dimensionality reduction can be performed by selecting the eigenvectors corresponding to the largest singular values and transforming the original WFCI based on the selected eigenvectors. The number of components retained is typically decided by the variance explained by them. However, it is difficult to automatically decide the number of principal components that should be retained in a given dataset, and the appropriate number of components can vary widely across different datasets [15]. The components obtained by PCA (SVD) are able to capture most of the original information and have fewer dimensions. To summarize, the original WFCI is decomposed into spatial components a_k and temporal components c_k:

$$\tilde{X} = \sum_{k=1}^{K} a_k c_k \tag{6}$$

$$\text{subject to } \tilde{X} \approx X \tag{7}$$

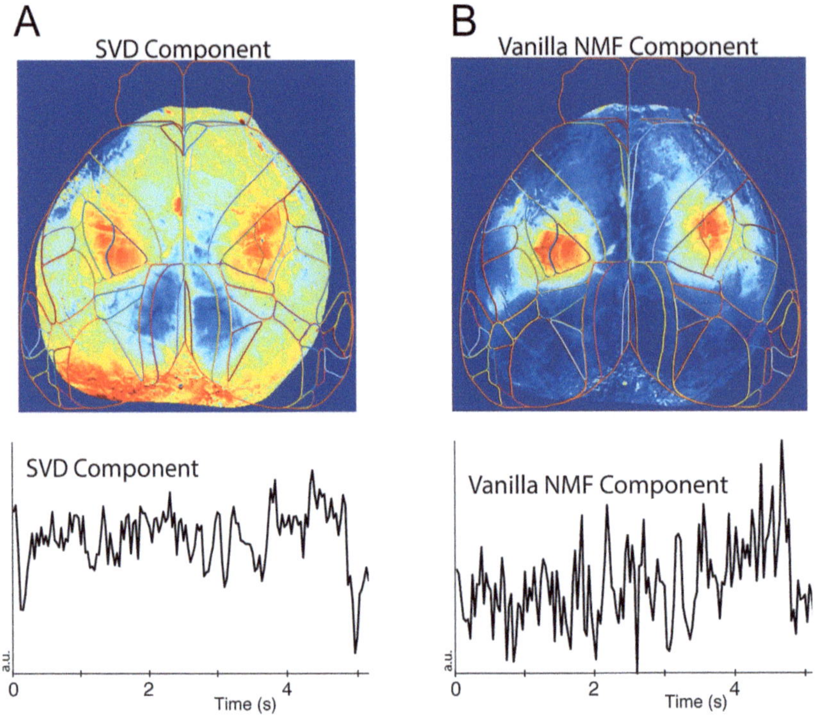

Fig. 3 (**a**) Example spatial and temporal components of SVD for WFCI in [2] and (**b**) example spatial and temporal components of NMF for WFCI in [2]

where K is the total number of selected components as described above. In addition, PCA (SVD) generates a linear transformation where the principal components are orthogonal. Other methodologies, such as Convolutional AutoEncoder (AE), may also exhibit similar capabilities to PCA. However, the transformation of Convolutional AE has nonlinearity, and it may be necessary to impose constraints on the orthogonality of the weights in the AE.

As one of the most concise spatiotemporal analysis methods, PCA (SVD) has a relatively easy application and can gather important information within several components. These components, however, are typically delocalized. Instead of exhibiting localization within well-defined brain regions, they contain information across the entire field of view, here, the dorsal cortex. For example, Fig. 3a shows an example of the spatial component from the SVD method in [2]; this component shows significant activation in multiple brain regions.

3.4 Nonnegative Matrix Factorization

Nonnegative matrix factorization (NMF) is a decomposition approach that optimizes a cost function that is similar to that used in SVD [2], with the addition of nonnegativity constraints on the components [16].

$$\tilde{X} = \sum_{k=1}^{K} a_k c_k \tag{8}$$

$$\text{subject to } \tilde{X} \approx X; \; a_k \geq 0; \; c_k \geq 0 \tag{9}$$

Unfortunately, numerous similar criticisms of PCA also apply to NMF, i.e., delocalization; hence, the NMF components are typically not interpretable. An example of NMF application is in Fig. 3b.

3.5 Localized Semi-nonnegative Matrix Factorization

Due to the delocalization of NMF and non-interpretability of its spatial components, a novel method was developed by Saxena et al. [2] where a localization constrain was added to NMF, termed "Localized Semi-nonnegative Matrix Factorization (LocaNMF)." It matches each of the spatial components to a well-defined brain region (an atlas), thereby imposing a constraint on each component a_k to exhibit sparsity in a highly specific manner, that is, to exhibit sparsity exclusively beyond the functional confines of the corresponding region. The authors used the Allen CCF atlas [17] to initialize the location of NMF spatial components and constrain the spatial components to be located close to their corresponding regions by including an appropriate penalization aimed at minimizing the sum of the square residual of the factorization. Moreover, in LocaNMF, the spatial components are limited to be nonnegative as in NMF. On the other hand, the temporal components can be negative since the mean-adjusted WFCI may be negative; thus, the method is termed semi-NMF. The equations of LocaNMF are as below:

$$min_{A,C} \; ||X - AC||^2 \tag{10}$$

$$\text{subject to } A \geq 0, \; ||a_k||_\infty = 1 \;\; \forall k \in [1, K] \tag{11}$$

$$\sum_{n=1}^{N} |d_k(n) a_k(n)|^2 \leq L_k \;\; \forall k \in [1, K] \tag{12}$$

where $||a_k||_\infty$ means $\max_n |a_k(n)|$, and $d_k(n)$ quantifies the smallest Euclidean distance from pixel n to the atlas region corresponding to component k. L_k is the threshold used to enforce localization.

Finally, the LocaNMF spatial components are constrained to their corresponding regions of the atlas and, therefore, have the property of localization. Examples of LocaNMF in [2] are in Fig. 4. As discussed above, LocaNMF results show more localized and interpretable spatial components; each spatial component can be

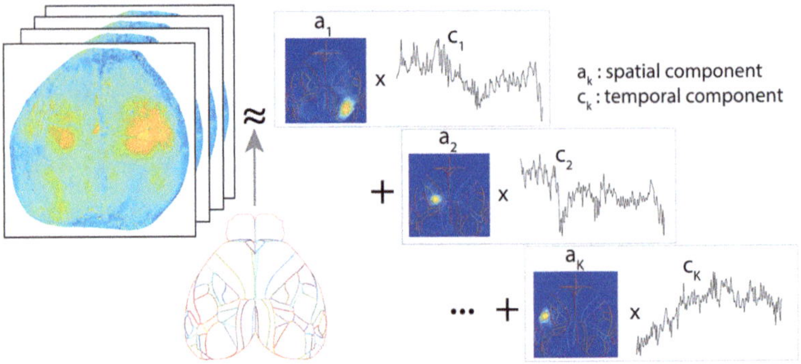

Fig. 4 LocaNMF decomposition of a WFCI into spatial components and temporal components, with the spatial components soft-aligned to an atlas. Figure source: [2]

explicitly explained as a component capturing the information of one brain region.

3.6 Spatiotemporal Flow

The similarity of spatial flows observed in WFCI to flows of physical fluids provides an opportunity to apply methods for fluid flow analysis on WFCI analysis, such as Lagrangian coherent structures (LCSs) that act as transport barriers in the flow and finite-time Lyapunov exponent (FTLE) whose ridge is used to visualize LCSs [18–20]. Recently, a visualization framework, FLOW (flow lines in optical wide-field imaging) portraits, was developed to capture the spatiotemporal dynamics of neural activity by extracting field lines in WFCI [21]. FLOW portraits are created by analyzing the temporal changes of frame-by-frame dynamics as time-varying optical flow vector fields. As a result, these portraits provide precise and easily understandable visual representations of WFCI. FLOW portraits directly transform the image stack into time-varying vector fields to extract discernible patterns of activity propagation within the data instead of approximating the data and explaining variance in the recordings, which decomposition methods usually do.

Figure 5 shows five steps to compute FLOW portrait. Starting with WFCI data preprocessed as $\Delta F/F$ (see Normalization), optical flow is used to convert the frame-by-frame changes in pixel intensity to a vector field. Next, the FTLE fields are computed in forward and backward time using an integration length. Consequently, ridges of these fields highlight coherent structures of the flow, and these ridges are used to compute the final FLOW portrait. The forward-time FTLE ridges (orange curves in Fig. 5e) highlight regions that repel flow, while the backward-time ridges (purple curves in Fig. 5e) show regions that attract activity.

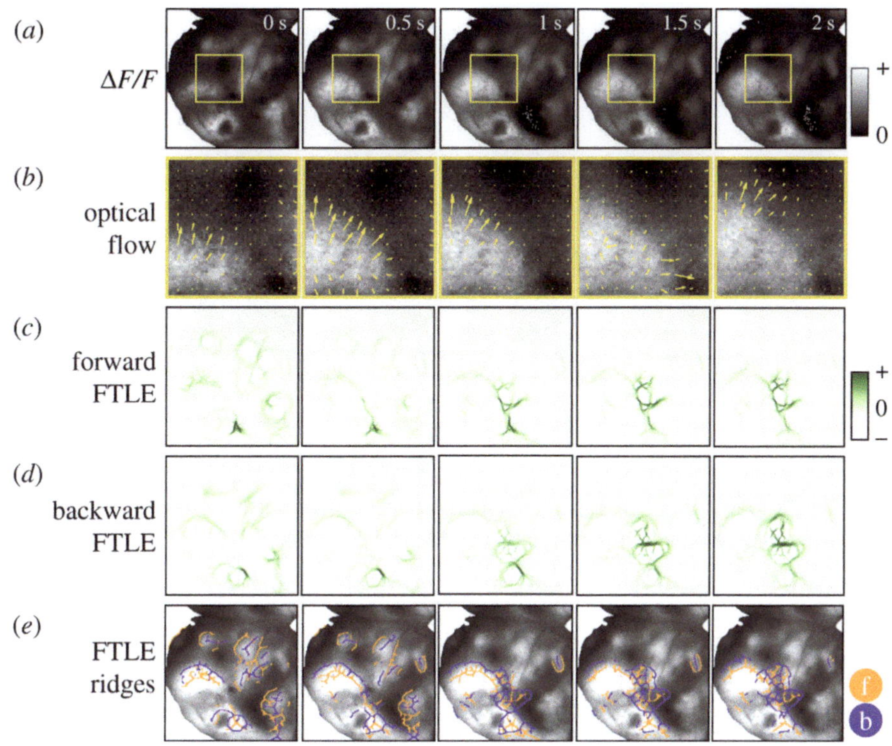

Fig. 5 Steps to compute a FLOW portrait. Figure source: [21]

4 Functional Connectivity

The exploration of functional connectivity in WFCI helps to understand how different regions of the brain communicate and interact with each other and uncover the underlying network dynamics and information flow on fundamental brain processes such as perception, cognition, and behavior. Once the data is decomposed into spatial and temporal components, the functional connectivity between different brain regions can be generated by the relationship between activity in different regions.

4.1 Correlation

Typically, pair-wise measure of the correlation between Ca^{2+} activity in regions is used to depict the functional connectivity map. For instance, Pearson correlation coefficient is commonly used. In practice, the temporal components of the data are used to compute the pair-wise correlation, such as ROI. The Pearson correlation coefficient $r_{k,k'}$ of ROI $c_k(t)$ and $c_{k'}(t)$ is

$$r_{k,k'} = \frac{\sum_{t=1}^{T}(c_k(t)-\bar{c_k})(c_{k'}(t)-\bar{c_{k'}})}{\sqrt{\sum_{t=1}^{T}(c_k(t)-\bar{c_k})^2\sum_{t=1}^{T}(c_{k\prime}(t)-\bar{c_{k\prime}})^2}} \quad (13)$$

where $\bar{c_k}$ and $\bar{c_{k'}}$ are mean values across time of c_k and $c_{k'}$.

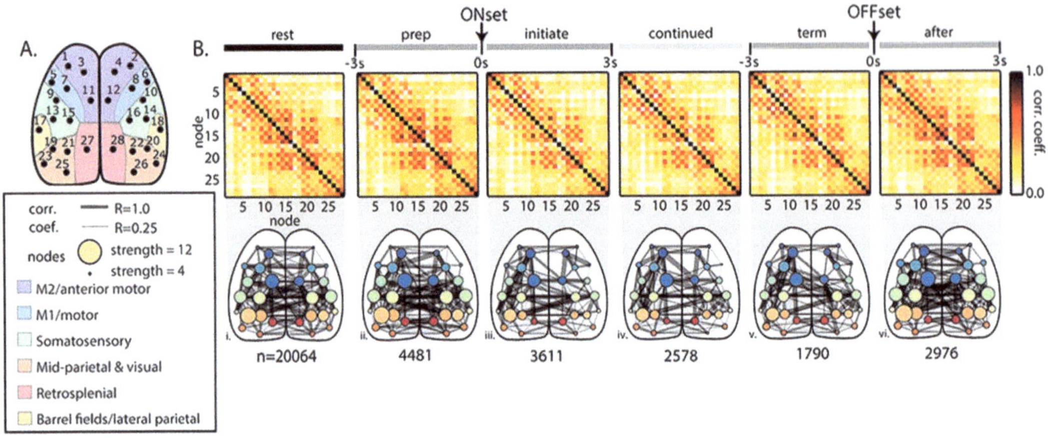

Fig. 6 (**a**) Diagram of nodes common to the catalog independent components across mice mapped onto the Common Coordinate Framework. (**b**) Top: Averaged correlation matrices between nodes across all mice. Bottom: Graphical representation of the correlation matrices. Figure source: [23]

Subsequently, graph theory analysis can be used to quantitatively assess the functional connectivity characteristics among distinct regions [22]. Brain regions are denoted as nodes within the graph, while functional connections are represented as lines or edges; an example is shown in Fig. 6. The graph uncovers the functional topology and structure of the network, where the statistical comparisons can be computed at both local and global levels.

Correlation-based measurement of functional connectivity is also used at different time periods to reveal the temporal dynamics of functional connectivity, such as [23] (*see* Fig. 6). Specifically, the correlation between independent temporal components of WFCI is computed within each 3-second period, thereby revealing distinct patterns of functional activity and their dissimilarity or similarity across time.

4.2 Centrality

In graph analytics, centrality is used to measure the importance of various nodes in a graph and, thus, identify the significant nodes. In functional connectivity analysis of WFCI, centrality provides a measure of how tightly connected the behavior of a component/node is to all other components/nodes in the network. One basic method of centrality calculation is

$$L_i^{-1} = \frac{n-1}{\sum_{j \in N, j \neq i} d_{ij}} \tag{14}$$

where d_{ij} is the shortest connection between node i and node j, and L is the characteristic path length which is the reciprocal of centrality [24]. More methods to compute the centrality are introduced in [24] and [25]. The change of centrality illustrates the domination of brain activity that is moved from one region/node to another [23].

4.3 Granger Causality

The Granger causality test is a statistical hypothesis test that aims to determine whether one time series is useful in forecasting another, thereby assessing the presence of causal relationships between them. In the analysis of WFCI, the Granger causalities between all nodes were calculated to estimate the directionality of the functional connectivity during different periods. There might be a tendency for specific nodes to exert a causal influence on others. Conversely, some nodes may receive input from many others. Granger causality analysis is usually performed by fitting a vector autoregressive (VAR) model to the time series. Specifically, we have temporal components $C(t)$ fitted to a VAR with L time lags:

$$C_j(t) = \sum_{\tau=1}^{L} B_\tau C_i(t-\tau) + \varepsilon \qquad (15)$$

where ε is a Gaussian random vector. If at least one of the elements B is significantly larger than zero in absolute value, the time series C_j and C_i have Granger causality. More measurement of Granger causality and examples can be found in [23] and [26].

The metrics discussed above are widely used; however, many other metrics of functional connectivity are also useful. For example, the measures of resilience reflect network vulnerability to insult. More metrics and their measurement can be found in [24]; most of them are inspired by functional magnetic resonance imaging (fMRI) and can also be applied on WFCI.

5 Relationship Between Mesoscope Imaging and Behavior

The phenomenon of behavior can be characterized as a dynamic and intricate process that entails a diversity of postural changes exhibited by organisms over time. Recent studies in the field of behavior decoding have demonstrated that behavior can be represented as a simplified construct. For example, certain decoding tasks involve classifying behaviors, such as determining right or left decisions in [27] or detecting the occurrence of a lever pull in [1] and [28]. In such studies, behavior is reduced to a discrete class that serves as a representation of the observed behavior. Moreover, the behavior also manifests as a sequence of movements, e.g., reach, grasp, and return movements in [29], thereby constituting a time-series signal.

5.1 Behavior Decoding

Decoding behavior through multiregional neural activity is helpful to uncover the corresponding relationships, and with this quantitative relationship, we can gain insights into how the brain processes information [30]. Behavior decoding from neural data can help us identify the ability of the data to explain different types of task-related and spontaneous activity. Practically, recorded brain neural activity is mapped to behavior using a combination of linear and nonlinear methods. Furthermore, predicting future behavior from neural recordings has the potential to perform corrective neural stimulation before the onset of behavior. The development of machine learning and deep learning approaches provides an opportunity to decode behavior from a large amount of complex neural activity using regression- and classification-based methods; the existing well-developed machine learning and deep learning Python-based libraries exhibit remarkable efficiency, e.g., scikit-learn, TensorFlow, and PyTorch.

5.1.1 Regression-Based Methods

A simple yet efficient and widely used approach is to use linear regression to decode temporally varying behavioral signals using neural imaging data recorded during the experiment. Specifically, this method connects the sequential behavior and neural activity by finding the linear relationship between them (*see* Eq. 16, where Υ is the behavior). Moreover, to increase expressive power in this relationship, f can also accommodate nonlinear transformations of input variables. Through behavior regression from neural activity, how changes in neural activity are associated with changes in behavior can be uncovered and neural correlates of behavior can be identified. Furthermore, regression analyses can predict behavior based on neural activity, which provides insights into the neural representations and computations involved in generating behavior.

$$\Upsilon(t) = f(C(t)) \tag{16}$$

Here, the behavior is typically continuous behavior of the subjects, such as body movements tracked via DeepLabCut (DLC) [31] and video frames of behavior or their low-dimensional latent [30]. An example of regression in behavior decoding is shown in Fig. 7 from [2], where sequences of paw positions recorded by DLC are linked to WFCI components.

5.1.2 Classification-Based Methods

Behavior classification using neural activity aims to find the difference between neural activities associated with distinct behaviors. By examining which neural patterns correspond to specific behaviors, we are able to discern the neural code underlying different behavioral states. This helps uncover the neural mechanisms that give rise to complex behaviors and cognitive processes, leading to insights into the underlying principles of brain operation. Meanwhile, decoding the ongoing behavioral states allows for behavior monitoring or prediction and understanding of the individual's actions, intentions, or cognitive states.

Machine Learning Methods

Models such as logistic decoders [2] are able to identify the difference between distinct neural activities. Logistic regression is a supervised learning algorithm that aims to predict the probability

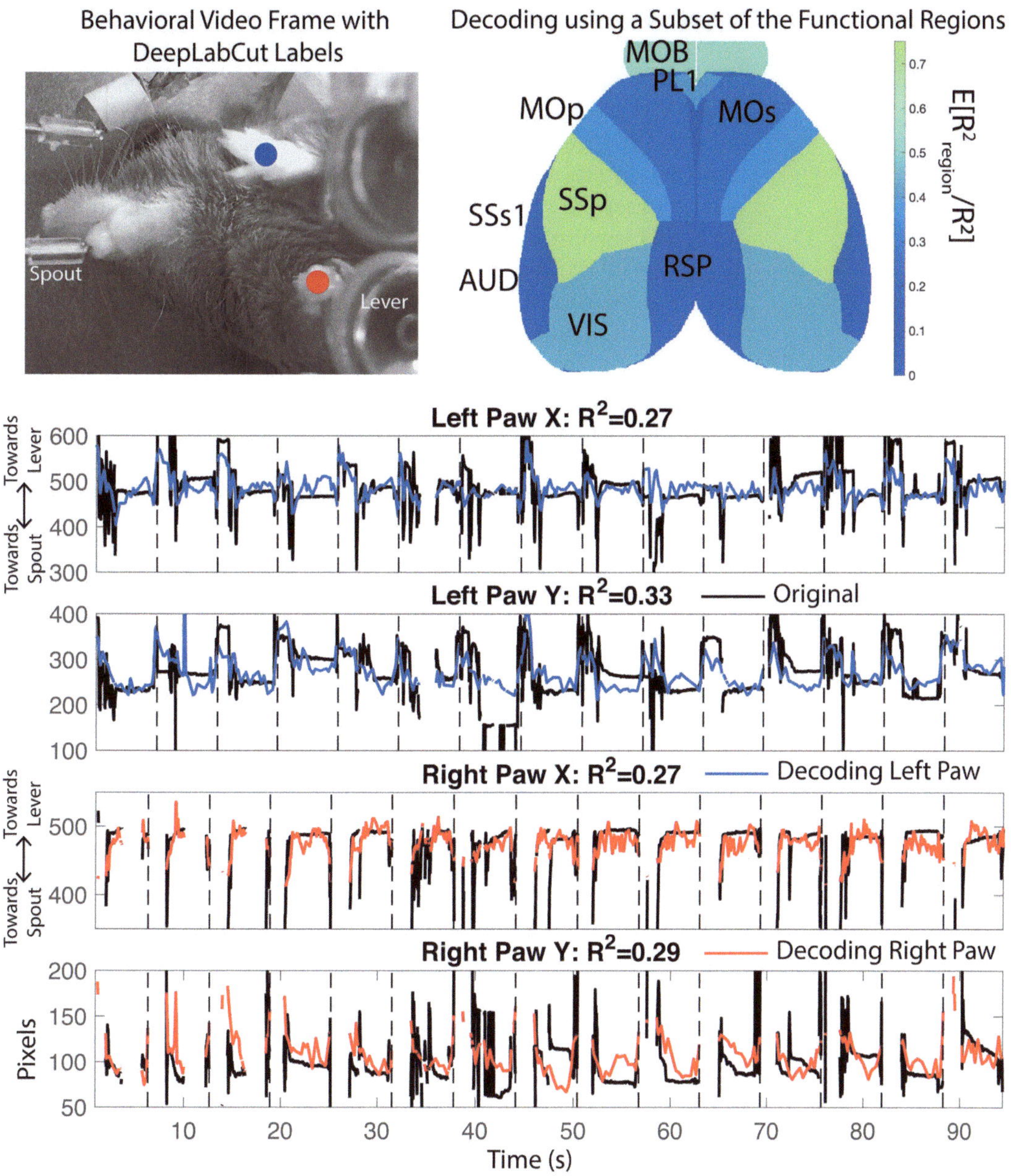

Fig. 7 Decoding paw position from WFCI signals. Figure source: [2]

of an instance belonging to a particular class or category. In the context of behavior classification, logistic regression is applied to neural activity data to determine the likelihood of a given behavioral state based on the observed neural patterns. In behavior classification, the neural activity may be complex, and thus the logistic decoder and other simple machine learning methods may not be able to identify the different neural activity efficiently. In this case, more powerful classifiers can be used to explore classification.

Support vector machine (SVM) is a widely adopted machine learning technique that is leveraged for the purposes of classifying;

it is a supervised learning model that separates the data into different classes by finding the best possible boundary or hyperplane that maximally separates the data points belonging to different classes. The SVM algorithm is particularly useful when dealing with complex and high-dimensional data which is challenging to be visualized and interpreted. Moreover, the nonlinearity of the data makes it hard to separate different classes, in which case, SVM is known for its kernel version which is able to handle nonlinear data classification. The SVM kernels first transform the nonlinear data into a higher dimensional space and then find a linear boundary that separates the classes.

Neural Networks

Recurrent neural networks (RNNs) are another set of common models for modeling sequence data. They process sequential inputs and generate predictions for the sequence class by leveraging recurrent hidden states that are able to retain a memory of previous inputs. Moreover, RNNs can convert their representations across time to adapt to the task, and thus, they perform well in classifying sequential data [32]. Typically, the decoding model of RNNs was built with time-series neural data $x \in^{R \times T}$ which is recorded and processed from R different brain regions and T time points as the input, with the outputs as the different types of behavior, e.g., a class of distinct behaviors and movement trajectories. Commonly, the activation function of the RNN layer is given by *tanh*, yet the activation function of the output layer depends on the behavior, e.g., *sigmoid* for classification. The following are the equations of the RNN network:

$$h_t = \tanh\left(W_h h_{t-1} + W_x x_t + b_h\right) \ \forall t \in [1, T] \tag{17}$$

$$y_t = \sigma(W_y h_t + b_y) \ \forall t \in [1, T] \tag{18}$$

where $W_x \in^{N \times R}$ means the input weights, $W_h \in^{N \times N}$ is the recurrent weights, and $W_y \in^{1 \times N}$ represents the output weights. $x_t \in^{R \times 1}$ is the neural data at time point t, $h_t \in^{N \times 1}$ is the value for the hidden units at time point t, y_t is the output of the dense layer, R and N are the number of brain regions and the number of RNN hidden units, respectively. Figure 8 shows the specific structure of unfolded

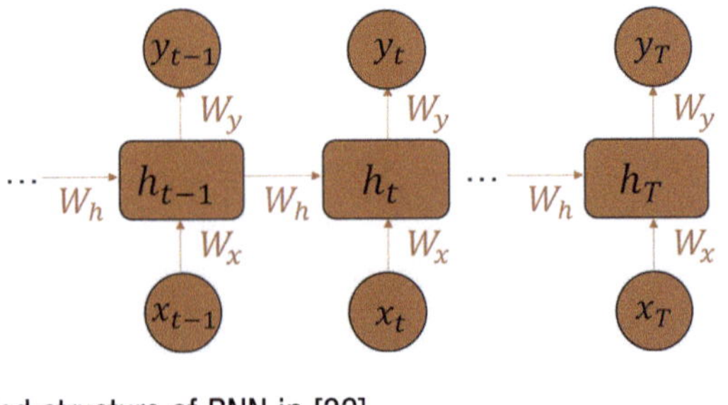

Fig. 8 Unfolded structure of RNN in [33]

RNNs. Backpropagation-through-time (BPTT) is widely used to train the RNNs with two commonly used loss functions: (a) the loss at the last output of RNNs (y_T) in order to focus on the prediction of the entire sequence and (b) the loss at all time steps of RNNs sequence ($\sum_t y_t$), which focuses on the aggregate performance of the RNNs.

5.2 Neural Encoding

We introduced behavior decoding in the previous section; however, the relationship between behavior and neural activity is not unidirectional, and neural activity can also be encoded by behavior, which has the potential to reveal how behavior affects neural activity. With this formulation, we can gain insights into the neural representations of different cognitive processes and their corresponding behavioral outcomes. Furthermore, neural encoding provides a means to explore the relationship between external stimuli, behavioral responses, and neural activity. By examining the patterns of neural responses associated with different behaviors, we can elucidate the neural codes that mediate the translation of outside stimulus into behavioral outputs.

$$X(t) = f(\Upsilon(t)) \tag{19}$$

Encoding models are able to connect the behavior or stimulus Υ to neural activity X.

By encoding the neural activity of different components/brain regions using behavioral inputs, researchers have recently found that movements dominate cortical activity in mice, and this effect is greater in some brain areas than others. An example of neural encoding is shown in Fig. 9. Here, neural activity is encoded by behavior of mice under different behavioral events through ridge regression, a form of regularized linear regression.

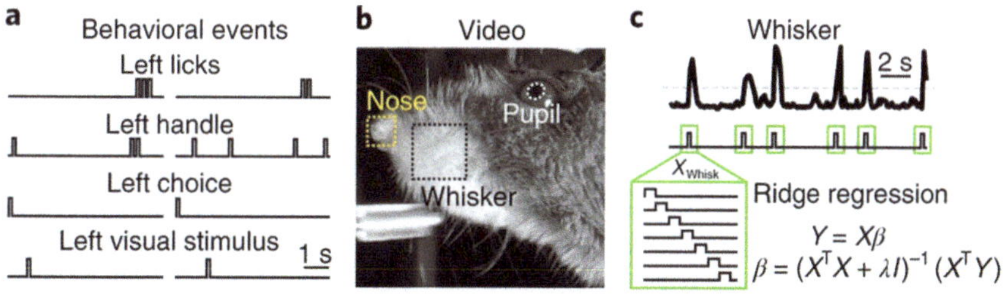

Fig. 9 Example of neural encoding using regression models. Figure source [12]

6 Future

6.1 Brain Connectivity

As based on correlation among spatial–temporal components of brain data, conventional approaches for measuring brain connectivity face challenges in adequately capturing its dynamic fluctuations, even when employing sliding window methodologies. This may cause difficulty in capturing underlying neural dynamics crucial to provide insights into behavior. Consequently, it is beneficial to develop methods to measure dynamic brain connectivity.

6.2 Relationship with Behavior

Cognitive and motor behavioral outputs are often under direct control of central nervous processing, and as such, it is possible to utilize neural stimulation techniques to regulate brain activity and modulate behavior. For example, undesired behaviors can be suppressed or redirected before occurring. To implement this strategy, it is necessary to measure the brain activity under distinct behavior and develop models of neural signals that correspond to different types of behaviors. Subsequently, stimuli and controllers can be designed to transform the neural signals associated with a particular behavior into those associated with a different behavior, effectively modifying the individual's behavior before it occurs. Moreover, the designed stimulus needs to be monitored and evaluated, and the behavior needs to be identified prior to the action of itself and stimulus. Mesoscope imaging has the potential to achieve identification of behavior and design of stimulus with its evaluation.

7 Summary

This chapter introduces and summarizes methods to analyze mesoscope imaging data and reveal the relationship with behavior. Mesoscope imaging data has emerged as a powerful approach to studying behavior because sequential neural activity stores the dynamic characteristic of the brain and multiregional neural activity measures different brain regions simultaneously. The utilization of single-photon excitation and fluorescent calcium indicators in the form of the wide-field calcium imaging (WFCI) technique has emerged as a robust and effective method to investigate neural functioning at the mesoscopic level during animal behaviors. This method enables the simultaneous measurement of the activity across extensive regions of the brain. However, because of its indirect measurement of neural activity, WFCI may not accurately reflect the underlying neural activity in brain, and the calcium indicators may cause pathological activity in the brain. Additionally, WFCI detects the signal from the cortical surface and may not be able to show the neural activity in the deeper brain. Decomposition-based methods have become an important step and have been widely studied: Methods such as ROI, SVD (PCA), NMF, LocaNMF, and ICA were developed to separate the spatial and temporal components, and the temporal components were then used to decode the behavior. Functional connectivity of WFCI enables the examination of interconnections and associations among distinct brain regions. Various methodologies, including correlation analysis, causality analysis, and centrality measures, have been designed to quantify diverse types of relationships between these brain regions.

Behavior and neural activity can be connected through machine learning models such as linear or logistic regression, recurrent neural networks, latent variable models, and others. Classical methods such as linear and logistic regression that map neural activity to behavior, while easy to implement, regard all temporal and spatial features as a huge space of features and may not be able to perform temporal decoding which is crucial in studying the dynamical characteristic of neural activity and behaviors (although training by sliding window offers a way to have temporal output). Dynamical systems such as recurrent neural networks are a promising approach toward behavior decoding; these models can reveal linear and nonlinear connections between neural activity and behavior, while also succinctly capturing neural dynamics.

Acknowledgements

We gratefully acknowledge the support from the NSF NCS 2219876.

References

1. Mitelut C, Zhang Y, Sekino Y, Boyd JD, Bollanos F, Swindale NV, Silasi G, Saxena S, Murphy TH (2022) Mesoscale cortex-wide neural dynamics predict self-initiated actions in mice several seconds prior to movement. Elife 11:e76506
2. Saxena S, Kinsella I, Musall S, Kim SH, Meszaros J, Thibodeaux DN, Kim C, Cunningham J, Hillman EM, Churchland A et al (2020) Localized semi-nonnegative matrix factorization (LocaNMF) of widefield calcium imaging data. PLoS Comput Biol 16(4):e1007791
3. Malik WQ, Schummers J, Sur M, Brown EN (2011) Denoising two-photon calcium imaging data. PLoS One6(6):e20490
4. Mukamel EA, Nimmerjahn A, Schnitzer MJ (2009) Automated analysis of cellular signals from large-scale calcium imaging data. Neuron 63(6):747–760
5. Nietz AK, Popa LS, Streng ML, Carter RE (2022) Wide-field calcium imaging of neuronal network dynamics in vivo. Biology 11(11): 1601
6. Pnevmatikakis EA, Soudry D, Gao Y, Machado TA, Merel J, Pfau D, Reardon T, Mu Y, Lacefield C, Yang W et al (2016) Simultaneous denoising, deconvolution, and demixing of calcium imaging data. Neuron 89(2):285–299
7. Cardin JA, Crair MC, Higley MJ (2020) Mesoscopic imaging: shining a wide light on large-scale neural dynamics. Neuron 108(1):33–43
8. Pinto L, Rajan K, DePasquale B, Thiberge SY, Tank DW, Brody CD (2019) Task-dependent changes in the large-scale dynamics and necessity of cortical regions. Neuron 104(4): 810–824
9. Chaudhary S, Moon S, Lu H (2022) Fast, efficient, and accurate neuro-imaging

denoising via supervised deep learning. Nat Commun 13(1):5165
10. Ma Y, Shaik MA, Kim SH, Kozberg MG, Thibodeaux DN, Zhao HT, Yu H, Hillman EM (2016) Wide-field optical mapping of neural activity and brain haemodynamics: considerations and novel approaches. Philos Trans Roy Soc B: Biol Sci 371(1705):20150360
11. Wang Q, Ding S-L, Li Y, Royall J, Feng D, Lesnar P, Graddis N, Naeemi M, Facer B, Ho A et al (2020) The Allen mouse brain common coordinate framework: a 3d reference atlas/ Cell 181(4):936–953
12. Musall S, Kaufman MT, Juavinett AL, Gluf S, Churchland AK (2019) Single-trial neural dynamics are dominated by richly varied movements. Nat Neurosci 22(10):1677–1686
13. Vanni MP, Chan AW, Balbi M, Silasi G, Murphy TH (2017) Mesoscale mapping of mouse cortex reveals frequency-dependent cycling between distinct macroscale functional modules. J Neurosci 37(31):7513–7533
14. Calhoun VD, Adali T, Pearlson G, Pekar JJ (2001) Spatial and temporal independent component analysis of functional MRI data containing a pair of task-related waveforms. Hum Brain Mapping 13(1):43–53
15. Buchanan EK, Kinsella I, Zhou D, Zhu R, Zhou P, Gerhard F, Ferrante J, Ma Y, Kim SH, Shaik MA et al (2018) Penalized matrix decomposition for denoising, compression, and improved demixing of functional imaging data. BioRxiv 334706
16. Liu J, Whiteway MR, Sheikhattar A, Butts DA, Babadi B, Kanold PO (2019) Parallel processing of sound dynamics across mouse auditory cortex via spatially patterned thalamic inputs and distinct areal intracortical circuits. Cell Rep 27(3):872–885
17. Lein ES, Hawrylycz MJ, Ao N, Ayres M, Bensinger A, Bernard A, Boe AF, Boguski MS, Brockway KS, Byrnes EJ et al (2007) Genome-wide atlas of gene expression in the adult mouse brain. Nature 445(7124): 168–176
18. Haller G (2002) Lagrangian coherent structures from approximate velocity data. Phys Fluids 14(6):1851–1861
19. Rossi V, Ser-Giacomi E, Monroy P, López C, Hernández-Garcıa E (2016) Lagrangian flow network: theory and applications. Transport 25(3):036404
20. Shadden SC, Lekien F, Marsden JE (2005) Definition and properties of Lagrangian coherent structures from finite-time Lyapunov exponents in two-dimensional aperiodic flows. Phys D: Nonlinear Phenom 212(3–4): 271–304
21. Linden NJ, Tabuena DR, Steinmetz NA, Moody WJ, Brunton SL, Brunton BW (2021) Go with the flow: visualizing spatiotemporal dynamics in optical widefield calcium imaging. J Roy Soc Interface 18(181):20210523
22. Bullmore E, Sporns O (2009) Complex brain networks: graph theoretical analysis of structural and functional systems. Nat Rev Neurosci 10(3):186–198
23. West SL, Aronson JD, Popa LS, Feller KD, Carter RE, Chiesl WM, Gerhart ML, Shekhar AC, Ghanbari L, Kodandaramaiah SB et al (2022) Wide-field calcium imaging of dynamic cortical networks during locomotion. Cerebral Cortex 32(12):2668–2687
24. Rubinov M, Sporns O (2010) Complex network measures of brain connectivity: uses and interpretations. Neuroimage 52(3): 1059–1069
25. Freeman LC et al (2002) Centrality in social networks: conceptual clarification. In: Social network: critical concepts in sociology, vol 1. Routledge, Londres, pp. 238–263
26. Barnett L, Seth AK (2014) The MVGC multivariate granger causality toolbox: a new approach to granger-causal inference. J Neurosc Methods 223:50–68
27. Soon CS, Brass M, Heinze H-J, Haynes J-D (2008) Unconscious determinants of free decisions in the human brain. Nat Neurosci 11(5): 543–545
28. Zhang Y, Mitelut C, Silasi G, Bolanos F, Swindale N, Murphy T, Saxena S (2021) Uncovering the effect of different brain regions on behavioral classification using recurrent neural networks. In: 2021 43rd Annual International Conference of the IEEE Engineering in Medicine & Biology Society (EMBC). IEEE, Piscataway, pp 6602–6607
29. Sani OG, Abbaspourazad H, Wong YT, Pesaran B, Shanechi MM (2021) Modeling behaviorally relevant neural dynamics enabled by preferential subspace identification. Nat Neurosci 24(1):140–149
30. Batty E, Whiteway M, Saxena S, Biderman D, Abe T, Musall S, Gillis W, Markowitz J, Churchland A, Cunningham JP et al (2019) BehaveNet: nonlinear embedding and Bayesian neural decoding of behavioral videos. In: Advances in neural information processing systems, vol 32
31. Mathis A, Mamidanna P, Cury KM, Abe T, Murthy VN, Mathis MW, Bethge M (2018) DeepLabCut: markerless pose estimation of

user-defined body parts with deep learning. Nat Neurosci 21(9):1281–1289

32. Farrell M, Recanatesi S, Moore T, Lajoie G, Shea-Brown E (2022) Gradient-based learning drives robust representations in recurrent neural networks by balancing compression and expansion. Nat Mach Intell 4(6):564–573

33. Zhang Y, Saxena S (2022) Behavioral classification of sequential neural activity using time varying recurrent neural networks. In: NeurIPS 2022 Workshop on Robustness in Sequence Modeling

Chapter 3

Real-Time Ultra-Large-Scale Imaging with High-Resolution Microscopy

Qionghai Dai, Hao Xie, and Jiamin Wu

Abstract

This chapter presents a novel microscopy technique, real-time ultra-large scale imaging with high-resolution microscopy (RUSH), which enables imaging of biological systems with unprecedented space–bandwidth product and data throughput. We describe the design and implementation of the RUSH system, which comprises a customized objective lens, a camera array, and a computer cluster. We also demonstrate the applications of the RUSH system in imaging neural activity in mouse brains and other biological samples with curved surfaces. Further, we introduce algorithms for image denoising and neural signal detection, and discuss the challenges and opportunities for large-scale imaging. The chapter aims to provide readers with an overview of the RUSH system and its potential for advancing system biology studies.

Key words Mesoscale, Computational imaging, Camera array, Large-scale neural recording, Denoising, Light-field

1 Introduction

Due to the intricate and constantly changing nature of living systems, predicting systematic behaviors solely from the properties of individual components is often challenging [1]. To investigate the biology of a system, such as neural network activity across cortical regions [2], leucocyte trafficking dynamics [3], or tumor metastasis [4], an appropriate microscope is required, which possesses a field of view (FOV) of at least 1-mm, sub-cellular resolution, and the ability to record biological processes in real-time.

Attaining such a system presents some fundamental challenges. Firstly, it is necessary to develop an optical system with a high space-bandwidth product (SBP), which combines high resolution and field of view (FOV) to capture the optical signal [5, 6]. Additionally, a data acquisition system with a high data throughput—characterized by a large pixel count and frame rate—must be implemented [7]. To manage the high data throughput, a data transfer and

James E. Niemeyer et al. (eds.), *Awake Behaving Mesoscopic Brain Imaging*, Neuromethods, vol. 214,
https://doi.org/10.1007/978-1-0716-4120-0_3,

storage system with adequate capacity is required. Moreover, when the FOV is extensive, the depth of field must be expanded to accommodate the curvature of the mouse brain [8]. Developing high-efficiency signal processing algorithms is also critical for managing the enormous data throughput, thus fast and highly accurate algorithms are required [9]. Finally, it is crucial to maintain low light exposure during long-term recordings to avoid photobleaching. Therefore, the signal-to-noise level of the images becomes a severe issue, and it is necessary to develop a universal denoise algorithm that can accurately reconstruct the raw image in various scenarios.

This chapter is structured into six sections. The first section serves as an introduction to the topic. The second section presents the real-time ultra-large-scale imaging with high-resolution microscopy (RUSH) approach [10], which involves using a customized microscope and a camera array to achieve ultra-large scale and high-resolution imaging. The third section describes the attempt to simultaneously monitor neural activity in the superficial cortex and hippocampus, and then the meso-scale imaging is extended to a curved surface. The fourth and fifth sections introduce algorithms for image denoising and neural detection, respectively. Finally, the limitations and future directions for large-scale imaging are discussed.

2 Real-Time Ultra-Large-Scale Imaging with High-Resolution Microscopy (RUSH)

The SBP of microscopes is fundamentally restricted by scale-dependent geometric aberrations of optical systems, leading to a compromise between achievable resolution and FOV. To overcome this limitation, two approaches have been developed: image stitching, in which the whole FOV is divided into patches and imaged sequentially [11–14], and Fourier ptychography, in which the specimen is imaged under various illumination conditions under a low-resolution system, and the final image is obtained via Fourier-domain post-processing [15, 16]. Unfortunately, both approaches suffer from low speed and are incapable of supporting single-shot, whole-FOV information acquisition, which is essential for imaging biological dynamics. Additionally, these methods require the samples to remain static during the entire imaging process, making them challenging to apply to in vivo imaging of awake, behaving animals. To address the above challenges, we have proposed the real-time ultra-large-scale imaging with high-resolution microcopy (RUSH), as shown in Fig. 1.

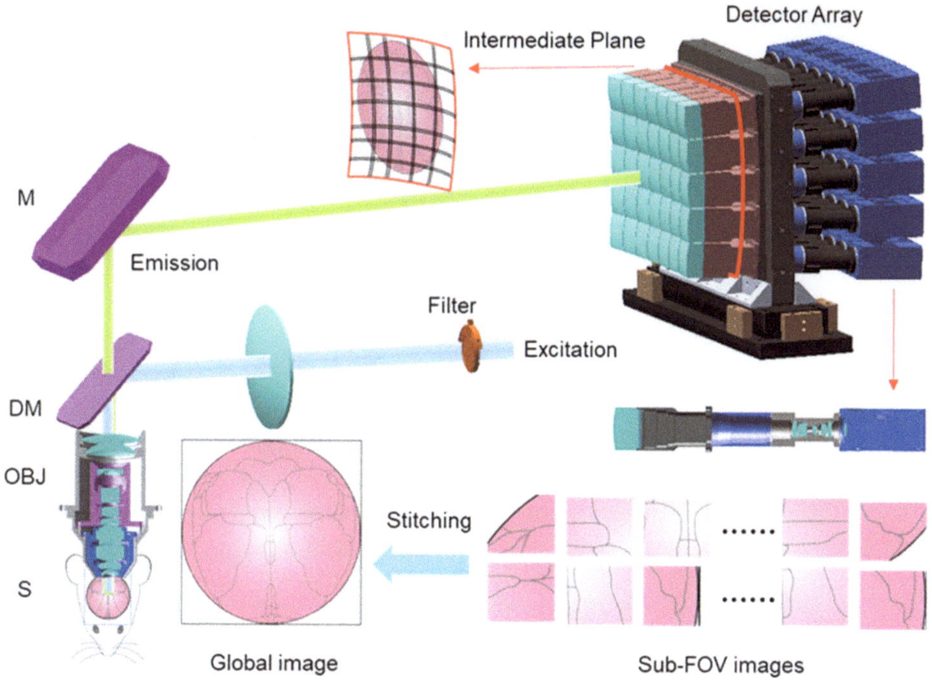

Fig. 1 Schematic of the RUSH system. In fluorescence imaging mode, the excitation beam from the light source is filtered with an excitation filter and reflected by the dichroic mirror (DM), before it passes through the customized objective and is projected onto biological samples. The fluorescence is collected by the same objective, filtered by the DM and an emission filter, reflected by a mirror (M), and forms a spherical intermediate image. The spherical field is divided into 5 × 7 sub-FOVs and imaged with corresponding collection units composed of the relay lens array and 35 sCMOS cameras. The customized objective lens is designed with 0.35 NA and 10 × 12 mm^2 FOV; the collection units provide a data throughput of up to 5.1 gigapixels per second. (The figure is reproduced from Fan et al. [10])

2.1 Optical Design

The large FOV of RUSH is realized by a customized objective and a camera array. The customized objective used in the RUSH system was composed of 14 optical elements in 11 groups, and was 161 mm in diameter and 280 mm in length. It was designed to achieve a 10 mm × 12 mm FOV and 0.35 NA over the wavelength range 420–680 nm. The entire optical system had a magnification factor of ×8 and an off-axis and spacing tolerance of ~2 μm. For wide-field fluorescence imaging, a light-emitting diode light source was used for excitation, with a bandpass excitation filter at 470 ± 20 nm and an emission filter at 525 ± 20 nm, chosen based on the fluorescence dyes or proteins used in the imaging. The excitation beam from the light source is filtered with an excitation filter and reflected by the dichroic mirror (DM), before it passes through the customized objective and is projected onto biological samples. The fluorescence is collected by the same objective, filtered by the DM and an emission filter, reflected by a mirror

(M), and forms a spherical intermediate image. The spherical field is divided into 5 × 7 sub-FOVs and imaged with corresponding collection units composed of the relay lens array and 35 water-cooled sCMOS cameras with 2560 × 2160 pixels and 16-bit depth. The customized objective lens is designed with 0.35 NA and 10 × 12 mm^2 FOV; the collection units provide a data throughput of up to 5.1 gigapixels per second.

2.2 Image Stitching

The alignment of sub-FOVs is crucial for seamless stitching and requires high-accuracy calibration. We have proposed a fast calibration approach using spatiotemporal structured illumination to build homographic matrices that map images from 35 cameras to the 10 × 12 mm^2 FOV, which provides sub-pixel calibration precision and feedback for mounting the camera array. During mounting, the displacement between the captured and expected mappings is calculated, and mounting parameters are fine-tuned accordingly. The proposed approach is computationally efficient and can elegantly address the challenges of assembly and calibration, as shown in Fig. 2b.

First, we set the global physical coordinates using the temporal-division coding of different cells on a high-density LCD module. Specifically, 12-bit temporal codes are used for x and y coordinates,

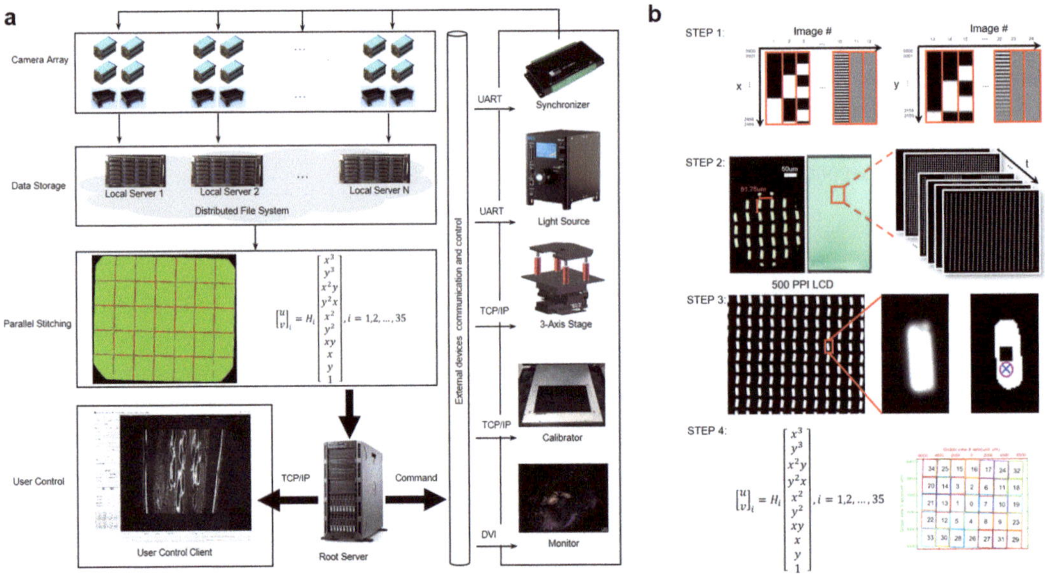

Fig. 2 Diagram for the data transmission, storage, and image calibration. (**a**) Our system involves a root server that controls cameras, computing cluster, user client, synchronizer, light source, 3-axis stage, calibrator, and monitor. Firstly, 35 cameras capture subFOV images in sync using a customized protocol. The high-bandwidth data is then sent through a DFS for processing. A parallel stitching algorithm is applied using calibrated homographic matrices for real-time display. Users can adjust exposure time and illumination intensity through the user control client. We also calibrate the local-to-global map table to enable seamless sub-FOV stitching. (**b**) Steps to coordinate calibration. (Figures are adapted from Fan et al. [10])

respectively. Second, each sub-FOV camera captures an image sequence of one specific small LCD region. Then the corresponding global position of each LCD cell can be decoded from its temporal pattern. For each cell in the sub-FOV image, we further extract its precise local image coordinates using center-of-area method, including binarization and weighted averaging. After retrieving the global and local coordinates of all the sub-FOVs, we build a set of homographic matrices for successive stitching.

2.3 Data Transfer and Storage

We have designed a computer cluster with nine nodes to handle the large amount of data generated by the RUSH system, as shown in Fig. 2a. The data from the 35 cameras were captured synchronously and continuously with a maximum frame rate of 30 frames per second. The data were then transmitted via dual-camera links to optical cables, with a bandwidth of 5.1 gigapixels per second. The system used a distributed file system with 324 hard disks to schedule local data writing requests to each node and to provide cross-node data exchange capability. This facilitated a highly parallel distributed stitching algorithm with adjustable granularity, which allowed real-time online stitching with adaptive resolution adjustment according to the regions of interest. The entire gigapixel video could be stitched and stored offline at up to 1.1 frames per second. A graphical user interface was developed to provide online flexible visualization of the multi-scale data with two high-definition displays.

We use a root server to control the cameras, the computing cluster, the user client and other external devices including the synchronizer, light source, 3-axis stage, calibrator, and the monitor. Firstly, 35 cameras are synchronized to capture the sub-FOV images based on our customized protocol. The high-bandwidth data is then transmitted through a DFS for data processing. Parallel stitching algorithm is conducted based on the calibrated homographic matrices for real-time display. Users can specify the parameters such as exposure time and illumination intensity by the user control client.

2.4 Ex Vivo Cellular Imaging

In order to demonstrate the efficacy of our RUSH macroscope for large-scale, high-resolution, and high-throughput imaging of biological dynamics, we conducted functional imaging of large cell ensembles. While fluorescence imaging of the cardiac cellular network is essential for gaining detailed insights into the cellular processes that are fundamental to cardiac function, it has previously been limited to small fields of view (FOVs) or low resolution. We cultivated rat primary cardiomyocytes and captured intercellular calcium dynamics at a rate of 30 Hz. Figure 3a presents the spatio-temporal propagation of a calcium wave across the cardiac cell ensemble, with wave phases depicted using different colors that correspond to different temporal delays. Enlarged views of

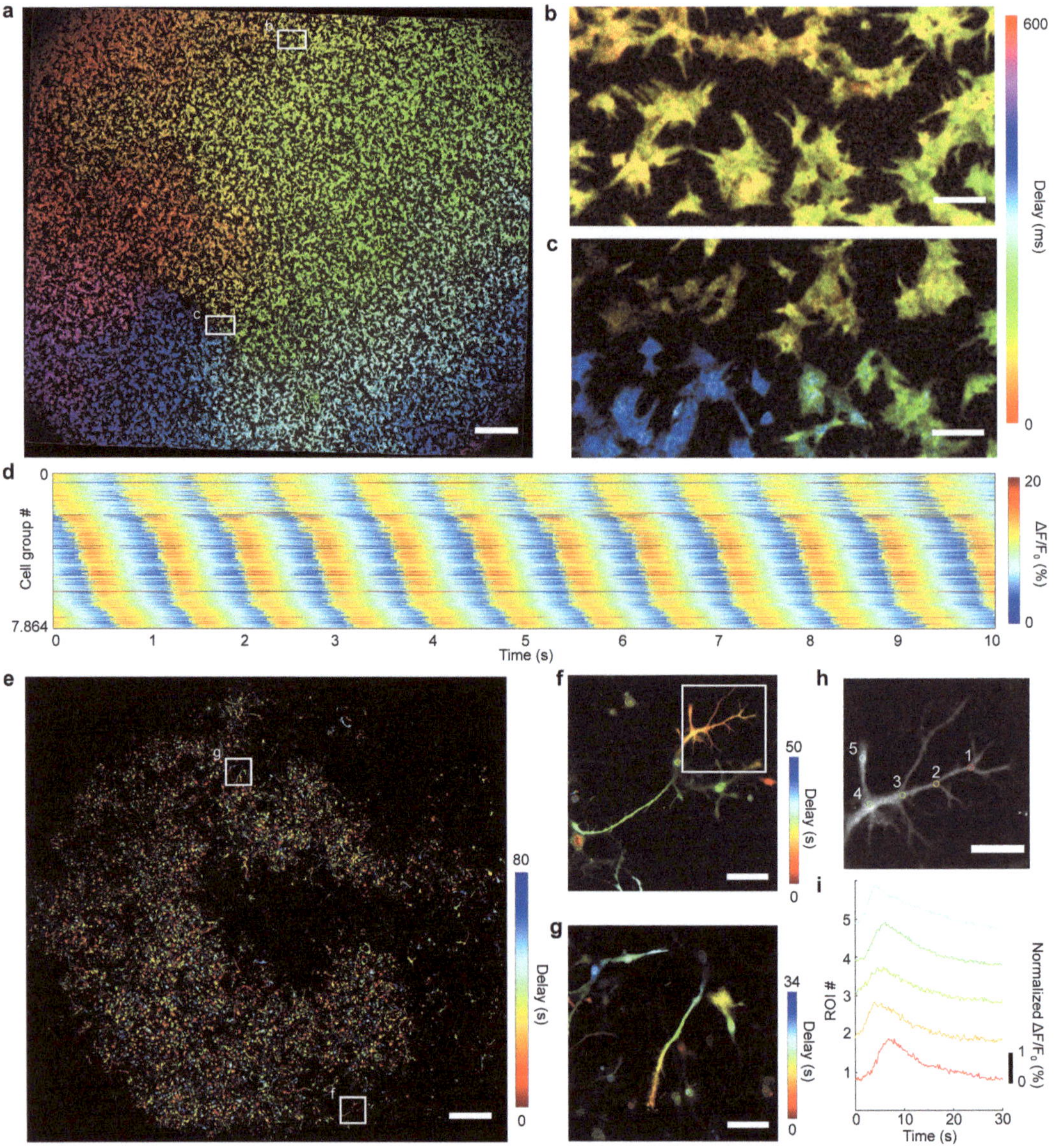

Fig. 3 High-throughput calcium imaging of neuron ensembles. (**a**–**c**) Color-coded spatiotemporal projection of calcium signals from cultured rat cardiomyocytes (**a**), with enlarged views shown in (**b**) and (**c**). Different colors visualize the peak instants of the calcium signal in one period, and the intensity corresponds to the standard deviation of the fluorescence signal. (**d**) Calcium intensity traces ($\Delta F/F_0$) of all segmented cell groups, indexed by their temporal delays. (**e**–**g**) Color-coded spatiotemporal projection of the calcium signals from cultured neuron ensembles (**e**), with enlarged views shown in (**f**) and (**g**). Colors and intensities are defined in the same way as in (**a**). (**h**) Maximum intensity projection image of the outlined region in (**f**). (**i**) Fluorescence signals of the numbered regions in (**h**). Scale bars, 1000 μm (**a**, **e**), 100 μm (**b**, **c**, **f**, **g**), 50 μm (**h**). (The figure is reproduced from Fan et al. [10])

subcellular structures are presented in Fig. 3b, c. Furthermore, we show the fluorescence signals ($\Delta F/F_0$) of all segmented cells in Fig. 3d, indexed by their wave phases. It can be observed that the calcium signal propagates in a periodically spiral pattern across the entire FOV. The RUSH system's high throughput allows us to capture non-periodical signal propagation in cardiac ensembles, either within a single cell or among multiple cells.

Compared to cardiac cellular ensembles, neural ensembles have the ability to form long-range connections. As a result, high-resolution and large-scale functional imaging is necessary to study neural network activity effectively. In this regard, we conducted a proof-of-concept experiment using cultured rat neuron ensembles by performing calcium imaging, which was prepared in a manner similar to the cardiac cells. Figure 3d depicts the spatiotemporal alterations of calcium signals across the neuron ensemble, while Fig. 3g–h illustrate the propagation of calcium signals along dendrites. Our RUSH macroscope enables the recording of fluorescence dynamics in somas and the observation of signal propagation along fine structures, such as single dendrites.

2.5 In Vivo Cellular Imaging

To show the proficiency of the RUSH macroscope in large-scale, high-resolution, and high-throughput imaging of biological dynamics, we performed calcium imaging of neurons and dendrites in awake mice in vivo. In this study, we utilized virus-infected adult C57BL/6 mice. First, we performed the craniotomy as described in a previous report, with a window size of ~8 mm × 8 mm, then installed flat optical windows and cemented the custom-made coverslips (D-shape) and aluminum head posts to the skulls. The neurons of adult C57BL/6 mice were sparsely labelled with a mixture of diluted AAV2-9-hSyn-cre and AAV2-9-Ef1a-DIO-GCaMP6f viruses (from BrainVTA Technology, China).

Calcium dynamics were recorded at 14 f.p.s., and the results are presented in Fig. 4. Our experiments enabled us to record calcium signals from neural somas (Fig. 4b–e), as well as to observe calcium propagation along dendrites (Fig. 4f–h). These investigations can provide the multi-scale data necessary for studying brain-wide correlations among neuron responses, including the response delay of calcium signals at different branches of a single neuron.

3 Mesoscale Multi-planar Imaging

As the imaging field of view increases, curvature of the mouse brain has become a crucial issue that need to be considered, because the lateral areas of the brain can be hundreds of micrometers to millimeters lower than the central part [8]. Another challenge is that many cognitive processes involve the neural interaction of multiple brain regions at different depths. For instance, the formation of

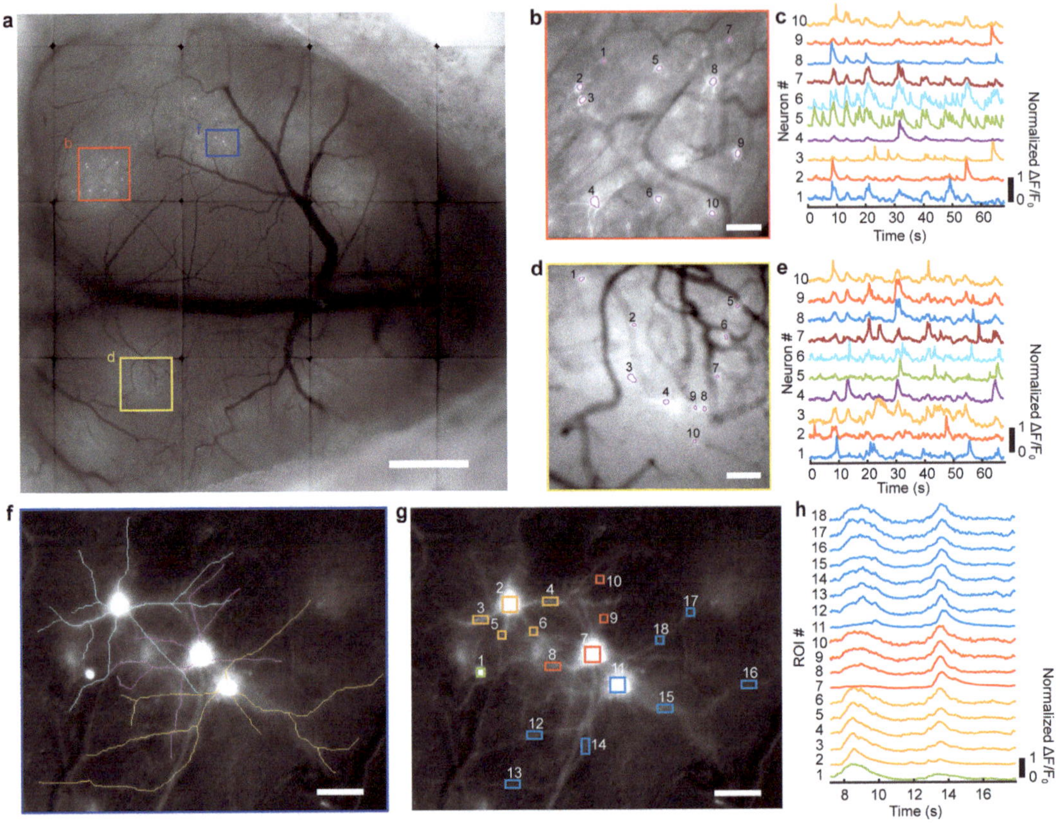

Fig. 4 In vivo brain-wide calcium imaging of adult C57BL/6 mice at dendritic resolution. (**a**) Large-FOV, high-resolution image of the mouse brain cortex. (**b**, **d**) Enlarged views of the areas labelled in (**a**). The contours of the segmented neuron cells are labelled. (**c**, **e**) Temporal traces of the segmented cells labelled in (**b**) and (**d**), respectively. (**f**) Enlarged view (s.d. projection) showing the clear structures of dendrites, highlighted with semi-automatic tracing of neurons. (**g**) The same enlarged view with labelled ROIs. Each color encodes an individual neuron. (**h**) Temporal traces of the ROIs labelled in g. Scale bars, 1000 μm (**a**), 100 μm (**b**, **d**), 50 μm (**f**, **g**). (The figure is reproduced from Fan et al. [10])

episodic memory is dependent on the interplay between the hippocampus and prefrontal cortex [17–19], while spatial navigation is dominated by hippocampal-parietal cortical interactions [20–22]. These nuclei are located more than 1 mm beneath the dura in adult mice and are spread across the whole brain. To investigate the dynamic behavior of different brain modules at a fine scale, it is essential to have an optical system with cellular resolution, brain-wide field of view, video-rate acquisition, and multi-depth imaging capabilities.

In this section, we firstly introduce a novel wide-field fluorescence microscope called the Spinning-disk Mesoscopic ARbitrarily shaped-surface imaging Technique (SMART, or previously termed MFIAS for multifocal fluorescence imaging of arbitrary surfaces) [23, 24]. This microscope has the ability to image on nonplanar surfaces while maintaining high spatial resolution across a large field of view. To achieve this, we have developed an active imaging framework that includes automatic detection of the surface profile, active control of illumination, high-speed spinning-disk scanning, and multiplexed detection. In the second part, we introduce a method for multi-planar imaging, which enables simultaneous recording of the superficial cortical and hippocampus dynamics with cellular resolution [25].

3.1 System Setup of SMART

The operation of SMART is shown in Fig. 5. The imaging process begins with the estimation of surface profiles, which is used to design a spatial-temporal selective illumination sequence. This sequence is then applied to the sample, while the spinning disk is used for fast axial scanning within a single frame exposure time. This allows for the acquisition of surface features with depth encoding. The obtained images, along with the illumination sequence, are then processed to decode cellular information with high accuracy. Overall, this approach provides an effective means for achieving high-resolution imaging of dynamics on complex surfaces. Key elements in this setup include:

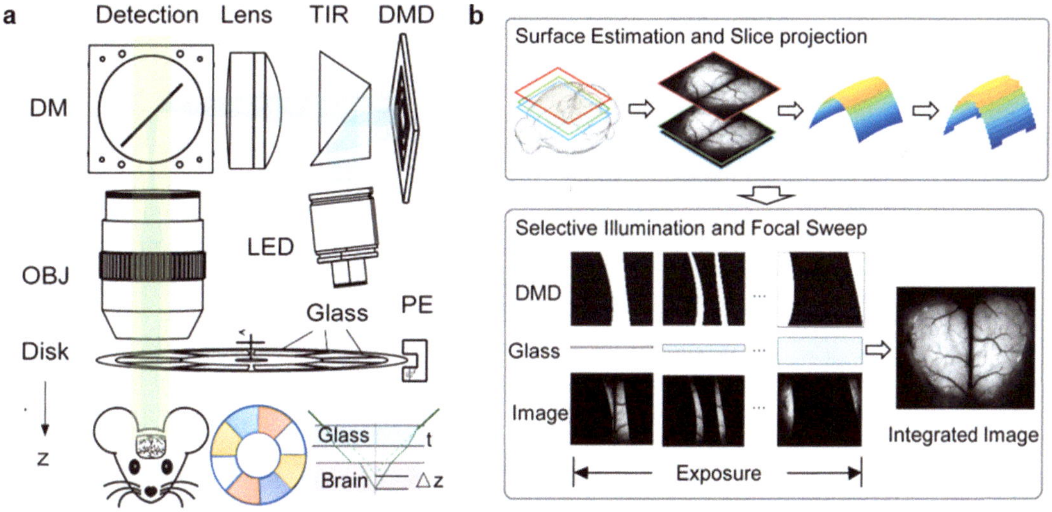

Fig. 5 Schematic diagram of the SMART system. (**a**) The SMART system comprises two components: selective illumination and focal modulation. The selective illumination component incorporates a DMD that is conjugated with the image plane and modulates the illumination patterns. The focal modulation component is a spinning disc with varying thicknesses of cover glass, which shifts the focal image to different depths. (**b**) Before frame acquisition, a depth map is generated and converted into a series of illumination patterns displayed on the DMD. During each exposure period, the spinning disc completes one full rotation, while the DMD displays one pattern at each glass thickness. The resulting image is an integration of focal areas from different depths. TIR, total internal reflection prism; DM, dichroic mirror; OBJ, objective; LED, light-emitting diode; PE, photon emitter. (Figures are adapted from Xie et al. [23])

1. A spinning disk for fast focal shifting. The spinning disk, composed of glasses with varying thicknesses, was positioned between the two surfaces to produce a focus shift. The use of flat optical elements ensures that all regions of the field of view are treated equally for a telecentric optical system as RUSH, allowing for high-resolution focal length changes across the entire field of view.
2. Selective illumination. The illumination patterns were used to modulate the illumination of the sample through the DMD, which enabled the bright regions to correspond to in-focus areas. The spinning disk was synchronized with the DMD and camera, and the acquisition of images was done in a single frame exposure time. The captured images were then processed to decode cellular information with high accuracy. The axial positions of the surface were also extracted from the illumination patterns and used to reconstruct the surface profile.

3.2 Results of SMART

This ability to image the entire superficial brain allowed us to study the spatiotemporal dynamics of neural activity across a large field of view, which is important for understanding the mechanisms underlying brain function. We were able to detect calcium signals from neurons across the mouse cortex in response to visual stimulation. This demonstrates the ability of SMART microscopy to provide high-quality, high-resolution images of neural activity in vivo, which is essential for advancing our understanding of brain function and disease.

The SMART microscopy approach enabled the acquisition of high-resolution images of neural activity with a large field of view and high imaging speed. This allowed for the detection of fast, micrometer-resolution spontaneous activity over the superficial dorsal cortex of the mouse brain, which was difficult to achieve with existing wide-field techniques. By using morphological methods, the time series of images at each depth were used to detect the three-dimensional positions of neurons. A polynomial function was then applied to fit the surface, as shown in the inset of Fig. 3a. The resulting images and surface profile provided a comprehensive view of the neural activity across the superficial layer of the cortex.

SMART's ability to capture an image of the entire superficial cortex in one snapshot and identify thousands of neurons with high resolution and efficiency represents a significant advance over existing wide-field techniques. By detecting 5400 neurons using the CNMF-E algorithm, SMART was able to evenly distribute neurons across the entire field of view, outperforming a conventional microscope that could only detect 1000 neurons within its depth of field. Additionally, by zooming in on four selected regions in detail, SMART was able to reveal the contours of all somata at high resolution. While the imaging speed of SMART was currently limited by the sCMOS camera speed of up to 50 frames per second, setting the speed to 10 Hz allowed it to match the dynamics of the calcium indicator.

3.3 Hippocampal Imaging

The hippocampus is exposed by removing the mice's skull and some cortical tissue. The hippocampus is 1 mm under the dura, while the cortex's layer 2/3 neurons are only 100 ~ 300 μm under the dura. A substance with a higher refractive index is added above the hippocampus to bring the hippocampal neurons closer to the cortical ones. The apparent depth Ha of the hippocampus in the substance depends on its real depth and the refractive indices of the imaging plane and the substance, respectively, according to Snell's Law. For a dry objective at around 515 nm wavelength, a glass column of 0.9 mm height and 2 mm diameter is attached to the cover glass above the mouse hippocampus, and an additional glass column of 1.8 mm thickness is put on top to further compensate the depth difference. A chronic craniotomy was performed by creating a 6 mm diameter window and aspirating the cortical tissue with a 0.9 mm diameter blunt needle connected to a vacuum pump. The center of the aspirated cortex was about 1.5 mm from the sagittal suture and 2 mm from the lambdoid suture. A chronic window was made by sticking a ϕ9 mm glass coverslip to a glass column with tissue adhesive. The window was implanted above the mouse cortex, and an aluminum head post was attached to the skull and secured with dental cement.

We have demonstrated the capability of our system to perform in vivo multi-planar neuron imaging in Thy1-YFP mice (JAX No. 003782). To assess the viability of our technology for chronic mouse imaging, we allowed a recovery period of 2 weeks following optical window implantation. We present brain images of Thy1-YFP mice captured on Day 14, 30, and 60 post-implantation. Neurons in all regions, including the hippocampus and superficial cortex, are visible, indicating that our microwindow implantation approach is suitable for chronic imaging [25].

4 Denoising

Fluorescence imaging is photon limited, which means that it is usually not possible to collect enough fluorescent photons to obtain a satisfactory imaging signal-to-noise ratio (SNR). The problem of low imaging sensitivity is the fundamental challenge of fluorescence imaging. The causes of the photon-limited challenge can be the low quantum yield of fluorescent indicators [26], vulnerability of biological tissue [27–32], and the quantum nature of photon detection [33–36].

Using deep learning to restore fluorescence images can compensate for the innate deficiencies of imaging systems and effectively improve microscopic image quality [35, 37–40]. The application of deep learning consists of two steps. Firstly, data is collected to make a training set, and deep neural networks are trained on the dataset using back-propagation and gradient descent algorithms. Then pretrained models are deployed and used to process unseen data. Those well-trained models still maintain good generalization on new test data. Such a data-driven paradigm dictates the strong data-dependent nature of deep-learning-based methods, where the dataset has a much greater impact on the performance than the network architecture. Thus, making a representative and category-balanced dataset is crucial for these methods.

Conventional deep-learning-based denoising methods for fluorescence imaging are based on supervised learning, which means that the training of the deep neural network relies on the supervision of paired ground-truth images (i.e., clean images without noise contamination or high-SNR images with the same underlying scene as the low-SNR images). For static samples, it is not difficult to obtain the ground-truth images by extending the exposure time of the camera. However, biological phenomena are often highly dynamic, non-repetitive activities where the same scene cannot be captured twice, and schemes to obtain training truth values by averaging are no longer feasible. The lack of ground-truth images is very common in fluorescence imaging of living organisms, such as neural calcium imaging, membrane voltage imaging, cell interaction, organelle dynamics, immune response, and hemodynamic processes. There is an urgent demand to develop new denoising methods to break the dependence of supervised denoising on truth values.

We proposed a self-supervised framework for fluorescence imaging denoising named DeepCAD [41], which can be used to train denoising models without requiring any ground-truth images and can achieve performance as good as the latest supervised denoising methods, as shown in Fig. 6. The original low-SNR image stack acquired by the imaging system is split into two sub-sequences (*xy–t* image sequences) of odd and even frames, which are used as the input and target of the deep neural network, respectively. Then, the neural network is trained with stochastic gradient descent. In brief, traditional supervised learning methods use ground-truth images as the supervised data, while DeepCAD uses adjacent frames of the input noisy images as the supervised data. Although the supervised data also contains the same kind of noise as the input data, the network parameters can still converge to values similar to those trained by supervised learning. After training, the parameters of the pretrained denoised network will be stored in a model file. When new low-SNR data arrives, a three-dimensional (*xy–t*) window will traverse the whole video and input these data blocks into the pretrained model. The output of the

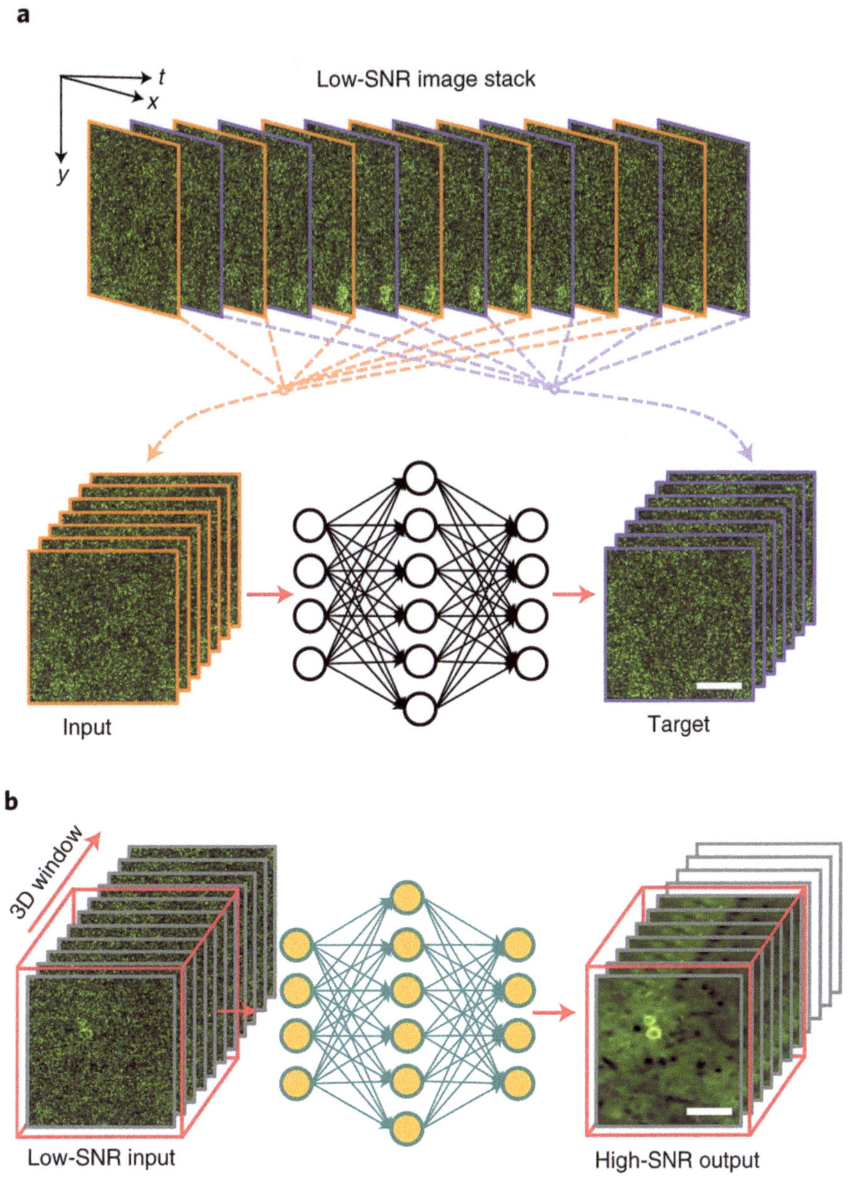

Fig. 6 General principle and validation of DeepCAD. (**a**) Self-supervised training strategy of DeepCAD. Consecutive frames in the original low-SNR stack are divided into two sub-stacks, which are then used as the input volume and corresponding target volume to train the deep neural network (3D U-Net20). After training, a denoising model can be established and memorized in network parameters. Scale bar, 50 μm. (**b**) Deployment of the DeepCAD model. For subsequent acquisitions, a 3D (*x*–*y*–*t*) window traverses the entire stack, and 3D tiles are sequentially fed into the pretrained model. Denoised recordings will be obtained after processing by the model. Scale bar, 50 μm. (Figures are adapted from Li et al. [41])

model is corresponding denoised blocks, and the complete denoised results will be obtained after stitching these output blocks together.

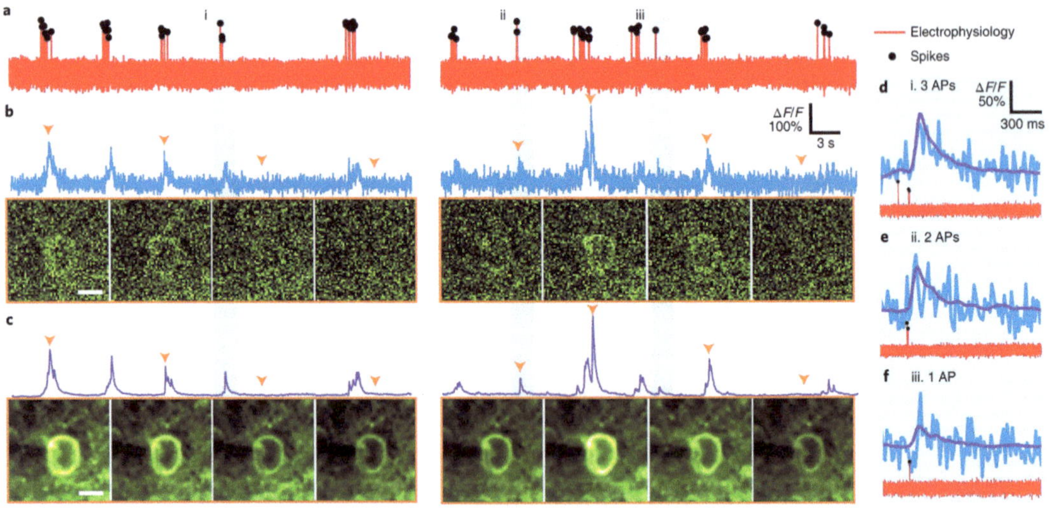

Fig. 7 Spatiotemporal enhancement with DeepCAD. (**a**) Single-neuron electrophysiology. Detected spikes are marked with black dots. (**b**) Two-photon calcium imaging data of the same neuron synchronized with cell-attached electrophysiology. Representative frames indicated with orange triangles are presented below the trace. Scale bar, 10 μm. (**c**) Fluorescence traces and representative frames after the enhancement with DeepCAD. Scale bar, 10 μm. (**d**–**f**) Examples of calcium transients evoked by three APs (**d**), two APs (**e**), and one AP (**f**) before (blue) and after (purple) noise removal. (Figures are adapted from Li et al. [41])

To verify the effectiveness and reliability of DeepCAD on experimentally obtained data, we then demonstrated its performance on previously released two-photon calcium imaging data [42]. In this dataset, simultaneous cell-attached electrophysiological recordings (Fig. 7a) are synchronized with two-photon imaging and serve as reference for calcium transients and as ground truth for spike inference. Contaminated by detection noise, both the spatial footprint and temporal traces of neurons were severely corrupted in the original data (Fig. 7b). After we applied DeepCAD to enhance these data, the annular cytoplasms became recognizable and calcium traces were liberated from noise (Fig. 7c). Even barely perceptible calcium transients evoked by one AP, two APs, or three APs were distinguished and maintained their original dynamics (Fig. 7d–g), which otherwise would be overwhelmed by noise.

Furthermore, to meet the growing demand of large-scale image processing and real-time high-sensitivity observation, we comprehensively optimized DeepCAD to improve its processing speed and performance [43]. We also designed a multithreaded processing pipeline, as well as an optimal hardware deployment scheme, to integrate the proposed denoising into the imaging system. We finally implemented real-time noise suppression on a two-photon fluorescence microscope for high-sensitivity imaging. With this high-sensitivity imaging technique, we observed various biological events, including neural calcium activity in multiple model organisms (Drosophila, zebrafish, and mice), the 3D migration of immune cells after acute brain injury, and the 3D dynamics of adenosine triphosphate (ATP) release in the mouse cortex after laser-induced injury [44]. The implementation of real-time high-sensitivity fluorescence imaging will help biologists solve the photon-limited challenge in fluorescence imaging and facilitate the revelation of underlying biological mechanisms in a wide range of applications.

5 Rapid Calcium Signal Extraction

Widefield microscopy can image multi-millimeter fields of view and thousands of neurons in mammalian brains at a video rate. But extracting neuronal activity signals from calcium imaging data at a cellular resolution is hard because of tissue scattering [45] and background signals [46], which makes it slow and difficult [47–50]. We propose a deep learning method called DeepWonder, which uses simulated calcium recordings but works well on experimental data [51]. DeepWonder achieves a processing speed improvement of nearly tenfolds compared to the widely used CNMF-E technique in widefield calcium imaging analysis. In only 17 h with workstation-grade computing resources, DeepWonder found over 14,000 neurons, which is much better than the nearly week-long processing time with previous methods. This new method can help with massive data processing in widefield neuronal imaging, avoiding the need for many computational resources.

The DeepWonder framework consists of two neural networks: The Removing Background Network (RB-Net) and the Neuron Segmentation Network (NS-Net). RB-Net removes background contaminations from the imaging data, while NS-Net segments neurons from the background-removed data, as shown in Fig. 8.

The effectiveness of widefield microscope in detecting neurons is limited by background contamination, which hinders signal extraction and detection quality. In DeepWonder, we use a neural network to map images with background contamination to background-free data by generating synthetic widefield calcium imaging data. We create synthetic data by modeling vessels, neurons, and background dendrites and axons based on a specific widefield microscope model to produce realistic virtual recordings with accurate pixel, $\Delta F/F$, and spatial frequency distributions. Paired virtual recordings are fed to the removing background network (RB-Net) to learn the mapping between contaminated and background-free captures. Trained RB-Net learns interpretable features and outputs high-contrast images and vivid neuronal activities without contaminations. DeepWonder enhances correlation scores to the ground truth signals and signal-to-background ratios significantly compared to raw data. RB-Net-driven by virtual data in DeepWonder is effectively applied to remove backgrounds of real

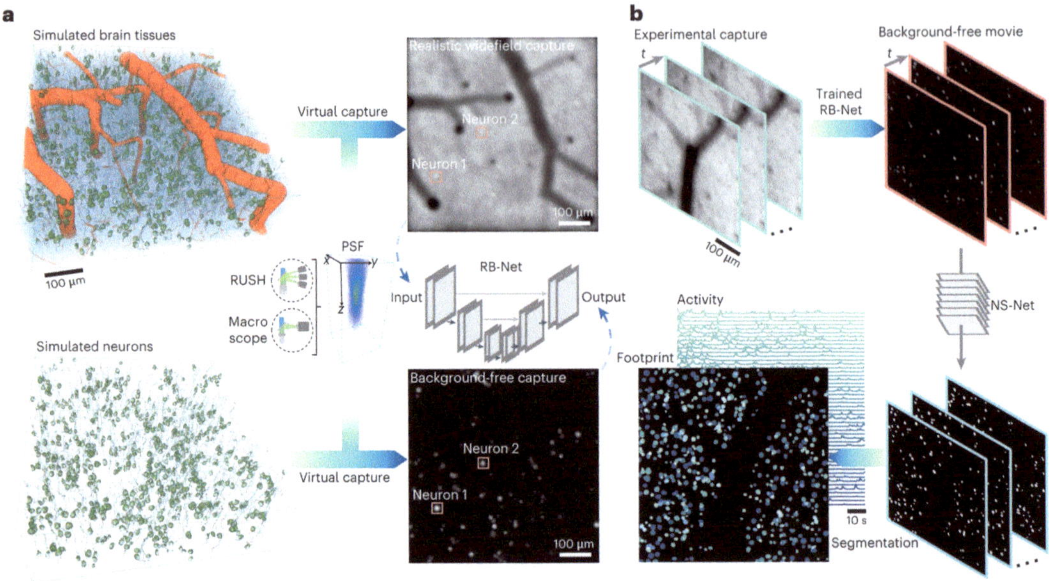

Fig. 8 Principle of deep widefield calcium finder (DeepWonder). (**a**) The deep widefield calcium finder (DeepWonder) operates on the principle of removing background noise from widefield calcium imaging. In the training stage, the removing background network (RB-Net) is trained using a realistic widefield simulator to generate virtual captures with similar parameters to experimental captures. These captures are used as inputs along with corresponding labels that represent the same neuron distributions without background contamination. The RB-Net is trained to remove the background from the captured images and restore background-free neuronal images. (**b**) In the testing stage, DeepWonder is used on new recordings to remove the background noise from the experimental captures. The neuron segmentation network (NS-Net) is then applied to segment the neurons and extract their signals from the background-removed movies. (Figure reproduced from Zhang et al. [51])

recordings. The RB-Net achieves superior performance in SBR, correlation score, and neuron finding scores compared to other state-of-the-art background removal methods while spending almost sevenfold shorter time in removing background. The similarity between virtual generations and real recordings guarantees the high effectiveness of RB-Net in real recordings. We illustrate an SBR improvement in real recordings by RB-Net compared to raw data across 1543 neurons.

NS-Net, the neuron segmentation network proposed in DeepWonder, is designed to segment neurons from background-removed data. The NS-Net employs a lightweight CNN that segments neurons from the output of the RB-Net at high speed. The network then semantically segments roughly isolated neurons based on their spatiotemporal connectivity and yields mostly exclusive segmentations. This approach enables the direct readout of the temporal activities of individual neurons since there is no inter-neuron crosstalk. Neurons that are tiled and overlapped are further demixed by a local nonnegative matrix factorization (NMF) algorithm, which eliminates activity crosstalk. NS-Net reliably demixes neurons as close as 0.3 of the neuron diameter, resulting in a temporal similarity of over 0.9 and a spatial similarity of over 0.85.

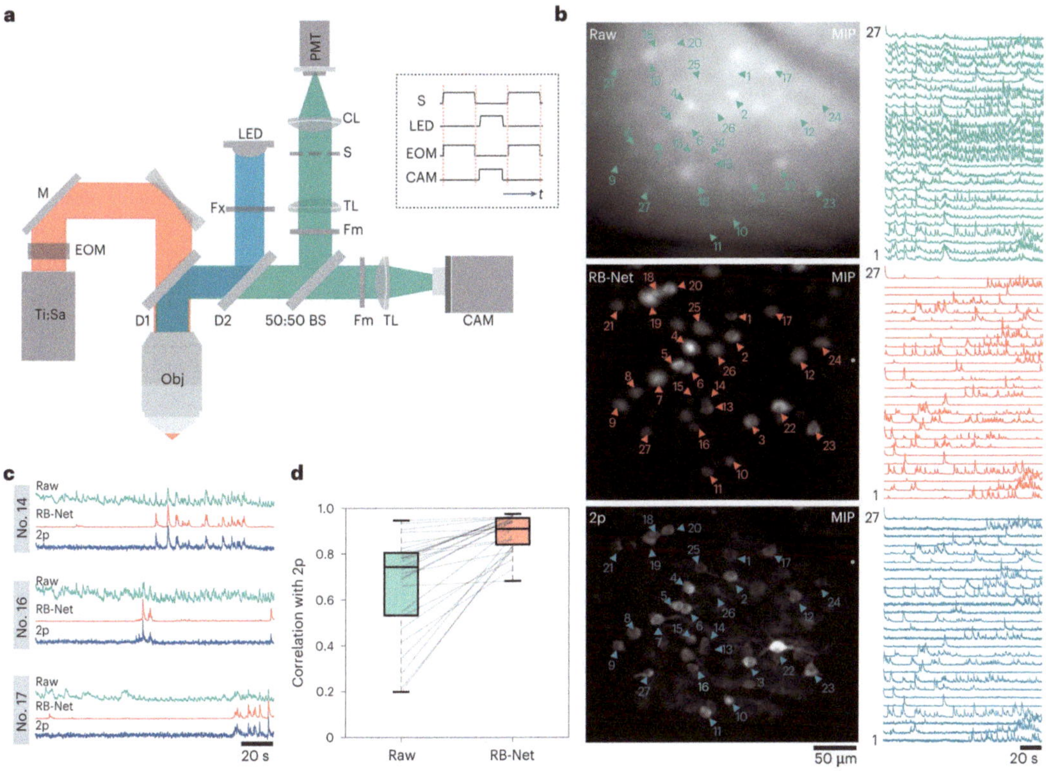

Fig. 9 A validation of single-photon microscopy accuracy using a two-photon microscope. (**a**) The hybrid 1p–2p microscope setup. LED, light-emitting diode light source; Ti:Sa, titanium:sapphire laser; M, mirror; DM, dichroic mirror; BS, beam splitter; Fm, emission filter; Fx, excitation filter; CL, collection lens; TL, tube lens; S, triggerable shutter; CAM, sCMOS camera; Obj, objective. Right box: control signals of the shutter, LED, EOM and camera exposure. (**b**) Maximum intensity projection (MIP) of widefield (top), RB-Net in DeepWonder processed widefield movie (middle) and two-photon movie (bottom). Triangles mark neurons, and the corresponding temporal activities are plotted on the right side. The mean background value for widefield, DeepWonder and 2p movies are 0.64, 0.001 and 0.12, respectively (normalized by maximum value). (**c**) Zoom-in plots of temporal activities of neurons 14, 16 and 17 in the widefield raw movie (green), RB-Net processed movie (red) and 2p movie (blue.) (**d**) Temporal correlations of 27 representative neurons, from raw data imaged with with 2p (0.68 ± 0.22, mean ± s.d.) and after processing with RB-Net (0.89 ± 0.08, mean ± s.d.). $^{***}P = 3.55 \times 10^{-6}$, two-sided Wilcoxon signed-rank test. Central black mark: median. Bottom and top edges: 25th and 75th percentiles. Whiskers extend to extreme points excluding outliers (1.5 times above or below the interquartile range). (Figure reproduced from Zhang et al. [51])

DeepWonder is able to significantly improve the processing speed and accuracy of calcium imaging analysis compared to current techniques, such as CNMF-E. Additionally, the hybrid microscopy device and comparison with two-photon recordings suggest that DeepWonder provides accurate neuronal segmentation and activity inference in mouse recordings, with high precision scores and signal correlations with two-photon ground truth, as shown in Fig. 9. The results show that DeepWonder has advantages in both speed and performance compared to widely used CNMF-E.

6 Notes

The development of a large field-of-view high-resolution computational microscope has overcome the contradiction between field-of-view and resolution, making it possible to observe a large number of mesoscale life science phenomena. However, other factors still limit the further development of mesoscopic imaging. One of the core challenges is low signal-to-noise ratio in fluorescence microscopy imaging. Signal-to-noise ratio is a critical indicator of imaging sensitivity. Without sufficient signal-to-noise ratio, the observed phenomenon cannot be described qualitatively or quantitatively, and the data obtained will be meaningless. Another challenge is the large amount of data generated by mesoscale imaging. How to process and analyze these data to discover new phenomena and reveal new mechanisms is a worthwhile research direction. Moreover, the current instrument can only be used for fixed samples. Extending the instrument to freely moving animals will help reveal the correlation between animal behavior and neural activity and promote understanding of biological intelligence. These three problems are the main directions that we are tackling.

6.1 High-Sensitivity Fluorescence Imaging Beyond the Photon Noise Limit

Fluorescence imaging is a technique that allows us to see biological processes and structures in living organisms by using fluorescent molecules that emit light after absorbing radiation. Unlike conventional imaging of macroscopic scenes, fluorescence imaging faces the challenge of low photon counts, which means that the image signals are very weak and noisy. This affects the resolution, speed, sensitivity, and other aspects of imaging. Fluorescence imaging is a fundamental and difficult problem in biophysics and bioengineering. Breaking through the photon noise limit is a fundamental challenge and opportunity for fluorescence imaging. Different fluorescence imaging techniques have different excitation and detection mechanisms, which require different computational and physical methods to overcome the photon noise limit. So, a general theory and method for this problem is still lacking in the field of fluorescence imaging. Achieving fluorescence imaging that breaks through the photon noise limit will enable major breakthroughs in various fluorescence imaging applications, such as super-resolution microscopy, two-photon microscopy, light field microscopy, and structured light microscopy. We are aiming to break through the photon noise limit, a fundamental challenge for optical imaging sensitivity. Photon noise is caused by the quantum nature of light and cannot be suppressed by classical methods. Therefore, how to develop computational restoration and physical compensation methods that break through the photon noise limit of fluorescence imaging is a frontier and basic problem in life sciences, optics, information sciences, and other fields. It can promote the progress of the entire field of fluorescence imaging and is expected to bring a series of new discoveries in life sciences.

6.2 Mesoscale Data Analysis Platform

The mesoscale imaging platform generates large-scale data, where the experimental data of a single project sometimes reaches the PB level. It is necessary to build a supporting mesoscale data analysis platform to effectively store and process data in order to achieve efficient use of large-scale data. The main functions of the mesoscale data analysis platform include: (1) Data storage cluster: This is the basic component of a large-scale data analysis platform, which stores various types of large-scale data collected by instruments. The data storage cluster can use cloud storage services or distributed file systems to support efficient storage and access of data. (2) Data visualization: This is the process of presenting data in different forms, such as videos, charts or other forms, which can help researchers better understand and analyze data and discover patterns and trends in the data. (3) Quantitative life science analysis, which deeply processes and analyzes biomedical imaging data. Quantitative life science analysis can use various life science data analysis tools such as R and Python, and provides support for life science research, helping researchers better understand life science data. High-performance analysis requires strong hardware support, including high-performance CPU and GPU clusters, to shorten analysis time.

The main components of a large-scale data analysis platform include: (1) Storage devices: Used to store large amounts of data, including hard drives, solid-state drives, and network storage, with a large storage capacity. (2) Processors: Used for data processing and computation, mainly including CPU and GPU clusters. The former is for general computation while the latter is responsible for image processing and matrix operations. (3) Network equipment: Used for communication and data transmission between different parts, with each component connected by fiber optics. (4) Software system: Including operating system, distributed computing framework, database system, data processing, and analysis tools.

Based on this hardware platform, by deploying low-level processing modules such as denoising, deconvolution, and background removal, along with high-level processing modules such as segmentation, signal extraction, signal analysis, and loop construction, we can build a high-performance mesoscale data analysis platform. Further by coupling the analyzed data using self-supervised or unsupervised algorithms, we can achieve inductive fusion of massive amounts of data to effectively discover relevant biological conclusions promoting the development of frontier life sciences.

6.3 Miniaturized Robust Imaging Instrument

Optical macroscopes are a type of optical microscope that can achieve high-resolution and large-field-of-view imaging of small objects. However, existing optical macroscopes face several challenges that limit their performance and applicability. First, optical aberrations such as coma, astigmatism and field curvature degrade the image quality at the edges of large fields-of-view, forcing a trade-off between resolution and field-of-view. To overcome this problem, optical engineering design is required, which results in bulky and costly systems. Second, in vivo imaging of biological samples is hampered by tissue distortion and scattering, which affect the accuracy and contrast of the images. Traditional methods use adaptive optics to compensate for static samples, but dynamic samples pose great difficulties. Third, background noise from light scattering reduces the signal-to-noise ratio and obscures the structures and functions of interest. Nonlinear excitation or selective illumination can mitigate this issue, but at the expense of time resolution. Therefore, there is a need for novel and scalable solutions for three-dimensional high-resolution mesoscale imaging systems that can reveal multiscale physiological regulation and interaction in living organisms.

We are now working on the next-generation mesoscale intravital fluorescence imaging system based on a digital adaptive optics (DAO) imaging framework that uses microlens arrays and periodic scanning elements to acquire four-dimensional full-light information [52–55]. As a generally accessible mesoscope, this framework decouples signal acquisition from adaptive wavefront correction and allows post-processing correction of system aberrations and dynamic tissue aberrations. By implementing large-scale spatial block adaptive optics correction, we improve the spatial resolution to the optical diffraction limit. This instrument can record accurate multidimensional multi-scale data of living organisms, reflecting the biological characteristics of individual cells and studying pathological physiological processes and mutual influences under intact microenvironmental conditions. This instrument is expected to reveal important life phenomena such as structural-functional coupling of neural circuits under physiological and pathological conditions and occurrence and metastasis of malignant tumors.

References

1. Anderson PW (1972) More is different – broken symmetry and nature of hierarchical structure of science. Science 177(4047):393–396
2. Ji N, Freeman J, Smith SL (2016) Technologies for imaging neural activity in large volumes. Nat Neurosci 19(9):1154–1164
3. Jung S et al (2000) Analysis of fractalkine receptor CX(3)CR1 function by targeted deletion and green fluorescent protein reporter gene insertion. Mol Cell Biol 20(11): 4106–4114
4. Valastyan S, Weinberg RA (2011) Tumor metastasis: molecular insights and evolving paradigms. Cell 147(2):275–292

5. Lohmann AW et al (1996) Space-bandwidth product of optical signals and systems. J Opt Soc Am A 13(3):470–473
6. Lim D, Chu KK, Mertz J (2008) Wide-field fluorescence sectioning with hybrid speckle and uniform-illumination microscopy. Opt Lett 33(16):1819–1821
7. Goda K, Tsia KK, Jalali B (2009) Serial time-encoded amplified imaging for real-time observation of fast dynamic phenomena. Nature 458(7242):1145–1149
8. Kauvar IV et al (2020) Cortical observation by synchronous multifocal optical sampling reveals widespread population encoding of actions. Neuron 107(2):351–367
9. Zhou P et al (2018) Efficient and accurate extraction of in vivo calcium signals from microendoscopic video data. eLife 7:e28728
10. Fan J et al (2019) Video-rate imaging of biological dynamics at centimetre scale and micr ometre resolution. Nat Photonics 13(11):809–816
11. Potsaid B, Bellouard Y, Wen JT (2005) Adaptive Scanning Optical Microscope (ASOM): a multidisciplinary optical microscope design for large field of view and high resolution imaging. Opt Express 13(17):6504–6518
12. Li AA et al (2010) Micro-optical sectioning tomography to obtain a high-resolution atlas of the mouse brain. Science 330(6009): 1404–1408
13. Economon MN et al (2016) A platform for brain-wide imaging and reconstruction of individual neurons. eLife 5:e10566
14. Tsai PS et al (2015) Ultra-large field-of-view two-photon microscopy. Opt Express 23(11): 13833–13847
15. Luo W et al (2016) Pixel super-resolution using wavelength scanning. Light Sci Appl 5: e16060
16. Zheng GA, Horstmeyer R, Yang CH (2013) Wide-field, high-resolution Fourier ptychographic microscopy. Nat Photonics 7(9): 739–745
17. Wing EA et al (2020) Cortical overlap and cortical-hippocampal interactions predict subsequent true and false memory. J Neurosci 40(9):1920–1930
18. Moscovitch M et al (2016) Episodic memory and beyond: the hippocampus and neocortex in transformation. Annu Rev Psychol 67:105–134
19. Preston AR, Eichenbaum H (2013) Interplay of hippocampus and prefrontal cortex in memory. Curr Biol 23(17):R764–R773
20. Saleem AB et al (2018) Coherent encoding of subjective spatial position in visual cortex and hippocampus. Nature 562(7725):124–127
21. Wolbers T et al (2007) Differential recruitment of the hippocampus, medial prefrontal cortex, and the human motion complex during path integration in humans. J Neurosci 27(35): 9408–9416
22. Yu JY, Frank LM (2015) Hippocampal-cortical interaction in decision making. Neurobiol Learn Mem 117:34–41
23. Xie H et al (2023) Multifocal fluorescence video-rate imaging of centimetre-wide arbitrarily shaped brain surfaces at micrometric resolution. Nat Biomed Eng 8:740–753. https://doi.org/10.1038/s41551-023-01155-6
24. Xie H et al (2022) Spinning Disk Multifocal Microscopy for dynamic arbitrarily shaped surface imaging at centimetre scale and micrometre resolution. bioRxiv. https://doi.org/10.1101/2022.02.22.481397
25. Zhuang C et al (2021) Real-time brain-wide multi-planar microscopy for simultaneous cortex and hippocampus imaging at the cellular resolution in mice. Biomed Opt Express 12(4):1858–1868
26. Hirano M et al (2022) A highly photostable and bright green fluorescent protein. Nat Biotechnol 40(7):1132–1142
27. Hoebe RA et al (2007) Controlled light-exposure microscopy reduces photobleaching and phototoxicity in fluorescence live-cell imaging. Nat Biotechnol 25(2):249–253
28. Huang XS et al (2018) Fast, long-term, super-resolution imaging with Hessian structured illumination microscopy. Nat Biotechnol 36(5):451–459
29. Icha J et al (2017) Phototoxicity in live fluorescence microscopy, and how to avoid it. Bioessays 39(8). https://doi.org/10.1002/bies.201700003
30. Laissue PP et al (2017) Assessing phototoxicity in live fluorescence imaging. Nat Methods 14(7):657–661
31. Skylaki S, Hilsenbeck O, Schroeder T (2016) Challenges in long-term imaging and quantification of single-cell dynamics. Nat Biotechnol 34(11):1137–1144
32. Verweij FJ et al (2021) The power of imaging to understand extracellular vesicle biology in vivo. Nat Methods 18(9):1013–1026
33. Casacio CA et al (2021) Quantum-enhanced nonlinear microscopy. Nature 594(7862): 201–206
34. Taylor MA, Bowen WP (2016) Quantum metrology and its application in biology. Phys Rep 615:1–59

35. Meiniel W, Olivo-Marin JC, Angelini ED (2018) Denoising of microscopy images: a review of the state-of-the-art, and a new sparsity-based method. IEEE Trans Image Process 27(8):3842–3856
36. Chen JJ et al (2021) Three-dimensional residual channel attention networks denoise and sharpen fluorescence microscopy image volumes. Nat Methods 18(6):678–687
37. Weigert M et al (2018) Content-aware image restoration: pushing the limits of fluorescence microscopy. Nat Methods 15(12):1090–1097
38. Li XY et al (2021) Reinforcing neuron extraction and spike inference in calcium imaging using deep self-supervised denoising. Nat Methods 18(11):1395–1400
39. Lecoq J et al (2021) Removing independent noise in systems neuroscience data using DeepInterpolation. Nat Methods 18(11): 1401–1408
40. Li XY et al (2021) Unsupervised content-preserving transformation for optical microscopy. Light Sci Appl 10(1):44
41. Li X et al (2021) Reinforcing neuron extraction and spike inference in calcium imaging using deep self-supervised denoising. Nat Methods 18(11):1395–1400
42. Chen T-W et al (2013) Ultrasensitive fluorescent proteins for imaging neuronal activity. Nature 499(7458):295–300
43. Li X et al (2022) Real-time denoising enables high-sensitivity fluorescence time-lapse imaging beyond the shot-noise limit. Nat Biotechnol 41:1–11
44. Wu ZF et al (2022) A sensitive GRAB sensor for detecting extracellular ATP in vitro and in vivo. Neuron 110(5):770–782
45. Helmchen F, Denk W (2005) Deep tissue two-photon microscopy. Nat Methods 2(12): 932–940
46. Waters J (2020) Sources of widefield fluorescence from the brain. eLife 9:e59841
47. Lu J et al (2018) MIN1PIPE: a miniscope 1-photon-based calcium imaging signal extraction pipeline. Cell Rep 23(12):3673–3684
48. Friedrich J, Giovannucci A, Pnevmatikakis EA (2021) Online analysis of microendoscopic 1-photon calcium imaging data streams. PLoS Comput Biol 17(1):e1008565
49. Mukamel EA, Nimmerjahn A, Schnitzer MJ (2009) Automated analysis of cellular signals from large-scale calcium imaging data. Neuron 63(6):747–760
50. Inan H et al (2021) Fast and statistically robust cell extraction from large-scale neural calcium imaging datasets. bioRxiv. https://doi.org/10.1101/2021.03.24.436279
51. Zhang Y et al (2023) Rapid detection of neurons in widefield calcium imaging datasets after training with synthetic data. Nat Methods 20: 747–754
52. Mukati MU, Gunturk BK (2018) Light field super resolution through controlled microshifts of light field sensor. Signal Process Image Commun 67:71–78
53. Lu Z et al (2019) Phase-space deconvolution for light field microscopy. Opt Express 27(13): 18131–18145
54. Wu J et al (2022) An integrated imaging sensor for aberration-corrected 3D photography. Nature 612(7938):62–71
55. Wu J et al (2021) Iterative tomography with digital adaptive optics permits hour-long intravital observation of 3D subcellular dynamics at millisecond scale. Cell 184(12):3318–3332.e17

Part II

Mesoscale Imaging Techniques

Chapter 4

Mesoscale Calcium (Ca^{2+}) Imaging in Freely Behaving Mice

Suhasa Kodandaramaiah, Mathew L. Rynes, Daniel A. Surinach, Kapil Saxena, Zahra Navabi, Eunsong Ko, and Malachi Lehmann

Abstract

Mesoscale imaging of calcium dynamics has emerged as a powerful technique to study the coordination of activity of multiple cortical regions simultaneously during awake behavior. While much of the work in the field has focused on studying cortical dynamics during head-fixed preparations, we have recently developed the mini-mScope—a head mounted widefield imaging microscope capable of imaging large swathes of the dorsal cortex of freely behaving mice. In this chapter, we guide the reader with methodological details to build and use the mini-mScope for widefield Ca^{2+} imaging in freely behaving mice performing complex behaviors such as spatial navigation.

Key words Wide-field calcium imaging, Spatial navigation, Naturalistic behaviors, Neuroscience, Fluorescence microscopy

1 Introduction

Cognition and behavior are mediated by neural activity at multiple spatial scales, and the underlying computations are distributed across the brain. Interactions between regions are known to be important for behaviors and cognitive processes. The advent of genetically encoded calcium and voltage indicators and transgenic approaches for broadly expressing these indicators in the brain in a cell-type specific fashion has enabled mesoscale imaging of multiple cortical regions simultaneously [1–3]. These studies have revealed how neural activity across multiple regions of the cortex are coordinated in a variety of behaviors [4–6] such as cognitive and spatial navigation tasks [7–10]. Studying brief, stereotyped behaviors during head fixation has revealed much about cortex-wide calcium dynamics during locomotion [11–14], motor learning [4, 5], sensory discrimination [15], decision making [16, 17], and goal directed behavior [4]. Cortical dynamics are also altered in diseased states [18, 19].

James E. Niemeyer et al. (eds.), *Awake Behaving Mesoscopic Brain Imaging*, Neuromethods, vol. 214,
https://doi.org/10.1007/978-1-0716-4120-0_4,

Head fixation, however, poses numerous constraints to behavior. First, training animals to perform behavioral tasks, even when relatively simple, takes significant time when head-restrained [20]. To enable mesoscale imaging studies in freely behaving animals performing complex, ethologically relevant behaviors, we have recently developed a miniaturized imaging microscope, the mini-mScope, which allows imaging calcium dynamics across most of the dorsal cortex of mice [21]. The mini-mScope weighs 3.8 g and has a field of view (FOV) of 8 mm × 10 mm, which can be imaged at resolution ranging between 39.3 μm at the center of the FOV and 55.7 μm at the edge of the FOV (Fig. 1).

In this chapter, we will describe the methodology and techniques used for mesoscale calcium imaging in freely behaving mice, using the mini-mScope. We will first describe the construction and benchtop testing of the open-source mini-mScope system, followed by descriptions of surgical preparations, behavioral acclimatization, and the procedures for imaging mesoscale calcium activity in vivo. When followed, this procedure allows us to image across the cortex in transgenic animals broadly expressing calcium indicators in the cortex.

2 Materials

2.1 The Mini-mScope and See-Shells

The mini-mScope is a miniaturized imaging device developed by our group to be an open-source device. While there are no commercially available devices, we have made all the computer-aided design (CAD) files available in Rynes et al. [21]. The mini-mScope system can be built using commonly available laboratory tools and electronic testing equipment. Below, we will first describe the mini-mScope hardware followed by instructions for constructing the device. These sections have been adapted from our original publications [21, 22].

The mini-mScope is designed to dock onto transparent polymer skulls that are chronically implanted in mice [22]. The transparent polymer skull (See-Shell) was developed to enable optical access to most of the dorsal cortex of mice (~45 mm^2) by mimicking the morphology of the skull surface it replaces [22]. For mini-mScope imaging, the See-Shell incorporates a planer top surface and three tabs with attachment points in the three-dimensional (3D)-printed implant with the titanium head-plate eliminated. Two lateral tabs have two circular slots, where two circular magnets (B07C8ZZ2K9; Amazon) are affixed; the device bottom surface or a protective cap have two corresponding magnetic slots, allowing for easy attachment to the implant. A posterior tab is screwed to the short sleeve around the mini-mScope base using 18-8 flat head screws (catalog no. 96817A704; McMaster-Carr). This further secures the device position once inserted on the implant. The

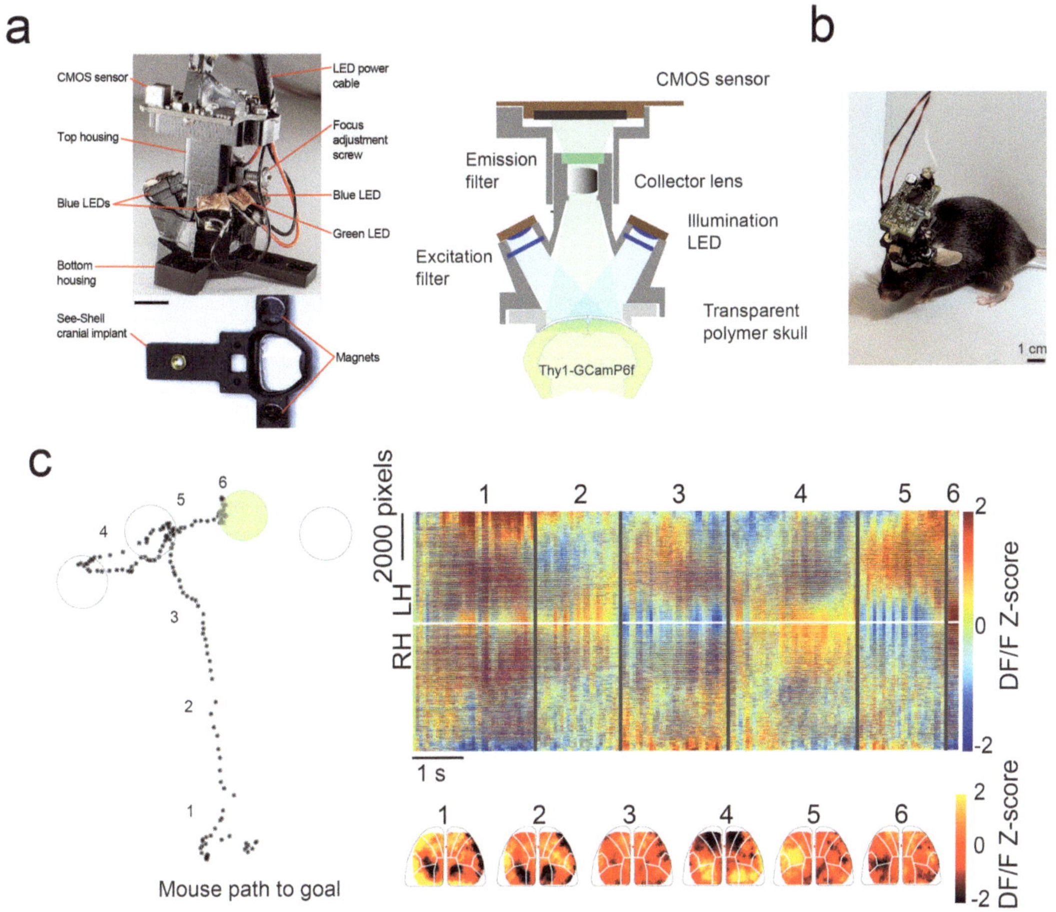

Fig. 1 The mini-mScope: a miniaturized head-mounted microscope for whole-cortex calcium imaging in freely behaving mice. (**a**) *Left:* A photograph of a fully assembled mini-mScope and the See-Shell that attaches to the device via interlocking magnets. *Right:* A cross-sectional view of the mini-mScope from a CAD model. Excitation light sources provide illumination to the cortex for calcium imaging. The green light-emitting diode (LED) on the posterior of the device is used for reflectance imaging for hemodynamic correction. The emitted light from the fluorophores is focused through a collector lens and filtered through an emission filter onto a CMOS sensor that is manually focused by moving the top housing vertically with respect to the bottom housing. (**b**) A still image of a mouse bearing the mini-mScope. (**c**) *Left:* Mouse pose tracking data from a trial completed during a spatial navigation task in the Barnes Maze. Numbers denote corresponding DF/F calcium spatial map time points shown in the right image. *Right:* A DF/F spatial map of the calcium data for all pixels across the field of view separated into left and right cortical hemispheres. The DF/F 2D spatial maps are shown below at various time points marked by the numbers on the pose plot on the left image

overall interlocking mechanisms of the mini-mScope allow for an easy and quick attachment in less than 5 s to an awake mouse.

The mini-mScope comprises two custom-milled Delrin housings that interlock (see Fig. 2). The bottom housing features illumination assemblies containing blue LEDs (LUXEON Rebel Color Blue, 470 nm; Digi-Key), along with excitation filters (450–490 nm, 3 × 3 × 1 mm^3, ET 470/40×; Chroma). For

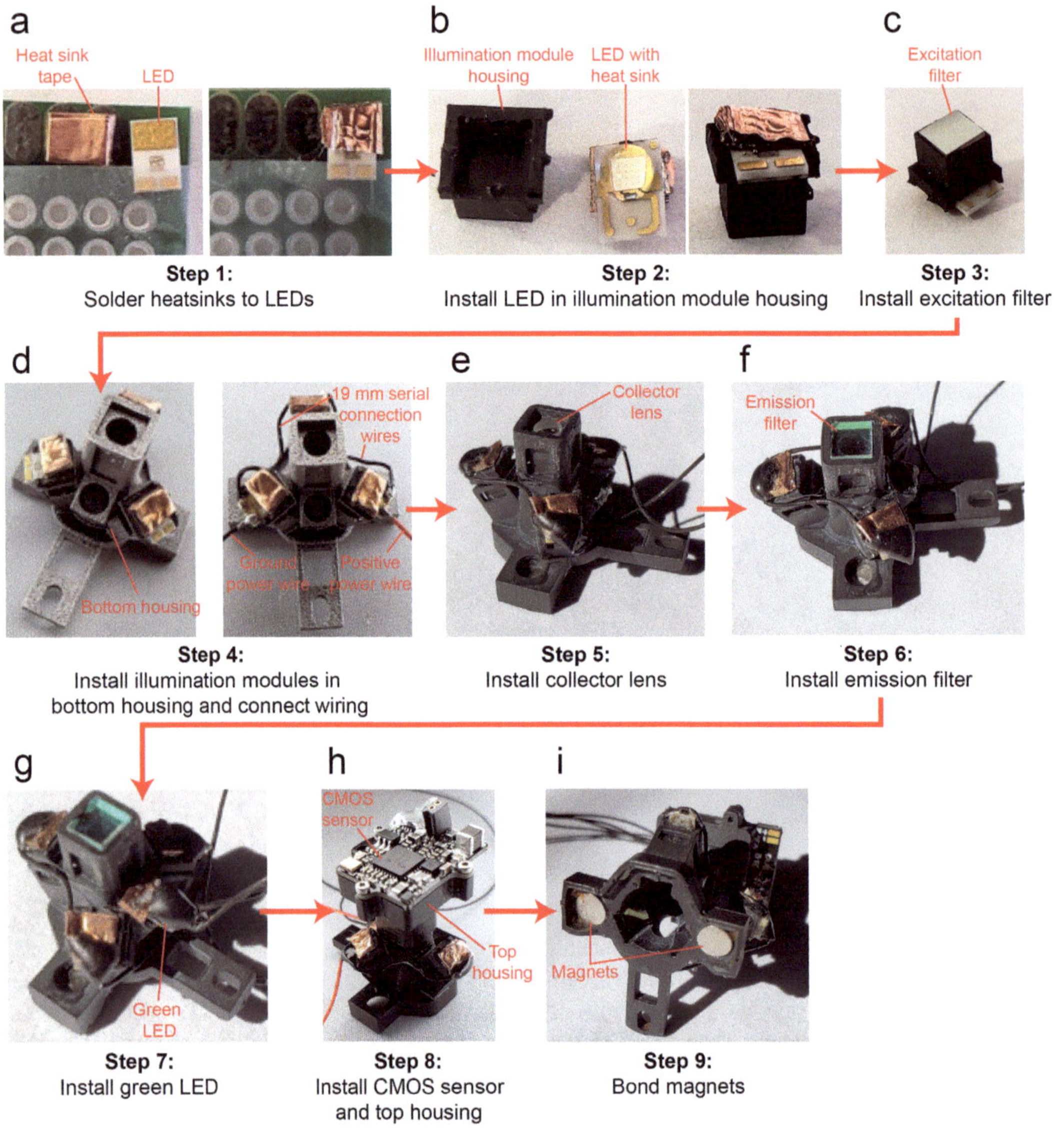

Fig. 2 Step-by-step instruction of the mini-mScope assembly: (**a**) *Left:* A photograph of a heat sink tape and the back of the LED secured with a tape on a prototyping board. *Right:* An image of a soldered LED with the heatsink. (**b**) *Left:* An illumination module sub-housing along with the soldered LED for the illumination module assembly. *Right:* The installation of LED in the illumination module sub-housing using a cyanoacrylate glue. (**c**) The installation of the excitation filter in the bottom side of the illumination module. (**d**) *Left:* An image of three illumination sub-housings inserted in the corresponding arms within the mini-mScope bottom housing. *Right:* Series wiring to connect each sub-housing. (**e**) A photograph of a collector lens embedded in the lens chute in the bottom housing's central shaft. (**f**) The insertion of the custom-diced emission filter on the top of the central shaft. (**g**) The insertion of a green LED to the circular slot on the remaining arm in the bottom housing. (**h**) The installation of the CMOS sensor and the top housing using 18-8 flat head screws. (**i**) Two disk magnets are glued to the slots on the bottom housing with the aligned magnet orientation

reflectance imaging, a green LED (LUXEON Rebel Color Green, 530 nm; Digi-Key) is installed in the circular shaft at the back of the bottom housing. The central shaft of the bottom housing aligns with the emission light paths and includes a biconvex lens and an excitation filter. The top housing facilitates a focusing adjustment by moving along the central shaft of the bottom housing. Once the optimal focal plane is found, the position is fixed with a 0–80 adjustment screw (catalog no. 91735A262; McMaster-Carr). Then, a complementary metal oxide semiconductor (CMOS) sensor (Miniscope CMOS v3 PCB; LabMaker) is attached to the top housing using the 18-8 flat head screws (*see* **Note 1**). The LED power wires are threaded through a circular hole in the top housing and then connected to a commutator (Carousel Commutator 1× DHST 2× LED; Plexon) for load reduction.

2.2 Transgenic Animals with Broad Expression of Calcium Indicators in the Cortex

Studies done in our group have used Thy1-GCaMP6f mice [23] (#024339, Jackson Laboratories), that broadly express GCamP6 calcium indicator in excitatory neurons in layers 2–3, 4 and 5 of the cortex. We have also used transgenic Emx-CaMKII-Ai85 expressing iGluSnFR in glutamatergic cortical neurons to visualize mesoscale glutamate neurotransmission [24]. Additionally, red shifted indicators such as jRGECO and its variants may be used [25]. *See* **Note 4**.

3 Methods

3.1 Mini-mScope Assembly

3.1.1 Pre-fabrication of Custom Parts of the Mini-mScope Assembly

Prior to constructing the mini-mScope, custom parts need to be prefabricated, which form the top and bottom housings of the scope and the illumination module sub-housings that host the illumination LEDs and excitation filters. CAD files for all the components can be accessed in Rynes et al. [21]. The top and bottom housings are fabricated out of Delrin using CAD files "*meso4.22_top_delrin.SLDPRT*" and "*meso4.27_bottom_delrin.SLDPRT*," respectively. We currently outsource the top and bottom housing manufacturing that uses a 3-axis computer numerical control (CNC) mill. Then, the illumination module sub-housings, the frames of the See-Shell implants, and the protective cap were 3D-printed using black poly(methyl methacrylate) (PMMA) resin (catalog no. RS-F2-GPBK-04; Formlabs) through a desktop stereolithography (SLA) printer (Form 2: Formlabs). The following CAD files are used: "*filter_cube_ver5.STL*," "*082019_mesoscope_frame.STL*," and "*10-15-19_frame_cap.STL*." Once the prints come out, they are cured under ultraviolet (UV) light.

3.1.2 Illumination Module Assembly

Each illumination module includes a blue LED, the 3D-printed illumination sub-housing, and a custom-diced emission filter (450–490 nm, 3 × 3 × 1 mm^3, ET 470/40×; Chroma). First, a mixture of solder paste and flux was applied to the LED heatsink pad. Then, a copper plate (~ 80 mm) (B071JKLFXX, Amazon) was folded 4–5 times and soldered on the LED heatsink with a heat gun at 400 °F (HG5520-US, Amazon). To mount the LED on the illumination sub-housing, a cyanoacrylate glue (1818A5, McMaster) was applied to the distal end of the LED. During this process, the LED bulb should remain intact from the glue. Similarly, a few drops of (UV)-curable optical glue (AA352 Light Cure Adhesive; LOCTITE) was applied to the distal end of the illumination sub-housing, where the excitation filter is inserted in the correct orientation as the blue light traverses in one direction. The light direction is typically indicated by the marking along the edge of the filter from the manufacturer. In a consistent manner, care was taken to ensure the glue did not contaminate any filter surface. This process needed to be repeated for all illumination modules.

3.1.3 Mini-mScope Bottom Housing Assembly

Each housing is fitted into the corresponding illumination arm within the bottom housing. Once it was installed, the blue LEDs were serially connected with 19 mm long wires (9510T2, low-voltage high-temperature wire with FEP insulation 29 wire gauge; McMaster-Carr). Likewise, two black (negative) and two red (positive) wires were soldered to the corresponding LED pad and the passive slip ring commutator (Carousel Commutator 1× DHST 2× LED; Plexon). During soldering, the device was inspected for any shorts of wires and LED functionality by using a multimeter open-loop test. Then, black epoxy (RS-F2-GPBK-04, Formlabs) was layered on the top of LED power pads to enhance the soldered wires as well as light shielding. This procedure is repeated for all the blue LEDs.

To mount the collector lens (47-721, Edmund Optics), a few drops of UV curable optical glue were applied to its cylindrical face, and then the lens was mounted into the lens chute in the central shaft of the bottom housing. Using the same glue, the custom-cut emission filter (ET525/50 m, Chroma) was attached to the square emission filter slot within the central shaft. Then, a green LED was glued to the circular slot on the remaining arm of the bottom housing with a few drops of super glue. As done previously for the blue LEDs, any shorts of wires and LED functionality were checked during soldering, and then the black epoxy was applied to reinforce the soldered wires. Finally, two circular neodymium magnets (B07C8ZZ2K9; Amazon) were attached to the bottom surface of the bottom housing using a cyanoacrylate glue. This completes the assembly of the bottom housing of the mini-mScope.

3.1.4 Mini-mScope Device Assembly

The CMOS sensor was tightened to the top surface of the top housing using 18-8 flat head screws. Then, the device assembly was completed by moving the top housing along the central shaft of the bottom housing and fastening it with a 0–80 set screw. To adjust focusing during operation, the set screw can be loosened to allow the top housing to slide along the central shaft. Lastly, the LED power wires in the wire enclosure of the top housing are fixed in place by applying a little epoxy.

While we used the UCLA Miniscope V3 CMOS sensor in the first version of the mini-mScope device, the subsequent versions of the mini-mScope [26] used the higher sensitivity Mini-FAST sensor developed by the Kemere group [27]. While the mini-FAST sensor adds an additional weight of 0.8 g to the device, we have found no significant effects on behavior of the mice during a spatial navigation task [26] compared to control mice, where they exhibit traditional behaviors commonly scored during the task.

3.1.5 Mini-mScope Electronic Circuitry and Wiring

The circuit used to connect the internal mini-mScope components with an external microcontroller (Teensy 3.5, PJRC), behavioral camera, and desktop computer is shown in Fig. 3. The objective of this circuit is to synchronize all components, both internal and external, with the image-acquisition unit of mini-mScope, the optical CMOS sensor collecting fluorescence data from the dorsal surface of the cortex. The pulsation of blue and green LEDs, behavioral camera frame-capture, and CMOS sensor frame-capture are controlled from the internal clock of the CMOS sensor data acquisition board (DAQ) (Miniscope DAQ PCB, Labmaker.org). The internal clock of the CMOS DAQ is therefore oriented as the "command clock," while all other components, including the Teensy 3.6 microcontroller, blue and green LEDs, and behavioral camera, are "subservient" to this clock.

To implement this control scheme, the microcontroller first sends a single trigger to initiate the CMOS DAQ clock and waits for instruction thereafter. The DAQ begins collecting frame-data from the CMOS sensor, and with each captured frame, delivers a 3.3V transistor-transistor logic (TTL) to the microcontroller. This initiates the microcontroller to deliver a 3.3V TTL pulse to power one of two metal oxide field effect transistor relays (power mosFET, IRL520, Digikey Inc.), connected to respective blue and green LEDs. For odd frames, the TTL is sent to the blue LED mosFET relay and powered for 20 ms, and for the even frames it is sent to the green LED mosFET relay and powered for 4 ms. Simultaneously, the same frame-by-frame TTL is delivered to a desktop computer and a mounted behavioral camera configured to write frames as the TTL is received. In this manner, CMOS frames containing fluorescence data, blue and green LED pulsation, and behavioral frames containing video feed of the experiment are synchronized by frame.

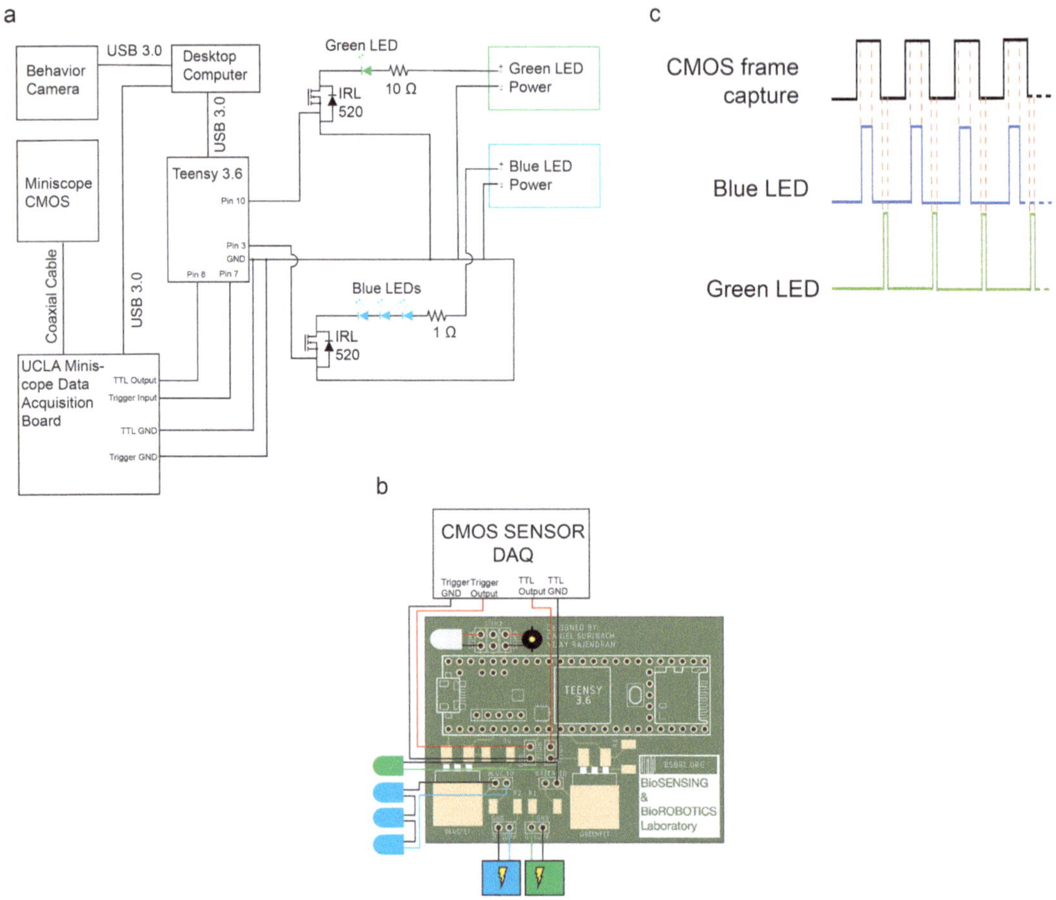

Fig. 3 mini-mScope illumination design and wiring. (**a**) Schematic of wiring used to synchronize both the internal and external components of a mini-mScope recording setup. The internal components (a blue LED array, a green LED, and a CMOS sensor) illuminate and record from the dorsal cortex, while the external component (a behavioral camera) records free behavior of a mouse. For every frame captured by the CMOS sensor, the CMOS data acquisition (DAQ) board sends a transistor-transistor-logic (TTL) pulse to a microcontroller, which commands power switches for three blue LEDs and one green LED. The TTL pulse is also delivered to the behavior camera to write frames synchronously with the CMOS camera. (**b**) Schematic of custom PCB to incorporate the wiring as shown in (**a**). (**c**) Representation of the synchronization of components in the mini-mScope array. The TTL pulses from the CMOS DAQ ensure that the blue LED array is switched on during only even frames, while the green LED is switched on during only odd frames. The TTL pulses also ensure synchronous even and odd frame capture from a behavioral camera

To wire the CMOS sensor to the CMOS DAQ for data transfer, a single coax cable (50 Ω coaxial silicone rubber-jacketed cable; Cooner Wire) is used. Then, the blue and green LEDs are powered through the 29-gauge wires that run from PCB ground and power slots to their respective LEDs. It is advisable to route both the power wires and the coax cable through a commutator (Carousel Commutator 1× DHST 2× LED; Plexon) to relieve strain on the system. Similar low-impedance solid-core wires should be used to connect the trigger and TTL communication lines from the CMOS DAQ to the PCB and the TTL ports of the behavioral camera directly. Finally, the microcontroller and behavioral cameras are connected to the desktop computer via a micro-USB and USB 3.0 cable respectively.

The overall schematic of the circuit is used to synchronize the data acquisition in the CMOS sensor, with data acquisition from the behavior camera, and the alternate pulsing of the blue and green LEDs is shown in Fig. 3. The Miniscope DAQ sends an external trigger to a microcontroller (Teensy 3.5, PJRC) for every frame captured. At each odd frame captured, the microcontroller sends a 3.3V transistor-transistor logic (TTL) pulse to the power metal oxide field effect transistor (power mosFET, IRL520, Digi-key Inc.) relay to power the three blue LEDs connected in series. Similarly, a second TTL pulse lasting 4 ms is sent to a mosFET relay to power the green LED. A single coax cable (50 Ω coaxial silicone rubber-jacketed cable; Cooner Wire) is used to connect the CMOS sensor to the main data acquisition board (Miniscope DAQ PCB, Labmaker.org) and the blue LEDs and the green LED are connected to their respective mosFET relays using the 29-guage wires. The wires powering the LEDs and the coax cable are routed through a commutator (Carousel Commutator 1× DHST 2× LED; Plexon) for strain relief.

3.2 See-Shell Implant Assembly

The mini-mScope attaches to the See-Shell implant, which is chronically implanted on the mouse (*see* **Note 2**). The assembly instructions for the See-Shell implants have been described extensively in our previous work [21, 22]. Briefly, once the 3D-printed See-Shell frames are cured, a polyethylene terephthalate film (Melinex 462, 50-μm-thick; Dupont) was bonded to the curved surface of the frame by applying a small amount of epoxy (ScotchWeld DP100 Plus Clear; 3M). Then, two disk magnets were adhered to its slots on the implant and protective cap with a few drops of cyanoacrylate glue. During this step, the magnet orientation must be aligned with the opposing magnetic polarity of the magnets on the mini-mScope bottom surface. Then, a 0–80 nut was screwed into the hole in the posterior tab of the implant. This nut is used to fasten a stabilizing screw to prevent motion artifacts when the mini-mScope is docked to the implant during experiments (*see* **Note 3**).

3.2.1 Benchtop Testing of the Mini-Scope Optics

To prepare for in vivo imaging experiments with the mini-mScope, the device needs to be tested in the benchtop to ensure optimal performance. To avoid scratching the surfaces, we carefully inserted the optical components into the mini-mScope bottom housing and LED filter cubes using rubber-tipped forceps.

The excitation and emission filters were pointed towards the field of view, as was the curved end of the lens. To test the resolution of the device, we attached a USAF 1951 test target strip to the

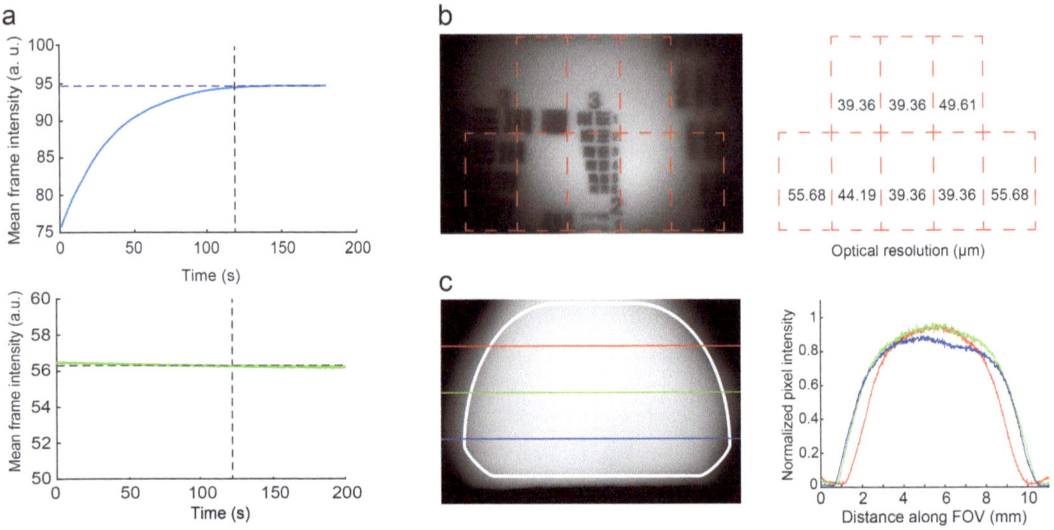

Fig. 4 Resolution and LED dynamics characterization: (**a**) Long-term LED stability during a controlled test with a fluorescein dye agar phantom with excitation and emission bands similar to GCaMP6f. Blue excitation and green reflectance LEDs were pulsed as described in Fig. 3 and emitted signals were collected onto the mini-mScope CMOS sensor. *Top:* Average frame intensity captured from blue excitation light illumination. The dashed line at approximately 120 s indicates stability in LED intensity dynamics as they heat up during the initial experiment duration. *Bottom:* Average frame intensity captured from green reflectance illumination. The dashed line at approximately 120 s indicates stability in LED intensity dynamics as they heat up during the initial experiment duration. (**b**) *Left:* 1951 USAF resolution test target overlaid onto the mini-mScope FOV. *Right:* Corresponding resolution values obtained within each grid in the left image. (**c**) *Left:* Fluorescein dye agar phantom illumination profile using the blue excitation LEDs and captured onto the mini-mScope CMOS sensor. The colored lines indicate illumination intensity profiles obtained along the mediolateral sections of the FOV. *Right:* Corresponding normalized illumination profiles obtained from the mediolateral sections of the left image

bottom of a fully assembled See-Shell implant and placed the mini-mScope on top of a dimmable microscope LED ring (B0BQBKVZG9, Amazon). Adjusting the height of the top housing containing the CMOS sensor relative to the bottom housing fixed on top of the LED ring allowed us to test the device's focus. We then connected the CMOS sensor and launched the Miniscope software, adjusting the LED intensity until we saw an image with no saturation spots in the live view. Finally, we checked the full range of focus possible to ensure that we could see the resolution test target's smallest feature, which corresponded to a resolution of approximately 35 μm.

Once the optics are assembled, the excitation and reflectance LEDs can be tested. The array of three blue LEDs need to deliver a cumulative ~31 mW of instantaneous power to the brain (Fig. 4). The green LED delivers ~0.22 mW of instantaneous power. Plug in the microcontroller and upload the "*LED pulse test.ino*" Arduino script. This script alternately pulses the LEDs without synchronization to the Miniscope CMOS sensor. Switch on the power supplies for the blue and green LEDs and set the blue LED supply to 7 V 0.8 A and the green LED supply to 5 V 0.1 A. Visually confirm that the LEDs are pulsing.

Once validated, the mini-mScope hardware and LED control software interface can be tested with the Miniscope DAQ software (see Github repository for software files). The testing phase encompassed several checks: (1) The LED control hardware triggers the Miniscope DAQ successfully to start and stop video recordings. (2) The DAQ sends TTL triggers when each frame is captured on the CMOS sensor. (3) The DAQ TTL pulses are registered via hardware interrupts on the LED control hardware. (4) The LEDs turn on/off while synchronized to the DAQ TTL pulses. To test these, upload the "*mini-mScope master LED.ino*" Arduino script. This will interface with the CMOS sensor DAQ and LEDs. Upload the software and open the Serial Monitor; there should be a user-input prompt waiting for a keyboard interrupt. In a separate window, open the Miniscope software and connect the CMOS sensor. Select a desired frame rate and tick the "trigger ext" feature, which allows the Miniscope DAQ to enter a video start/stop trigger mode. Ensure the blue and green LED power supplies are switched on. Navigate back to the Arduino Serial Monitor window and type "y" followed by enter. This will trigger the microcontroller to send a pulse to start video recordings for the CMOS. When a frame is captured and registered through the DAQ, it will send a TTL pulse back to the microcontroller that will be ready in a hardware interrupt mode. When the signal is received, the microcontroller will pulse either the blue or green LED depending on the incoming frame number.

3.3 Animal Preparation

Ensure all animal procedures are performed in accordance with approved protocols by the institutional animal care and use committee. Ensure proper aseptic procedure is followed during the surgery. All mini-mScope experiments in our group were conducted in adult Thy1-GCaMP6f [23] or Emx-CaMKII-Ai85 mice [24] with ages ranging between 8 and 30 weeks. These mice typically weigh 30–45 g and are housed in cages maintained at 14 h light/10 h dark cycle at 20–23 °C and 30–70% relative humidity with ad libitum access to food and water.

The implantation procedure was adapted for standard protocols for implanting glass coverslip cranial windows [28]. Prior to implant surgery, 2 mg/kg of sustained-release buprenorphine was preemptively administered (Buprenorphine SR-Lab; ZooPharm) and 2 mg/kg of meloxicam for analgesia and inflammation prevention was administered. Mice were introduced into an anesthesia induction chamber containing 1–5% isoflurane in pure oxygen, and once the animals were anesthetized with no response to toe-pinch stimulus, the animal was prepared for a craniotomy. The scalp was shaved over the dorsal cortex area and sterilized using standard aseptic techniques. Sterile eye ointment (Puralube; Dechra Veterinary Products) was applied to prevent the eyes from drying during the procedure.

Following surgical preparations, mice were immediately transferred to a stereotax (e.g., Model 900LS; Kopf). The scalp was excised using surgical scissors and the fascia was removed using a 0.5 mm microcurette (catalog no. 10080-05; Fine Science Tools). The See-Shell implant was sterilized with a blank 3D-printed See--Shell frame by soaking in 70% ethanol for 2 min, followed by thorough rinsing with sterile saline (*see* **Note 2**). The blank See-Shell frame was placed on the dorsal skull so most of the edge of the frame sits flush on the exposed skull. The See-Shell is designed to cover an area spanning 2 mm anterior and 6 mm posterior to bregma and should be placed symmetrically about the midline. A surgical marker (SKM-0005, AD Surgical) was used to mark the inside perimeter of the See-Shell frame. A craniotomy was drilled slightly outside the See-Shell frame outline using a high-speed pneumatic dental drill. As this is a challenging procedure, we have also used an automated craniotomy system to perform this step in many of the experiments conducted in our group [29, 30]. Fine forceps were used to gently remove the bone island in the craniotomy. The open craniotomy was covered with a sterile gauze pad soaked in sterile saline immediately after bone removal to keep the brain hydrated. Since the skull is vascularized, removal of the bone may cause bleeding. The mouse was left and observed in this state for 5–10 min to allow for the bleeding from the skull to stop. Heart rate and breathing rate of the mouse were continuously monitored to ensure the anesthetic plane is maintained throughout the procedure.

The gauze pad placed on the brain was removed, and sterilized cotton tip applicators were used to clean up remaining blood and skull tissue (catalog no. 823-WC, Puritan). The See-Shell implant was placed gently on the skull so that the frame corresponded to the craniotomy. Cotton tipped applicators were used again to dry the area surrounding the implant frame. A few drops of surgical grade cyanoacrylate glue (Vetbond, 3 M) were applied around the edge of the See-Shell to bond the implant to the skull surface. After the glue set, opaque dental cement (Metabond, Parkell) was applied to cement the implant to the skull. Once the cement was fully cured (~10 min), the protective cap was attached onto the implant, and the mouse was removed from the sterotax and placed onto a heated recovery pad (catalog # 72-0492, Harvard Apparatus). Care was taken to ensure the animal was fully ambulatory before transferring it to a clean cage for recovery. Post-operative analgesia and anti-inflammatory drugs were administered for up to 72 h after the procedure to ensure proper recovery from the surgery. In our experience, the implanted windows become clear 4–5 days after

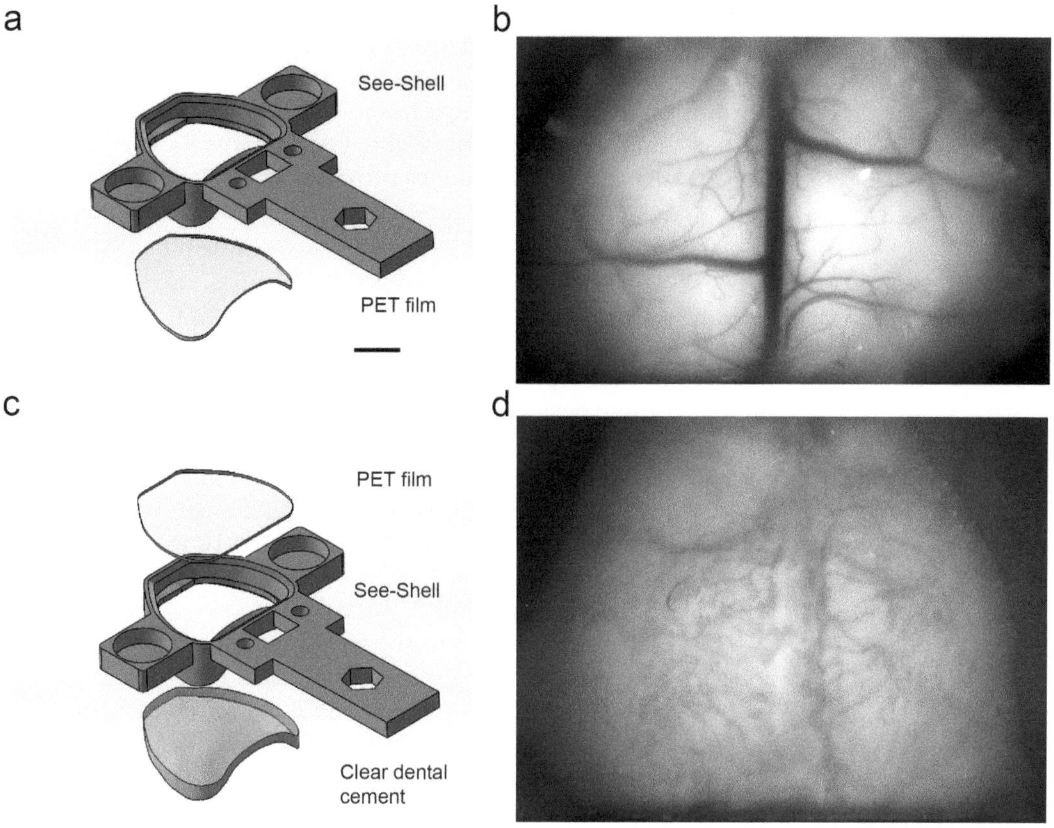

Fig. 5 Mesoscale imaging through an intact skull using the mini-mScope. (**a**) A CAD schematic illustrating the See-Shell for imaging through a craniotomy. (**b**) Image of mouse brain through fully transparent See-Shell implant. (**c**) A CAD schematic illustrating the modifications to the mini-mScope See-Shell for intact skull calcium imaging. Scale bar, 4 mm. (**d**) Cortex of a mouse through the intact skull imaged during the sensory evoked stimuli experiments. Scale bar, 2 mm. (Adapted from Rynes et al. [21])

implantation, and can remain clear over a year if the surgical procedure is carefully executed [22].

Alternatively, an intact skull preparation can be used. To do this, a See-Shell implant frame can be fixed directly onto the skull to create a chamber. The chamber should then be filled with a thin layer of clear dental cement to cover the skull and create a refractive index-matched surface by capping it with PET film cut to the profile of the outer perimeter of the frame. A CAD schematic demonstrating the modification to the See-Shell implant required to complete this procedure is shown in Fig. 5a, and a photograph of the resulting through skull imaging is shown in Fig. 5b.

3.4 Behavioral Acclimatization

In all our experiments, animals were allowed to fully recover from the surgery for up to 7 days prior to behavior experiments. Before starting the actual experiment, the mice were first handled by the experimenter in their home cage. Handling the animals for

3–5 days before the start of the experiment decreases their anxiety as they get used to the touch of the experimenter. On the first day, mice are picked up a few times (3–5 times), held briefly for a few seconds and released into their home cage. On the second day, mice are picked up and allowed to explore both arms of the experimenter for 20–30 s with gentle nudges provided as needed to correct the course of exploration. This process is repeated for 3–5 days until the animal voluntarily comes closer to the experimenter's hand when the hand is placed in the animal's cage.

Once the mice are fully acclimatized to the experiments, the animals are acclimatized to the process of mounting and unmounting the mini-mScope. We 3D-printed mini-mScope dummy bodies and filled these devices with silicone glue to match their weight to the functional mini-mScope. Animals were picked up, placed on the arm of the experimenter and the dummy mini-mScopes were mounted on the implanted See-Shells. The animals were released to their home cage and allowed to explore for 10 min after which they were picked up and the dummy device was then unmounted. This process was repeated for 2–3 days.

Once fully acclimatized to the weight and the processes of the mounting and unmounting the mini-mScope, mice can perform a range of behaviors and show comparable locomotion characteristics compared to untethered mice in the same arena. In our original study, we performed comparisons of the average speed, total distance traveled and max speed of mice bearing the mini-mScope to untethered mice. We found no significant differences in any of the metrics (Fig. 6a–d). In a second study, we have also demonstrated that mini-mScope bearing mice can navigate the Barnes maze, a complex 2D spatial navigation task where mice learn the location of an escape in the edge of the maze (Fig. 7e, f). The Barnes maze is a circular arena with holes in the perimeter; mice are presented with mildly aversive stimuli, typically light and sound, to encourage them to find which hole contains the goal, which contains cover from the aversive stimuli. Mice bearing the mini-mScope were able to learn to navigate to the goal via the utilization of a progression of search strategies which can be called random, serial, and spatial.

3.5 In Vivo Mini-mScope Imaging

One day prior to imaging during freely behaving experiments mice were lightly anesthetized (0.5–1% isoflurane in pure oxygen) and fixed it in a stereotactic apparatus. The mini-mScope was mounted on the anesthetized mouse, and the green LED was used to illuminate the surface for focusing. This was done by adjusting the top housing along the central shaft of the bottom housing until the area around the superior sagittal sinus was in focus. Figure 5b, shows an example image acquired from a Thy1-GCaMp6f mouse. As the mini-mScope images a curved surface using a single collector lens, it is not possible to have the entire FOV in focus at any time. At the best focus, the image resolution is ~40 μm, while at the edges of the

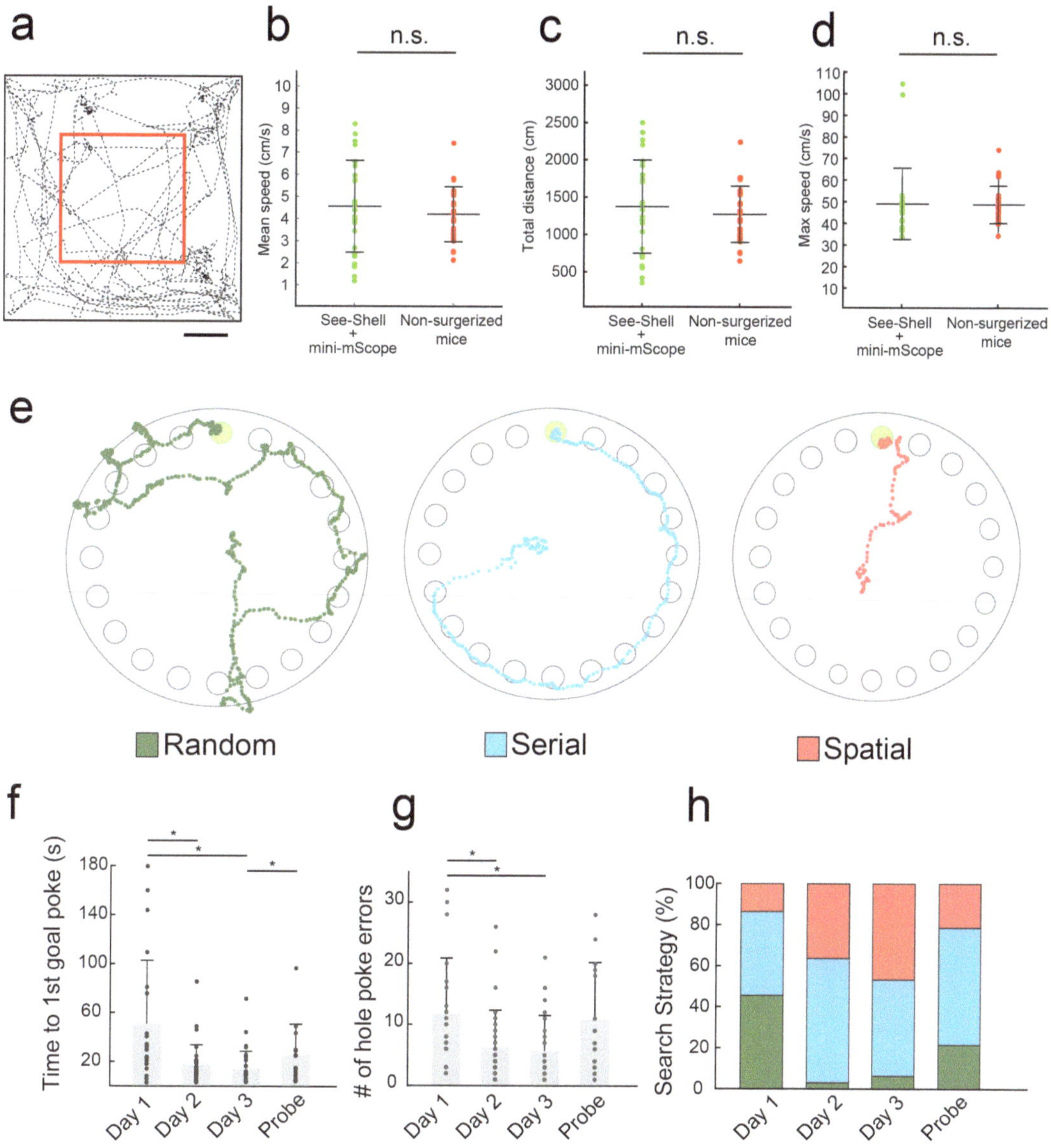

Fig. 6 Mouse behavior performance bearing mini-mScope. (**a**) Trace of mouse position in a 5-minute trial while implanted with a See-Shell window and fitted with the mini-mScope. (**b**) Plot of mean speed ± SD of mice with a See-Shell implanted, mice with both the See-Shell and head-mounted mini-mScope, and non-surgerized control mice (n = 5 mice per group, 5 trials per mouse). (**c**) Plot of total distance ± SD traveled by mice with a See-Shell implanted, mice with both the See-Shell and head-mounted mini-mScope, and non-surgerized control mice (n = 5 mice per group, 5 trials per mouse). (**d**) Plot of maximum speed ± SD of mice with a See-Shell implanted, mice with both the See-Shell and head-mounted mini-mScope, and non-surgerized control mice ($n = 5$ mice per group, 5 trials per mouse). (**e**) Traces obtained from tracking data of mice which utilized random, serial, and spatial search methods in the Barnes maze fitted with the mini-mScope. (**f**) A bar plot showing the mean primary latency, or time to first goal hole discovery, across days as mice learned to navigate the Barnes maze fitted with the mini-mScope. * indicates $p < 0.05$, Wilcoxon rank sum test. (**g**) A bar plot showing the mean number of primary errors, or the number of times the mouse checked an incorrect hole before reaching the goal. * indicates $p < 0.05$. (**f**) A bar plot depicting the percentage of search strategies utilized across trial days in the Barnes maze task

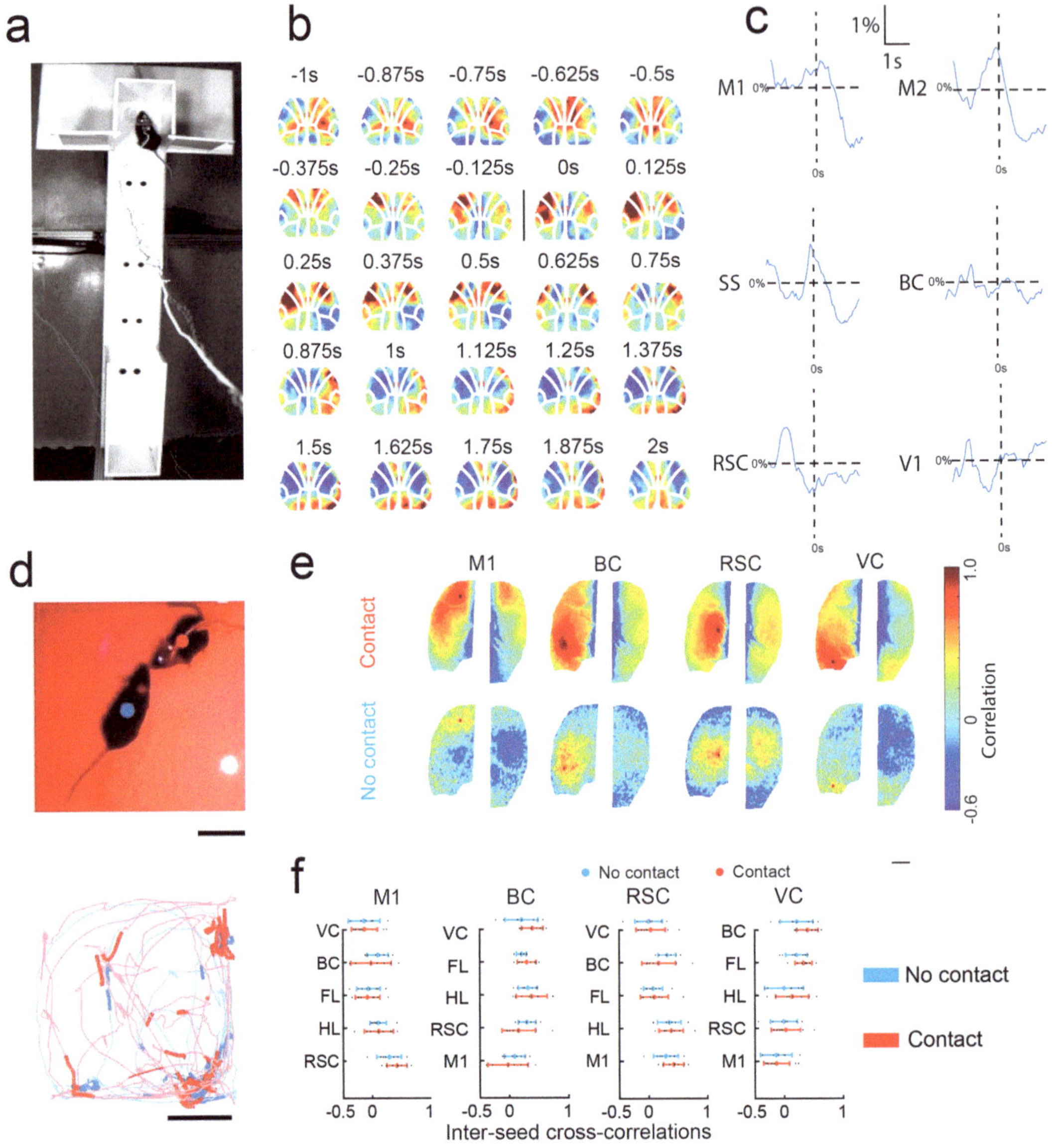

Fig. 7 Calcium imaging data analysis. (**a**) A behavioral experiment using a mouse bearing a mini-mScope during a water lick port task. The mouse crosses from a safe zone into a water reward zone to collect a water reward while the mini-mScope records calcium dynamics. (**b**) Example montage of DF/F pseudo-color spatial maps during a water lick task from −2 s to 2 s. $t = 0$ denotes when the mouse reaches the lick port. (**c**) Average DF/F maps taken at various anatomical cortical ROIs. Averages were computed across all water lick port events across all trials for one mouse shown here. (**d**) *Top:* A different experimental paradigm showing a social behavior experiment. Two mice of the same sex are introduced and allowed to socially interact while the mini-mScope records dynamics from one of the mice. Scale bar is 5 cm. *Bottom:* The pose tracking data for each mouse is computed and plotted for one trial. Segments of darker colors indicate positions of the mice when they are within 5 cm of each other. Scale bar is 10 cm. (**e**) Representative seed pixel correlation maps for the social behavior experiment generated from manually scored behavioral data that segmented trials into periods of social contact and no social contact. Seed pixel maps were calculated for seed pixels within anatomically defined cortical ROIs. Scale bar 2 mm. (**f**) Average inter-seed correlations associated with the seeds in (**e**) but across trials and mice ($n = 6$ trials, $n = 6$ mice). The error bars denote the mean +− S.D

FOV, the image resolution is ~60 μm. Once the top surface of the brain was in focus, the imaging plane was moved slightly (~50–200 μm) below the dorsal surface of the brain to minimize blood vessel artifacts, and then the focusing adjustment screw was tightened. This was performed for each mouse as the plane of focus slightly differed. Once the FOV was focused, imaging was performed by alternately pulsing the three blue LEDs and the green LED. Power was adjusted to the set of LEDs so that at ideal imaging conditions, mean pixel intensity values across the FOV were ~180/255 for the blue LED image and ~100/255 for the green LED image while ensuring that there were no saturated regions of the FOV in either imaging wavelength. A protective cap was applied to the top of the implant to protect the brain and implant from damage and to cover the PET film.

On days when imaging, the protective cap was removed, and the surface of the PET was examined for debris. If the imaging FOV was found to be occluded, PBS or DI water was used to flush the surface of the implant and dried. The mini-mScope was then mounted to the implant. Before imaging, the LEDs were pulsed for up to 2 min prior to the experiment to ensure they are completely warmed up and the output illumination reaches an equilibrium (Fig. 4). mini-mScope imaging can be carried out for a tested limit of 1 h generating ~1 Gb of data per minute of data acquisition with the Miniscope V3 sensor and ~3 Gb of data per minute for the miniFAST sensor. We performed longitudinal behavioral experiments for up to 30 min and validated the mini-mScope did not have significant effects on locomotion compared to shorter experiment durations. Longer imaging sessions will likely require additional handling and training of the mice to acclimate them steadily to increasing loading over time.

3.6 Data Analysis

All custom scripts used for preprocessing and analysis of the data are available at the following code repository: https://github.com/bsbrl/mini-mScope. Much of the preprocessing of the data and the subsequent analyses are similar to pipelines developed for analyses of mesoscale data in head-fixed animals. In addition, we have created additional scripts made to account for the mini-mScope hardware and recording conditions unique to freely behaving animals.

3.6.1 Preprocessing

Analysis of mesoscale calcium imaging data acquired with the mini-mScope progresses in two stages. First, the data is run through a preprocessing stage where the data goes to a series of steps to prepare for further analysis. First, the data is acquired in two separate illumination conditions—at the excitation wavelength of the fluorophore and while illuminated at the isosbestic wavelength or at a wavelength used to capture reflectance images. Corrections need to account for fluctuations in illumination intensity as well as motion artifacts during the freely moving nature of the recordings. This is followed by correcting for hemodynamic artifacts.

The data from the mini-mScope CMOS sensor is captured in AVI format (RGB, 480 × 752 image size when using the Miniscope V3 sensor, 1800 × 1040 image when using the mini-Fast sensor). Frames are acquired under blue and green light illumination conditions. A custom script in MATLAB (see Github repository) can be used to convert the videos for RGB to grayscale, excluding the 2-minute period at the start of acquisition when the LEDs are warming up and segmenting the videos into the blue and green channel videos.

The green channel and blue channel videos are then motion corrected for lateral artifacts using rigid motion correction [31]. The parameters used for the Moco plugin were $w = 40$ and downsample value = 0.5. Log files of pixels are generated, and a corrected output video is generated for further analysis.

To remove both the heartbeat oscillation and incorporate the Nyquist frequency for calcium indicators' rate-limiting frequency, a zero-phase Chebyshev bandpass filter is applied, with a low-pass cutoff of 0.1 Hz to remove any slow dynamic changes that likely are not calcium-dependent, and a high-pass cutoff of 5–7 Hz depending on the imaging acquisition frame rate [32, 33]. These filters will remove sources of noise that aren't physiological. The calcium imaging data can then be passed to a correction algorithm that removes global calcium transients and excitation illumination fluctuations [34]. The correction algorithm removes intensity changes that arise from fluctuations in intensity of LED illumination. These fluctuations can be caused by heating of the LEDs during usage. Finally, to increase contrast and reduce the contribution of the imaging sensor's photon noise, a spatial smoothening filter is applied to the calcium imaging data [21, 26, 34, 35].

The last important step in the preprocessing stage of the data analysis is the correction for hemodynamic artifacts. Increases of neuronal activity are paired with and contaminated by hemodynamics, which is the neurovascular coupling and increased blood flow that occurs with neural activity [36]. The excitation and emission wavelengths of GCaMP are found in the same spectra as that of oxygenated and deoxygenated hemoglobin. As a result, changes in blood oxygenation influence the scattering of excitation and emission light, thus changing the raw pixel fluorescence that is captured during imaging [4, 37]. There are several methods available to correct for neurovascular coupling. A commonly employed method is to utilize dual wavelength imaging to capture hemodynamic changes during recording. Regression-based correction, usually by subtraction or division, is then implemented to the raw fluorescence data to remove the influence of the neurovascular coupling. A common wavelength of light used to measure hemodynamics without impeding GCaMP signals can be found in the green spectrum (~540 nm) [12], which measures reflectance of the surface tissue and broad changes in blood flow.

In the mini-mScope the green LED illuminates the dorsal cortical surface at a peak excitation wavelength of ~540 nm. We perform hemodynamic correction using the method described by Musall et al. [12]. Briefly, reflectance DF/F was calculated for each pixel in the green channel images. To capture the slow kinetics of hemodynamics, a zero-phase Chebyshev low-pass filter (cutoff frequency of 0.15) is applied. The DF/F calcium traces are divided by the DF/F reflectance traces to generate a hemodynamic corrected trace.

3.6.2 Image Registration to a Common Anatomical Reference Frame

Multi-brain region imaging relies on defining anatomically relevant regions of interest (ROIs). This is commonly performed by manually aligning an image of the brain to a commonly used anatomic reference atlas [35, 38–40]. ROIs are then manually drawn over each desired anatomic region. While a common reference atlas can be useful to compare results from different studies, manual ROI placement can suffer from individual variances in mouse-to-mouse anatomy, leading to imprecise parcellation and ROI definition during smaller sample size or complex tasks that have not been studied yet [11, 41]. An alternative to manual ROI curation is to use unsupervised platforms to discover pixels and regions that fire in groups or patterns by using methods such as clustering analyses [42, 43], nonnegative matrix factorization [44], and independent component analysis [5, 19]. Once defined, manually curated ROIs or sequences of pixels that are grouped with unsupervised methods can be observed for activation patterns and functional connectivity across behaviors or imaging sessions.

3.7 Calcium Data Analyses

The first step after preprocessing the data is to construct the percent change in fluorescence when compared to baseline (DF/F). In our analyses, we define the baseline of fluorescence pixel-wise as the average of each pixel's time series for each trial. For more comparable results across mice, a z-score is applied to each DF/F trace to normalize it by standard deviation (Fig. 7).

Several analysis methods can be applied to the data acquired using the mini-mScope, that provide insights into the activity of individual brain regions and the temporal relationship between them. While not exhaustive, we detail below examples of two such analysis techniques used on the mini-mScope data acquired in our group.

Seed-Pixel Correlation Maps Seed pixel correlation maps are a commonly used framework to investigate the functional connectivity between one brain ROI and others in a behavioral epoch. A seed pixel within an anatomical brain ROI is selected and the trace for that seed is extracted. The Pearson's correlation coefficient, a metric that shows how similar two vectors are temporally, is computed between the seed pixel selected and all the other pixels in the FOV. For the mini-mScope, we selected 1 seed pixel located within each of 6 anatomically defined ROIs via alignment to the Allen Brain Atlas. The seed pixel regions included motor cortex (M1), forelimb, hind limb, barrel cortex (BC), somatosensory cortex, retrosplenial cortex (RSC), and the visual cortex (VC). We manually scored behavioral data by watching behavior videos and extracting interesting periods, such as moving, walking, grooming, and rearing. Seed pixel correlation maps are constructed by comparing the correlation between the seed pixels and all other pixels in the FOV during each behavioral period time window.

Image Correlation and Unsupervised Clustering to Identify Cortical Activation Patterns Cortical activity during free behavior is characterized by repeated patterns of cortical activation that occur sporadically over time, each activation pattern lasting 50 ms to over 1 s. In freely behaving mice, where task structure is relatively unconstrained, representing the complex spatiotemporal dynamics of the mesoscale calcium activity in a low dimensional space allows tractable analyses of the contribution of cortical activity to such complex behaviors. This activity can even be compared with simultaneously recorded electrographic data (*see* **Note 5**). In recent work, we developed a novel unsupervised methodology to cluster spatial patterns of cortical activation into a set of commonly occurring patterns of cortical activation or "states." This methodology relies on using k-means clustering to discover these patterns (Fig. 8).

To perform the cortical activation detection, we first load all the trials for one mouse for the experiment set up into computer memory. A common mask was drawn for each trial and averaged to form a common mask for each mouse where the superior sagittal sinus (SSS) and background is omitted to reduce sources of noise and decrease computational time. Each frame was correlated with all other frames using a Pearson's correlation coefficient. The correlation indicates what frames share patterns of repeated cortical activity across multiple time scales. The image correlation matrix is then sorted using k-means clustering to search for the patterns of repeated activity. K-means is run 50 times with different cluster initialization points from 1 to 30 clusters. At each cluster number and initialization, the image correlation matrix was sorted into the iteration that minimizes the cluster centroid distance to each point placed in that cluster. A t-distance optimization algorithm was used to calculate the average and variance of the image correlation value within each cluster and across other clusters [45]. The optimization

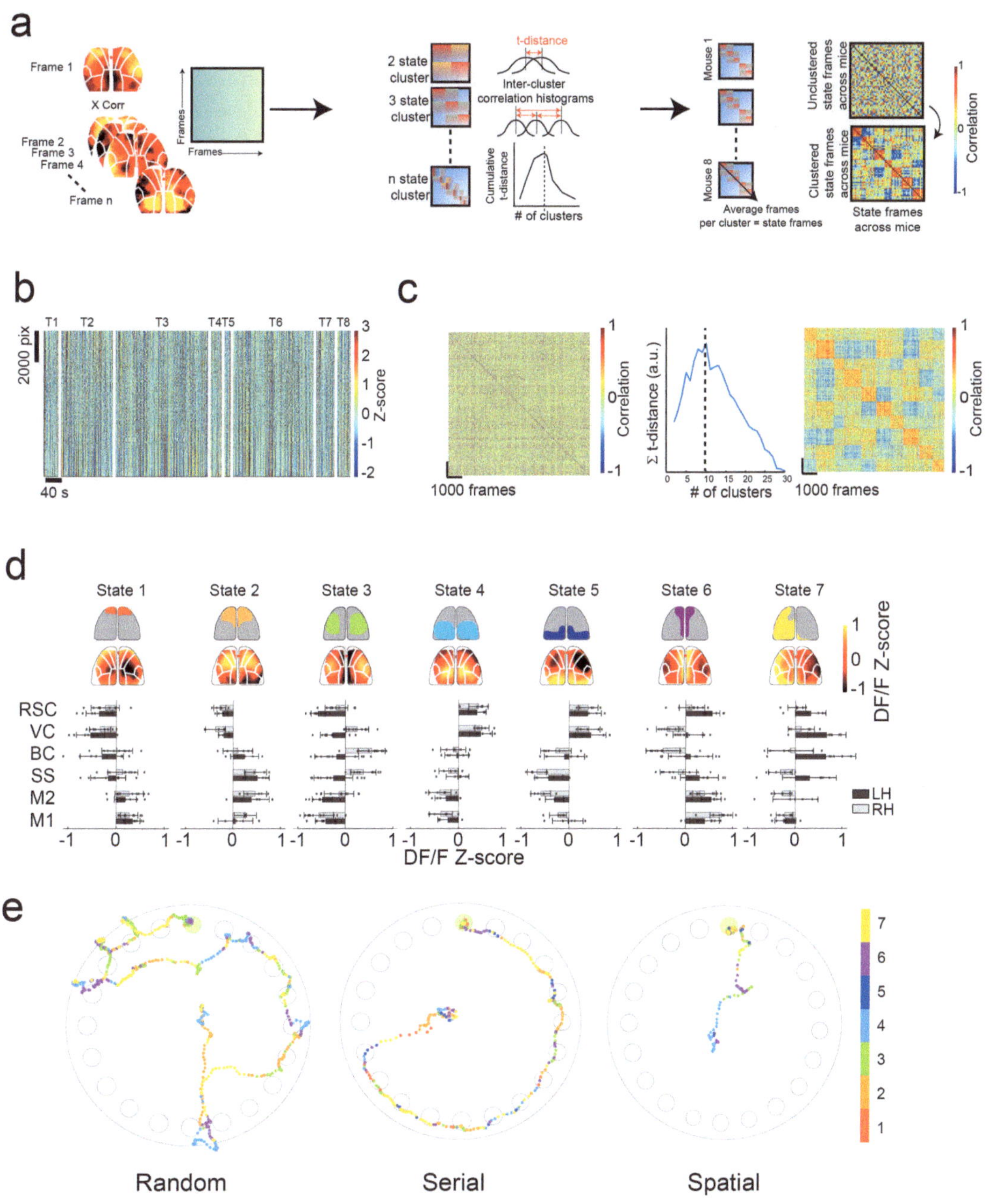

Fig. 8 Repeated patterns of cortical activity across large time scales during a Barnes Maze spatial navigation task. (**a**) *Left:* An image correlation matrix is constructed by using a Pearson's correlation coefficient for all frames and trials in a mouse. Images with high correlations share similar spatial activation patterns. *Center:* K-means clustering is used to cluster the image correlation matrix into n clusters. For each cluster, a t-distance optimization algorithm is used to compute the variance of the correlation within and across different clusters defined by k-means. Since the number of clusters (defined as states) is unknown a priori, the t-distance optimization will have a global maximum when the correlation is maximized within clusters and minimized across clusters. *Right:* Mean DF/F activity maps are constructed for each state and mouse. The average DF/F maps are grouped into a matrix and a second image correlation is computed. This will show the spatial patterns of DF/F that are similar across states found in different mice. The correlation matrix is sorted into 7 states of h highly correlated average DF/F activity maps across mice, forming a common state model.

will converge with a global maximum when the correlation is maximized within clusters and minimized across clusters. The maximum t-distance value will indicate the optimum number of states that exist across trials for each mouse. The number of states detected across $n = 8$ mice varied from 5–10 optimum states with an average of 6.5 states across mice.

To generate a brain state model across mice, a similar computation was performed. First, within each mouse's optimization, the index of all the frames that classified into each optimum cluster were stored. The frames in each cluster were then averaged to produce a mean representation of each state's peak activity patterns. The states' mean frames were stored and grouped into a matrix across mice. A Pearson's correlation was performed on this matrix, which formed a second image correlation matrix that shows the similarities of average activity patterns across mice. From here, a common set number of states can be selected to cluster the image correlation matrix, where for our experiments, we had an average of 6.5 states across mice, so we selected 7 states as the common framework. K-means clustering was used to cluster the second image correlation matrix unsupervised into 7 groups, forming the final state representation used for our experiments.

4 Notes

1. Our device borrows the imaging elements from the open source Miniscope (www.miniscope.org) environment. Thus, while custom parts are used, all components can be readily procured and a user can have a functioning mini-mScope in <4 weeks. Once assembled, the mini-mScope is a robust device that requires little maintenance.

Fig. 8 (continued) (**b**) Example of the clustering methodology used to identify cortex-wide brain states from widefield calcium data during the spatial navigation task for one mouse. DF/F data from all pixels in the FOV across all trials for one mouse are combined into one matrix and shown, where trials are indicated with T1-T8 labels. (**c**) *Left:* A correlation matrix is constructed by computing the Pearson's correlation coefficient between all frames. *Middle:* K-means clustering is used to sort the image correlation matrix into n clusters, calculating a t-distance value at each step. The maximum t-distance value indicates the optimal number of states for this mouse ($k = 10$ states). *Right:* The image correlation is re-sorted by the optimal number of clusters defined by the t-distance optimization algorithm, resulting in 10 states for this mouse. (**d**) *Top:* Common state space model across $n = 8$ mice for the spatial navigation task. Optimum number of states varied from 5–10 states across all mice, with an average of 6.4 +– 2 states. Seven states were used in the final model across mice. States were identified by cortical areas of high DF/F activation. The top row illustrates simplified activity maps with high DF/F z = score activity. The DF/F maps below the top row illustrate the average DF/F z-score maps for the mouse optimized in (**b, c**). *Bottom:* Bar graphs denote the average DF/F z-score of cortical regions across all mice using the Allen Brain Atlas. Error bars denote standard deviation

2. The imaging data is highly dependent on several aspects of experimentation with freely behaving animals that need to be carefully considered. First, while the implantation of the See-Shell procedure is adapted from chronic cranial window implantation procedures (see Subheading 3.3), the surgery involves opening a craniotomy over a significantly larger area of the brain than typical circular cranial windows. Thus, care must be taken to ensure the surgical procedure is conducted in sterile conditions and all aseptic techniques are followed. We have previously shown that the See-Shell can be chronically implanted on mice for several months, without any occlusion. In our hands, >90% of the mice survive the procedure, with skull clear for >340 days in over 60% of the mice implanted [29].
3. Care must also be taken to ensure the mini-mScope is attached firmly to the implant during imaging sessions, as any relative motion between the mini-mScope and the implant will result in motion artifacts in the acquired calcium data. In our experiments, we were able to limit the absolute maximum x and y displacement of the device by 21 ± 21.8 μm and 14.1 ± 17 μm, respectively, and we reduced the mean absolute displacement in x and y of the device by 2.0 ± 2.5 μm and 9.3 ± 4.9 μm, respectively.
4. We primarily use transgenic mice broadly expressed in excitatory populations of cells [23]. We have also shown the efficacy of mini-mScope imaging in transgenic animals expressing optical reporters of glutamate [21]. In the future, calcium indicators expressed in specific populations of cells can be used to investigate the role of these specific cell types to global coordination of activity [12, 46]. Recent advances in neural activity reporters [47, 48] and neurotransmitter reporters [24, 49, 50] can be leveraged in the near future to perform widefield imaging at higher temporal resolution and greeted dynamic range. Further, fitting mini-mScopes with ultra-fast imaging sensors [27] coupled with new voltage indicators optimized for voltage imaging [51] may allow pan-cortical voltage dynamics to be imaged in freely behaving mice.
5. Optical imaging can also be coupled with electrophysiological recording modalities to obtain complementary datasets at much high temporal resolution. For instance, we have previously shown that the See-Shells can be seamlessly integrated with patterned electrode arrays [52, 53] for recording surface field potentials. Inkjet-printed transparent arrays allow simultaneous ECoG and optical imaging across the same field of view without loss of signal fidelity [54]. Finally, mini-mScopes can also be modified to integrate miniature penetrating neural recording probes that can combine widefield optical imaging

with electrophysiological recordings from localized regions of the brain.

In conclusion, mini-mScope imaging will contribute to unraveling the mechanisms by which activity across the brain is coordinated during complex behaviors that will be difficult to replicate in head-fixed preparations.

References

1. Chen T-W et al (2013) Ultrasensitive fluorescent proteins for imaging neuronal activity. Nature 499:295–300
2. Adam Y et al (2019) Voltage imaging and optogenetics reveal behaviour-dependent changes in hippocampal dynamics. Nature 569:413. https://doi.org/10.1038/s41586-019-1166-7
3. Gong Y et al (2015) High-speed recording of neural spikes in awake mice and flies with a fluorescent voltage sensor. Science (80-) 350: 1361–1366
4. Allen WE et al (2017) Global representations of goal-directed behavior in distinct cell types of mouse neocortex. Neuron 94:891–907
5. Makino H et al (2017) Transformation of cortex-wide emergent properties during motor learning. Neuron 94:880–890
6. Wekselblatt JB, Flister ED, Piscopo DM, Niell CM (2016) Large-scale imaging of cortical dynamics during sensory perception and behavior. J Neurophysiol 115:2852–2866
7. Fischer LF, Soto-Albors RM, Buck F, Harnett MT (2020) Representation of visual landmarks in retrosplenial cortex. elife 9:e51458
8. Abadchi JK et al (2020) Spatiotemporal patterns of neocortical activity around hippocampal sharp-wave ripples. elife 9:e51972
9. Negrón-Oyarzo I et al (2018) Coordinated prefrontal–hippocampal activity and navigation strategy-related prefrontal firing during spatial memory formation. Proc Natl Acad Sci USA 115:7123–7128
10. Rothschild G, Eban E, Frank LM (2017) A cortical–hippocampal–cortical loop of information processing during memory consolidation. Nat Neurosci 20:251
11. Nietz AK et al (2022) Wide-field calcium imaging of neuronal network dynamics in vivo. Biology (Basel). 11:1601
12. Musall S, Kaufman MT, Juavinett AL, Gluf S, Churchland AK (2019) Single-trial neural dynamics are dominated by richly varied movements. Nat Neurosci 22:1677–1686
13. Mitelut C et al (2022) Mesoscale cortex-wide neural dynamics predict goal-directed, but not random actions in mice several seconds prior to movement. elife 11:e76506
14. West SL et al (2022) Wide-field calcium imaging of dynamic cortical networks during locomotion. Cereb Cortex 32:2668–2687
15. Gilad A, Helmchen F (2020) Spatiotemporal refinement of signal flow through association cortex during learning. Nat Commun 11:1744
16. Orsolic I, Rio M, Mrsic-Flogel TD, Znamenskiy P (2021) Mesoscale cortical dynamics reflect the interaction of sensory evidence and temporal expectation during perceptual decision-making. Neuron 109:1861–1875.e10
17. Pinto L et al (2019) Task-dependent changes in the large-scale dynamics and necessity of cortical regions in brief article task-dependent changes in the large-scale dynamics and necessity of cortical regions. Neuron 104:810–824
18. Sepers MD et al (2022) Altered cortical processing of sensory input in Huntington disease mouse models. Neurobiol Dis 169:105740
19. Cramer SW et al (2023) Wide-field calcium imaging reveals widespread changes in cortical functional connectivity following mild traumatic brain injury in the mouse. Neurobiol Dis 176:105943
20. Aguillon-Rodriguez V et al (2021) Standardized and reproducible measurement of decision-making in mice. elife 10:e63711
21. Rynes ML et al (2021) Miniaturized head-mounted microscope for whole-cortex mesoscale imaging in freely behaving mice. Nat Methods 18(4):417–425
22. Ghanbari L et al (2019) Cortex-wide neural interfacing via transparent polymer skulls. Nat Commun 10:1500. https://doi.org/10.1038/s41467-019-09488-0
23. Dana H et al (2014) Thy1-GCaMP6 transgenic mice for neuronal population imaging in vivo. PLoS One 9:e108697

24. Marvin JS et al (2013) An optimized fluorescent probe for visualizing glutamate neurotransmission. Nat Methods 10:162–170
25. Dana H et al (2016) Sensitive red protein calcium indicators for imaging neural activity. elife 5:e12727
26. Surinach D et al (2023) Distinct mesoscale cortical dynamics encode search strategies during spatial navigation. bioRxiv. https://doi.org/10.1101/2023.03.27.534480
27. Juneau J et al (2020) MiniFAST: a sensitive and fast miniaturized microscope for in vivo neural recording. bioRxiv:2020.11.03.367466. https://doi.org/10.1101/2020.11.03.367466
28. Holtmaat A et al (2009) Long-term, high-resolution imaging in the mouse neocortex through a chronic cranial window. Nat Protoc 4:1128
29. Rynes ML et al (2020) Assembly and operation of an open-source, computer numerical controlled (CNC) robot for performing cranial microsurgical procedures. Nat Protoc 15: 1992–2023
30. Ghanbari L et al (2019) Craniobot: A computer numerical controlled robot for cranial microsurgeries. Sci Rep 9:1023
31. Dubbs A, Guevara J, Yuste R (2016) moco: Fast motion correction for calcium imaging. Front Neuroinform 10. https://doi.org/10.3389/fninf.2016.00006
32. Mohajerani MH et al (2013) Spontaneous cortical activity alternates between motifs defined by regional axonal projections. Nat Neurosci 16:1426–1435
33. White BR et al (2011) Imaging of functional connectivity in the mouse brain. PLoS One 6: e16322
34. Vanni MP, Murphy TH (2014) Mesoscale transcranial spontaneous activity mapping in GCaMP3 transgenic mice reveals extensive reciprocal connections between areas of somatomotor cortex. J Neurosci 34:15931–15946
35. Murakami T, Yoshida T, Matsui T, Ohki K (2015) Wide-field Ca2+ imaging reveals visually evoked activity in the retrosplenial area. Front Mol Neurosci 8:20
36. Gu X et al (2018) Long-term optical imaging of neurovascular coupling in mouse cortex using GCaMP6f and intrinsic hemodynamic signals. NeuroImage 165:251
37. Ma Y et al (2016) Resting-state hemodynamics are spatiotemporally coupled to synchronized and symmetric neural activity in excitatory neurons. Proc Natl Acad Sci USA 113:E8463–E8471
38. Busche MA et al (2015) Rescue of long-range circuit dysfunction in Alzheimer's disease models. Nat Neurosci 18:1623–1630
39. Montagni E et al (2018) Wide-field imaging of cortical neuronal activity with red-shifted functional indicators during motor task execution. J Phys D Appl Phys 52:074001
40. Wang Q et al (2020) The Allen mouse brain common coordinate framework: a 3D reference atlas. Cell 181:936–953.e20
41. Ren C, Komiyama T (2021) Wide-field calcium imaging of cortex-wide activity in awake, head-fixed mice. STAR Protoc 2:100973
42. Barson D et al (2019) Simultaneous mesoscopic and two-photon imaging of neuronal activity in cortical circuits. Nat Methods 171(17):107–113
43. Lake EMR et al (2020) Simultaneous cortex-wide fluorescence Ca2+ imaging and whole-brain fMRI. Nat Methods 17:1262–1271
44. Saxena S et al (2020) Localized semi-nonnegative matrix factorization (LocaNMF) of widefield calcium imaging data. PLoS Comput Biol 16:e1007791
45. Geerligs L, van Gerven M, Güçlü U (2021) Detecting neural state transitions underlying event segmentation. NeuroImage 236:118085
46. Mohan H et al (2023) Cortical glutamatergic projection neuron types contribute to distinct functional subnetworks. Nat Neurosci 263(26):481–494
47. Dana H et al (2019) High-performance calcium sensors for imaging activity in neuronal populations and microcompartments. Nat Methods 16:649–657
48. Zhang Y et al (2021) Fast and sensitive GCaMP calcium indicators for imaging neural populations. bioRxiv:2021.11.08.467793. https://doi.org/10.1101/2021.11.08.467793
49. Madisen L et al (2015) Transgenic mice for intersectional targeting of neural sensors and effectors with high specificity and performance. Neuron 85:942–958
50. Jing M et al (2018) A genetically encoded fluorescent acetylcholine indicator for in vitro and in vivo studies. Nat Biotechnol 36:726–737
51. Lu X et al (2022) Detecting rapid pan-cortical voltage dynamics in vivo with a brighter and faster voltage indicator. bioRxiv:2022.08.29.505018. https://doi.org/10.1101/2022.08.29.505018
52. Hu J et al (2023) Fully desktop fabricated flexible graphene electrocorticography (ECoG) arrays. J Neural Eng 20:016019
53. Donaldson PD, Ghanbari L, Rynes ML, Kodandaramaiah SB, Swisher SL (2019)

Inkjet-printed silver electrode array for in-vivo electrocorticography. In: 2019 9th international IEEE/EMBS conference on neural engineering (NER), pp 774–777. https://doi.org/10.1109/NER.2019.8717083

54. Donaldson PD et al (2022) Polymer skulls with integrated transparent electrode arrays for cortex-wide opto-electrophysiological recordings. Adv Healthc Mater 11:2200626

Chapter 5

High-Density Multichannel Fiber Photometry

Yaroslav Sych and Fritjof Helmchen

Abstract

Animal behavior is governed by brain activity distributed and coordinated across multiple regions. Parallel progress in optical microscopy and the development of genetically encoded calcium indicators has advanced experimental approaches in systems neuroscience to assess brain activity simultaneously in multiple brain regions. Here, we provide a brief overview of optical recording techniques at this mesoscale level and of their combination with genetically encoded sensors and actuators. Next, we focus on the novel opportunities offered by extending fiber photometry to many brain regions, using compact arrays of optical fibers targeted to specific subnetworks of brain regions. We provide a detailed description and construction guidelines of high-density fiber array implants, including step-by-step surgical procedures for implantation and extraction. We describe several types of experiments enabled by this approach and highlight specific examples. We also discuss the challenges for analysis and interpretation of such mesoscopic recording data. Finally, we outline potential future improvements in terms of further functionalization of the multisite neural interfaces and of making them even less invasive.

Key words Calcium imaging, Fiber photometry, Fluorescence indicators, Neuronal population dynamics, Mesoscale brain networks, Sensorimotor processing, Neocortex, Subcortical regions

1 Introduction

Recording brain activity at different spatial and temporal scales is essential for understanding the neural basis of behavior. Such recordings allow neuroscientists to correlate neural activity and behavioral variables and to generate hypotheses regarding the neural circuit mechanisms underlying brain functions. In addition, by studying large-scale brain activity we can gain insights into how different regions of the brain interact to produce specific behaviors. This knowledge can help us to better understand the neural basis not only of normal but also abnormal behaviors and thus may inform the development of new treatments for behavioral disorders.

James E. Niemeyer et al. (eds.), *Awake Behaving Mesoscopic Brain Imaging*, Neuromethods, vol. 214,
https://doi.org/10.1007/978-1-0716-4120-0_5,

Several milestone studies have argued for distributed neural computations, i.e., the idea that different regions of the brain work together to perform complex cognitive functions. For example, functional magnetic resonance imaging (fMRI) studies have demonstrated that the brain can be divided into several large-scale networks (e.g., the "default mode network") and that these networks are functionally specialized and interact with each other in a highly dynamic manner [1–3]. The networks are formed by interconnected brain regions, each containing specialized and highly organized local microcircuits [4] of multiple cell types. In addition, many recent studies using electrophysiology or optical imaging have highlighted such multilevel organization and the distributed manner of neural processing [5–8], which are considered key features of how brains solve challenging tasks.

The *mesoscale* in the brain refers to the organizational level of neural structures and systems that are intermediate in size and link the microscale (individual neurons and synapses) to the macroscale (whole brain regions and brain-wide networks). The mesoscale finds itself at the transition of where cellular granularity becomes diffuse and where next-level organizational principles become apparent. For example, neocortical columns are thought to perform local computations, the results of which are embedded in brain-wide networks—through clusters of output neurons that give rise to specific long-range projections—to support specific computations for sensory integration, cognitive processing, and motor control. Mesoscale brain organization comprises *maps* that are anatomically or functionally defined cell *ensembles* usually composed of thousands of neurons with similar tuning properties. *Sensory maps* refer to specific regions of the brain that respond to sensory stimuli of a particular modality and that are spatially organized based on the characteristics of the sensory modality. For example, *somatotopic maps* are representations of the body surface that are present throughout the ascending pathway up to the primary somatosensory cortex for processing of tactile information. Different ensembles within these maps process distinct aspects of touch information, e.g., pressure vs. vibration stimuli or superficial skin receptor information vs. proprioceptive information. Another obvious example is the *retinotopic map* in the primary visual cortex that forms a representation of the visual field in the neocortex, based on highly preprocessed information from the retina relayed via the thalamus. Higher-level features such as orientation or motion direction of visual stimuli are also represented in maps. Next, association areas integrate the activity from these sensory maps to combine information across different modalities and to form more abstract representations, e.g. of objects, which are then fed into the larger-scale networks for attention, memory, movement preparation, and executive control. Thus, higher-order *cognitive maps* may contain representations of multimodal stimuli and movement plans that result from neural abstraction, for example, representations of learned context-dependent behavioral rules. Clearly, the functional representations and brain region interactions are shaped to a large extent by experience-dependent plasticity mechanisms. In the end, brain network activity patterns converge onto the neuromodulatory centers and the *motor maps* in the brainstem and the spinal cord, controlling both brain state as well as body dynamics (the latter through the skeletal muscle apparatus).

The mesoscale is thought to be a key level within this hierarchical brain organization because many neural computations are believed to engage this level. For example, neural activity synchronized at the level of *maps* and *functional neuronal ensembles* may correspond to specific global brain states or may trigger transitions of brain activity to a new state, which may also result in a change in behavior. Albeit not a new idea, the recognition that behaviorally relevant variables are represented by widely distributed networks of neurons and coordinated across many regions has triggered new research directions in recent years, promoting a more holistic description of brain dynamics. One reason for this development is that for decades experiments at the mesoscale level have been limited and hardly feasible, especially for deeper brain regions that are more difficult to access. This situation has fundamentally changed with the recent advances in electrophysiology, optical tools, and data analytics. New experimental approaches have emerged for distributed electrophysiology, large-scale in vivo imaging, temporally precise targeted perturbations of neural activity, and sophisticated behavior analysis. Modern analytical approaches often relate to network neuroscience and neuronal population dynamics described as trajectories on manifolds in high-dimensional neural state space. While network neuroscience focuses on understanding the communication between different interacting brain regions, the neural manifold perspective aims to describe activity patterns in terms of dynamic latent space variables of reduced dimensionality. These approaches regard the brain as a complex dynamic system rather than a simple collection of largely independently working subunits.

In this chapter, we present recent developments in the field of fluorescence measurements through optical fibers that enable simultaneous recordings of neural activity from many brain regions. We argue that such "multi-fiber photometry" approaches are well suited to acquire functional data on the mesoscale brain network level and are straightforward to be applied in behaving animals. They also can be combined with optogenetic perturbations as these approaches often are performed using optical fibers targeted to specific brain regions of interest, too. In Subheading 1, we first place the fiber photometry approach in the context of other methods, explain its principle and how to combine it with the expression of fluorescent indicators of activity, and discuss considerations regarding the interpretation of the recorded signals. In Subheading 2, we provide the details of the required materials, the surgical procedures for implantation, and the preparation of brains for histological verification. Subheading 3 then illustrates how a multi-fiber implant is used in a typical experiment, providing examples of recorded data and analysis methods. A final discussion and an outlook wrap up our chapter.

1.1 Techniques to Study In Vivo Neural Activity in Deep Brain Regions

While whole-brain imaging with cellular resolution has become possible in smaller organisms such as *C. elegans* and zebrafish larvae, it remains a challenge to study mesoscale neural activity in the mammalian brain during behavior. Neuroimaging methods such as fMRI [9–11] have limitations, including low spatial resolution as well as difficulties to apply them to behaving animals. Other methods like functional ultrasound (fUS) [12] and electrophysiology [13] also have their own limitations set by the microvasculature dynamics and difficulty in identifying specific cell types. Here, we focus on optical imaging techniques, which provide complementary possibilities by enabling approaches that are less feasible using other methods. For example, using genetic tools for cell-type specific expression of fluorescent indicators enables targeting of specific circuit components [14–17]. Moreover, single-cell and population signals can be longitudinally measured for days and weeks [18]. Recently, two-photon mesoscopes have been developed to enable cellular resolution imaging across large field of views (FOV) or in multiple distant regions [19–25]. The required experimental setups are expensive, however, and even with multiphoton excitation subcortical imaging is hardly possible unless invasive cannula implants are used. As an alternative, wide-field camera imaging using single-photon excitation is a simpler and less expensive method, but it usually sacrifices cellular resolution [26–30] and tissue depth penetration is very low. Consequently, in rodents both two-photon and single-photon imaging have been applied mainly to neocortical circuitry close to the brain surface and far less to subcortical regions.

Several experimental approaches were developed to extend large-scale optical recordings towards subcortical regions, such as fiber photometry [31–33] and imaging through gradient-index (GRIN) lenses [34–40] or implanted cannulas [41–43]. Fiber photometry is the simplest method that allows optical recording in deep brain regions. Here, a fluorescence signal is generated by delivering excitation light through an optical fiber and collecting emitted fluorescence photons back through the same fiber for remote photodetection. The fluorescence signal in this case lacks cellular resolution and represents an average signal of the stained neural structures in the vicinity of the fiber tip (bulk signal). Several optical approaches [44–47], often based on the reconstruction of the fiber "transmission" matrix, were developed to achieve cellular resolution; however, such approaches often have other limitations on the number of reconstructed fluorescent sources and usually flexible fiber bending is not allowed [44, 48, 49]. For bulk fiber photometry, the effective excitation and collection volumes depend on fiber parameters [50, 51] (diameter, numerical aperture) and also tissue parameters (e.g., scattering length). The fiber photometry approach is easy to use, it can be combined with the entire arsenal of fluorescent sensors, and optical fibers can be chronically implanted in the brain to monitor long-term changes of brain activity, such as those associated with sleep-state transitions [52], learning of a behavioral task [53], or disease progression [54, 55]. Furthermore, fiber photometry has been scaled up to allow for measurement and manipulation of multiple brain regions in parallel. With commonly used fiber ferrules, recording from 4 to 8 brain regions were possible [56], but a higher channel density was fundamentally limited by the bulky single-fiber implants. As an alternative approach, we introduced high-density fiber arrays that comprise 12–48 fibers mounted jointly in a compact neural interface [57]. This method reduced the footprint of the implant and partly alleviated geometric limitations on the channel density (Fig. 1a). To fully explore large-scale functional circuits, an even higher number of channels is desirable. We think that creating minimally invasive multisite optical probes of more than a hundred fiber channels should be feasible and would allow broad coverage of regions across the mesoscale network for functional interrogation.

1.2 Molecular Indicators and Viral and Transgenic Expression Methods

A variety of optical sensors exist to measure different aspects of brain activity, e.g., calcium concentration changes, voltage changes, or neuromodulator signals. Classic indicators are synthetic dyes, but most modern indicators are based on fluorescent proteins so that viral vectors can be used for genetic expression in specific cell types. Calcium indicators are the most popular sensors, measuring cellular activity by detecting changes in the intracellular calcium concentration in neurons (or glial cells). By binding to calcium ions, these sensors change (mostly increase) their fluorescence yield. Because in neurons action potentials lead to brief surges of calcium influx that cause transient elevations of the intracellular calcium concentration, the fluorescence changes of calcium indicators can be used to infer approximate changes in neural spiking activity. First-generation calcium indicators were synthetic dyes (e.g., Fura-2, Oregon Green BAPTA, or Fluo-4) that were mostly applied to cells or tissues in vitro but also injected into the brain of animals in vivo. Once inside the cells, these dyes can report intracellular calcium dynamics and neural activity patterns. The use of synthetic dyes was limited to short-term applications, however, due to wash-out or uptake inside the cells, preventing chronic usage and

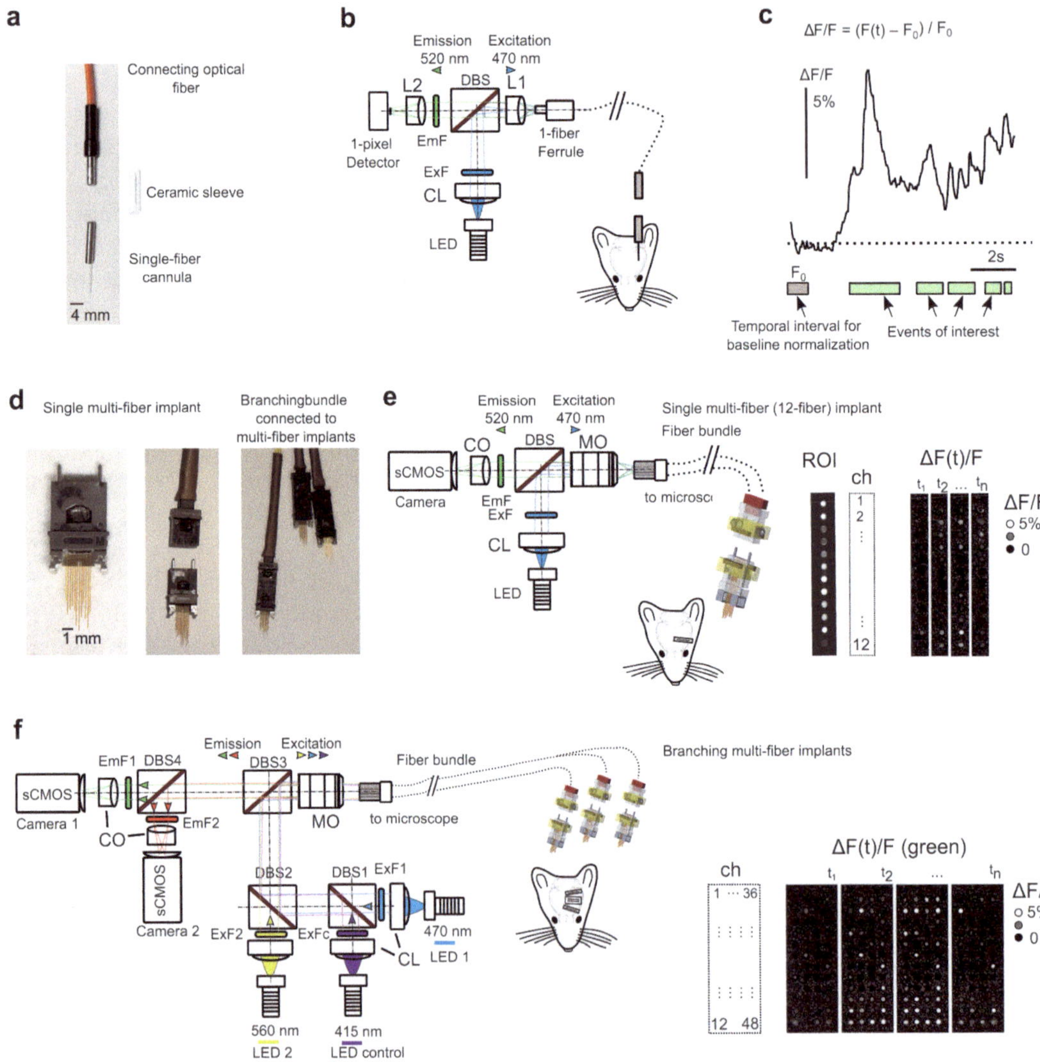

Fig. 1 High-density multichannel fiber photometry. (**a**) Photograph of single-fiber cannula (implant) connecting to the optical fiber via a ceramic sleeve. (**b**) Setup for single fiber photometry. The excitation light from the light source (here an LED) is collimated by a collimating lens (CL) and spectrally filtered with the excitation filter (ExF). The excitation light is coupled into the single-fiber with the lens (L1). Fluorescence emerging from the fiber is collimated with the same lens (L1), spectrally filtered with the emission filter (EmF) and focused with another lens (L2) on the photosensitive area of the photodetector. (**c**) Example of the $\Delta F(t)/F_0$ normalized fluorescence signal (in vivo GCaMP6 recording at 20 Hz averages, baseline fluorescence F_0 is calculated as mean over 1 s of pre-stimulus time, behavioral events of interest are indicated with green bars below the fluorescence signal, the first event of interest is a texture, stimulus predicting a water reward, that is followed by mouse licking on the waterspout). (**d**) Photograph of 12-fiber multi-fiber implant, fiber bundle connector and branching bundle connected to three multi-fiber implants. Multi-fiber implant dimensions, $7 \times 2.45 \times 8.05$ mm^3; weight, 0.32 g. (**e**) Left: schematic setup for multi-fiber photometry at one excitation wavelength. The excitation light from the light source (here an LED) is collimated by a collimating lens (CL) and spectrally filtered with the excitation filter (ExF). The microscope objective (MO) illuminated the whole field of view and, therefore, coupled excitation light into all fibers of the bundle. A dichroic beamsplitter (DBS) separates the fluorescence emission light (green ~520 nm) from the LED excitation light (blue ~470 nm). The

longitudinal experiments. Genetically encoded calcium indicators (GECIs) are the modern type of calcium sensors since 1997 [58]. The modification of green fluorescent protein (GFP) has led to the development of the GCaMP family of calcium indicators [59–61], among many other types, which can detect intracellular Ca^{2+} concentration changes [62] and track neural activity in a genetically targetable manner. GECIs cover a range of excitation and emission bands in the visible spectrum and some of them can optically resolve fluorescence changes evoked by single action potentials. GECIs are permanently expressed in neurons, either after viral infection or by incorporating the indicator's genetic code into the cell's genome in transgenic animals. GCaMP and RCaMP/R-GECO variants are some of the most widely used sensors based on green and red fluorescent proteins. Typically, sensors are expressed in the cytosol but they can also be targeted to subcellular structures such as somata [65, 66] or axons [67]. As a caveat, GECIs can be affected by other factors, such as gradually changing pH [68] or temperature. Protein engineering to further improve fluorescent sensors is continuing, focusing on the development of indicators with fast kinetics and large dynamic range to improve temporal fidelity of event detection [61, 69], and the design of red [63] or near-infrared optical indicator versions [70, 71] for increasing imaging depth in the brain or other scattering tissues.

The availability of transgenic mice that express GECIs in specific cell types under the control of distinct promoters has opened many further opportunities [64, 72–74]. There are several advantages of using transgenic animals as compared to delivering GECIs into the brain via viral injections. First, the expression profile in transgenic animals is controlled by the promoter, resulting in a highly specific labeling of subpopulations of cells. Second, expression is not only localized to the injection site but is present across all brain structures where cells express a particular target gene. Third, the use of transgenic animals may reduce variability by enabling

Fig. 1 (continued) emission filter (EmF) selects the optimal wavelength range for the respective fluorescent indicator (e.g., here exemplified for GCaMP). The camera objective (CO) creates an image of the fiber array on the CMOS camera sensor. Right: example image of the 12-fiber array including all channels. The temporal dynamics of normalized ΔF(t)/F signals is shown for four time periods (in vivo GCaMP6 recording at 20 frames per second; images are averages over Δt ~ 1 s) **(f)** Left: schematic setup for multi-fiber photometry at multiple wavelengths. The description is analogous to panels b and e. LED light is collimated (CL), spectrally filtered (ExF, ExF1, ExF2) and combined into one optical path with DBS1-2. DBS3 was designed to reflect spectral bands of LED1, LED2, and LED control and transmit fluorescence at emission spectral bands (in this example green and red). DBS4 separated green and red emission and the corresponding camera objectives (CO) that form images on two separate sCMOS cameras. The setup can be readily used with larger fiber arrays. We show the example of a 48-fiber bundle with multiple branching 12-fiber array implants. Bottom right inset shows total channel count (1–48). The temporal dynamics of normalized ΔF(t)/F signals is shown for four time periods separated by 1-s and accumulated in 50-ms intervals (in vivo GCaMP6 recording) more consistent GECI expression levels and thus provide more accurate and reproducible results. In contrast, following viral injections GECI expression may suffer from variable infection efficiency, viral titer, injected volume, etc., ending up less stable over time, potentially affecting cell health and limiting longitudinal experiments. Fourth, highly specific expression profiles can be achieved in double-transgenic animals. For example, mouse lines with layer-specific expression of GECIs in the cortex can be created by making use of the Cre-Lox system. This system can be used to generate transgenic mice that express Cre recombinase under the control of a promoter specific to a certain cell type or region of the brain (Cre driver line). Then, a second transgenic mouse is created that carries the GCaMP gene under the control of a "floxed" promoter (Cre reporter line). When these two transgenic mice are crossed, the Cre recombinase will excise the "floxed" region of the GCaMP gene, allowing for the expression of GCaMP only in cells that express Cre recombinase. Finally, GECI expression in transgenic mice may be switched on and off using an inducible gene expression system.

1.3 Fundamentals of Fluorescence Microscopy and Fiber Photometry

Fluorescence microscopy is a technique used to visualize specific structures or molecules within cells or tissues. It relies on the ability of fluorophores to absorb light within an excitation wavelength spectrum and subsequently emit light at longer wavelengths (within the emission spectrum). For example, for GCaMP calcium sensors the excitation wavelengths are in the blue part of the visible spectrum (470–490 nm), whereas fluorescence emission occurs in the green (500–540 nm). By selectively exciting and imaging fluorophores at appropriate wavelengths, specifically labeled cell structures or molecules can be visualized at high resolution. Fluorescence imaging can be performed using a multitude of microscopy techniques, including wide-field, confocal, multiphoton, and super-resolution microscopy, with each technique providing specific advantages for different applications.

Fiber photometry uses the same principle of fluorescence excitation but contrary to standard fluorescence microscopes, for which the sample (brain tissue) is positioned in the imaging plane of an objective, fiber photometry uses an optical fiber to deliver the excitation light to a region of interest and collect the emitted fluorescence as well (Fig. 1a, b). A fiber attached to the cannula is inserted into the brain tissue in vivo, and an optical fiber cable is connected to the cannula to deliver the excitation light and collect fluorescence from the target area. A sensitive detector, such as a photomultiplier tube or an avalanche photodiode is used to measure fluorescence at the distal end of the connecting fiber. Fiber photometry through a single fiber lacks cellular resolution as it measures a bulk signal from labeled cells (*see* **Note 1**), but it is a powerful method to study temporal dynamics of the activity of local cell populations by using fluorescent indicators sensitive to changes in calcium ion concentration, membrane voltage, second messenger levels, or activity of specific proteins. For time-dependent fluorescence recordings, the acquisition rate (or frame rate for 2D or 3D imaging methods) is an important parameter. However, in the case of calcium recordings, temporal fidelity of the recordings is also limited by the kinetics and response characteristics of the indicator itself. Changes in fluorescence over time, F(t), are often expressed as $\Delta F/F = (F(t) - F_0)/F_0$ to normalize for variations in indicator concentration, measurement volumes, and illumination (F_0 is the baseline fluorescence, and any background fluorescence has been subtracted beforehand for both F(t) and F_0). With fiber photometric calcium recordings, changes in neural activity can be monitored in real-time, for example in response to various stimuli or in association with particular behaviors (Fig. 1c). The relationship between neural spiking activity and photometric signals can be validated with simultaneous electrophysiology (*see* **Note 2**) Fiber photometry is complementary to wide-field fluorescence imaging from the brain surface, in that it allows measurements from deep brain regions and is suitable for chronic recordings in freely behaving rodents, which is more difficult to achieve with wide-field imaging.

Fiber photometry can be extended to multiregional recordings by using several fibers in parallel [56, 57]. Compared to single-fiber photometry, two key modifications are required, which are the use of a bundle of fibers with multiple parallel channels and fluorescence detection with a camera sensor (Fig. 1d–f). Excitation light from a light source (a laser or a light-emitting diode, LED) is coupled via an objective into all fibers at the proximal end of the fiber bundle. At the distal end, one either connects each fiber channel independently to a separate implant [56], or all fiber ends are integrated into one connector that attaches to a high-density multi-fiber implant, or one can also split the fiber bundle into several multi-fiber implants (Fig. 1d) [57]. For fluorescence detection from all fibers, the proximal fiber-bundle facet is imaged onto the camera sensor, typically using a collimating lens and a focusing objective. In this way, all fiber channel signals are detected simultaneously in specific regions of interest (ROIs) on the sensitive CMOS camera chip. For each camera frame, fluorescence intensities are averaged across all pixels for each fiber ROI (across all fiber modes) and time-dependent signals are expressed as $\Delta F/F$ changes.

1.4 Considerations on Crosstalk and Specificity

When using dense arrays of optical fibers, how much fluorescence light generated at one fiber tip can be seen by neighboring fibers? To estimate the amount of cross-talk between fiber channels, we sequentially illuminated single fiber channels (called "excited channel" below) while measuring fluorescence signals in all channels [57]. We normalized fluorescence signals from neighboring fibers to the fluorescence generated by direct illumination and analyzed their distance-dependence. We found that at the closest distance (0.25–1 mm), about 5–10% of the fluorescence came from the neighboring fiber channel, but this rapidly decreased to negligible levels with distance. There are two possible causes of crosstalk. First, propagation of the excitation light from the excited channel to neighboring fibers, which generates fluorescence underneath their tips. Second, propagation of fluorescence light excited at the tip of the excited fiber to neighboring channels. The first cause may only provide some additional excitation light to the neigboring channels, but does not contribute to the crosstalk during continuous LED illumination. Only the second cause, propagation of fluorescence, contributes to a crosstalk when neigboring fibers are positioned as close as possible to each other (*i.e.,* 250 micron apart). When neighboring fibers on the front-piece are designed to target different depths (>1 mm of relative distance), our multi-fiber arrays allow to record local signals in selected subsets of brain regions with high signal-to-noise and good regional specificity.

1.5 In Vivo Studies Enabled by Mesoscale Imaging of Fluorescent Activity Indicators

Mesoscale wide-field fluorescence imaging of brain activity has been used to study dynamic properties of large-scale networks of neuronal populations, revealing patterns of spontaneous activity [29, 75] and distributed patterns of sensory-evoked [76, 77] and motor-related [78] activity that spread over large regions of cortex. For example, brief deflections of a single whisker were found to elicit widespread depolarizing responses in the sensory cortex, starting in the primary somatosensory barrel cortex and subsequently activating the whisker motor cortex [77]. The flow of sensory information to the motor cortex was dependent on the animal's behavior and correlated with the whisker movements in response to the sensory stimulus. This suggests that sensorimotor transformations in the motor cortex play a significant role in the perception of active touch. Although the flow of activity is constrained by the anatomical organization of the brain [79], multiple studies reported widespread activation of the dorsal cortex during behavioral tasks [80, 81]. The observation that neurons in different cortical areas can exhibit coordinated activity, even when those areas are not directly connected, may be explained by the involvement of subcortical structures [53, 82]. Interaction between the cerebral cortex and subcortical structures can occur through multiple parallel pathways, including thalamo-cortical loops [83, 84] or the cortico-basal ganglia-thalamo-cortical networks [53, 84, 85].

Another key observation is that mesoscale patterns of brain activity reorganize depending on the behavioral task [86, 87]. This suggests that the brain may have a more distributed and flexible system for processing information. However, careful examination of animal behavior is necessary as those seemingly task-related activity patterns might also be explained by a richly varied movement repertoire of the animal [78, 88]. Considering the animal's movement, distinct patterns of mesoscale activity have been observed that varied depending on the behavioral strategy adopted by the animal [30]. In general, one has to distinguish between movement-related, state-related, and strategy-dependent patterns. Such distinctions might be accounted for by mathematical models to show that such global patterns of activity not only correlate to movement but also to the task performance [81] and the level of task expertise [53]. Additionally, mesoscale imaging has revealed that neurons in different brain regions can exhibit activity patterns that are specific to certain cognitive states, such as wakefulness [89] and sleep [90].

2 Materials

2.1 Multi-fiber Photometry Setup

The photometry setup for multi-fiber measurements can be built using a variety of devices and optical components. Here we describe a setup that was simplified compared to our first design for multi-fiber photometry [57] but is capable of dual-color measurements (Fig. 1f). Instead of lasers, this setup uses a 470-nm LED (M470L5; Thorlabs) for excitation of GCaMPs and a 561-nm LED (M565L3; Thorlabs) for excitation of RCaMPs. A third LED with the central wavelength at 415-nm (M415L4; Thorlabs) is used as reference channel, i.e., for exciting calcium-independent GCaMP fluorescence for correction of intrinsic and hemodynamic fluorescence changes (*see* **Note 3**) [56, 91–93]. To achieve stable CW operation, LEDs are operated at output powers as required during the experiment for 30 min before the start of recording. We use an LED driver controlling the current of the LED module (LEDD1B, T-Cube LED Driver; Thorlabs) to set the fluorescence excitation power for a GCaMP sensor to ~1.3 mW/mm^2 per fiber channel at the fiber tip. Please note that new versions of LED drivers exist with lower noise and ripple amplitude (e.g., UPLED, USB-Controlled LED Driver; Thorlabs). To collimate the light emitted by each LED we use aspheric condenser lenses (ACL25416U-A; Thorlabs, Ø1", f = 16 mm, NA = 0.79, ARC: 350–700 nm). All three collimated beams of LED colors are combined and aligned to the optical axis of the system with the help of dichroic beamsplitter/combiners consisting of a chromatic mirror and a cage cube with dichroic filter mount (CM1-DCH/M; Thorlabs). To combine 415-nm and 470-nm LED light we selected the AHF F48–442 beamsplitter with >90% reflection in the 350–442 nm wavelength range and >90% transmission between 460 and 650 nm. To combine both "blue" 415-nm/470-nm LEDs with the "yellow" 561-nm LED light we use an AHF F48–507-nm beamsplitter with >95% reflection between 350 and 495 nm and >95% transmission between 510 and 850 nm. In general, different types of dichroic beamsplitters can be chosen, with reflection and transmission bands according to the emission spectra of the set of excitation LEDs.

To couple excitation light into several optical fibers in parallel, the proximal end of the fiber array is positioned in the object plane of the setup and a wide-illumination pattern is created after the microscope objective (TL4x-SAP; Thorlabs) from partially collimated excitation light. A dichroic beamsplitter (F68-409 quadruple line, AHF) directs the excitation light for all wavelengths into the objective while transmitting fluorescence in the emission spectral windows of GCaMP and RCaMP to the detector. To separate fluorescence signals from residuals of the excitation light and to minimize autofluorescence generated in a broader spectral range, we use emission filters (either F67-432 or a separate filter F37-516, AHF, transmission 525-nm center wavelength, 50-nm range for GCaMP; and F47-605, AHF, transmission 605-nm center wavelength, 70-nm range for RCaMP). To image the proximal end of the fiber array onto the camera we use a 100-mm or 200-mm focal length tube lens (TTL100-A or TTL200, Thorlabs; depending on the magnification required). The image is created at the back focal plane of the tube lens on the CMOS sensor of the camera (ORCA Flash4.0, Hamamatsu).

2.2 Preparation of Multi-fiber Front Piece and Connecting Bundle

We use components of MT fiber connector technology, specifically ferrules that can fit different number of fibers (8, 12, 24, and 48 fibers; part numbers 17185, 7413, 12599; US Conec, Hickory, NC). Optical fibers with polyimide coating (100-μm core diameter, 124-μm outer diameter, NA 0.22; UM 22-100, Thorlabs) fit into the ferrule (Fig. 2a). To position all fibers for targeting specific brain regions, we align their length to a custom template that defines the required implantation depths (Fig. 2b). After alignment, we glue the fibers inside the ferrule with a two-component epoxy (302-3M 1LB Kit Part A and B; Epoxy Technology). An additional drop of epoxy is applied externally to the array side to be optically connected to the fiber bundle (Fig. 2c). We let the epoxy cure over 24 h and then polish the connecting surface in consecutive steps, starting with a rough diamond lapping sheet of 30-μm grit, to a silicon carbide lapping sheet of 5-μm grit, and then gradually moving to a calcined alumina lapping sheet of 0.3-μm grit (from LF30D, LF5P, LF3P, LF1P to LF03P; Thorlabs). We use a custom-designed bare ferrule polishing puck to hold the ferrule orthogonal to the polishing pad (NRS913A, Thorlabs). Ethanol or distilled water is added on the polishing pad to reduce friction. After the polishing procedure we add alignment pins to the implant (part numbers 16741, 16742; US Conec).

The connector for the fiber bundle is identical to the MT fiber connector (8, 12, 24, and 48 fibers; part numbers 17185, 7413, 12599; US Conec, Hickory, NC). To allow connection with the

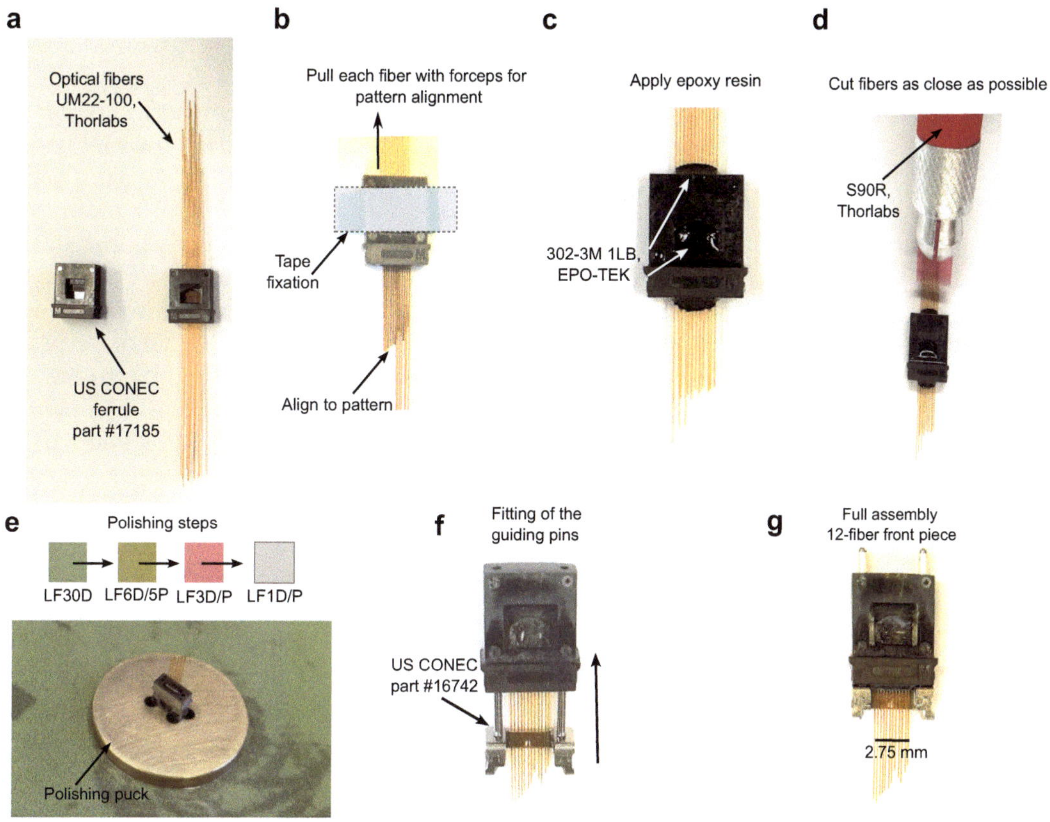

Fig. 2 Preparation of a multi-fiber front piece for implantation. (**a**) Insertion of the optical fibers into a 12-fiber connector ferrule. (**b**) Alignment of the fiber ends to a printed template of the intended pattern of fiber lengths. The ferrule is fixated with tape on the printout of the pattern. (**c**) Application of epoxy resin (glue) in the middle of the implant and on the distal surface for further polishing. (**d**) Cutting the fibers, prior to polishing, with a fiber optic scribe as short as possible. (**e**) Polishing with the help of a custom-designed bare ferrule polishing puck and polishing sheets of 30, 6, 3, and 1 μm grit size. (**f**) Fitting of the guiding pins. (**g**) Full assembly of a 12-fiber implant. (Modified with permission from Sych et al. [57])

front piece containing guiding pins, the multi-fiber bundle should not have pins. The fiber bundle can either be custom-made using the polishing procedure as described above. Or, alternatively, multi-fiber bundles based on these connectors are commercially available from the company Sylex (Bratislava, Slovak Republic). For example, we purchased a 48-channel fiber bundle splitting into four separate strands containing 12 fibers each and terminating with a linear 12-fiber ferrule connector (001-187284-85447000 EU, 7-m length, Sylex). Such a bundle can be connected to four independent implantation sites or to four independent animals with 12-fiber channels on each connector. Other configurations of fiber bundles can be easily specified by the customer. On average, we achieve a coupling efficiency close to 75% across the fibers of a 12-fiber implant and the connecting bundle. During the first uses of the bundle the coupling efficiency is higher; however, with frequent implant-bundle contact the coupling efficiency is decreasing mostly due to scratches on the surfaces. To further improve the coupling, use one of the last polishing steps with the lap LF1P or LF03P.

2.3 Surgical Procedures and Implantation

All procedures for animal experiments must be approved by the local authorities. For our experiments, all experimental procedures for animal experiments were carried out according to the guidelines of the Veterinary Office of Switzerland and approved by the Cantonal Veterinary Office in Zurich. We induced GCaMP6m expression in neurons by injecting AAV2.9-hSyn-GCaMP6m into the set of targeted brain regions in C57BL/6 mice and subsequently implanted a multi-fiber array (exemplified for a 12-fiber array front piece in Fig. 3).

Preparation Adult mice (2–6 months old) were anesthetized with 2% isoflurane (mixed in pure oxygen, Fig. 3a). Body temperature was maintained at 37 °C. To prevent inflammation and pain during anesthesia and postoperatively, 0.1 μL/g bw of Metacam was injected subcutaneously. Prior to implantation, connective tissue was removed from the skull bone, which was additionally polished and dried. iBond (Kulzer, Total Etch) was applied to ensure the best adhesion of the skull to the connective dental cement (Fig. 3b). To further stabilize the implant on the skull we used Charisma (Kulzer, A1) to produce a thin ring on the skull rim. Both iBond and Charisma require UV light curing. Small slit-like craniotomies were made at the appropriate stereotactic coordinates to allow for virus injections and implantation of the fiber array (carried out on the same day).

Viral Injection A selected viral construct is injected (e.g., ~200 nL of AAV2.9-hSyn-GCaMP6m) into each of the selected brain regions at a rate of 100 nL/min (Fig. 3c). A custom setup with a syringe and a barometer is used to control injection pressure and volume. To allow for local diffusion and avoid possible reflux, the glass injection pipette (10–20 μm diameter) is kept in place after each injection for at least 5 min.

Implantation Prior to the insertion of the fiber array, we mounted the implant on a manipulator, with the fibers oriented vertically, and then aligned the longest fiber tip relative to several anatomical landmarks, e.g., midline, bregma, and the surface of the brain (Fig. 3d). This procedure ensures that all fibers inserted will be oriented towards the selected regions according to the brain atlas. Prior to fiber implantation, we use the fine needle of an insulin syringe to slightly scratch the dura at the brain surface in several

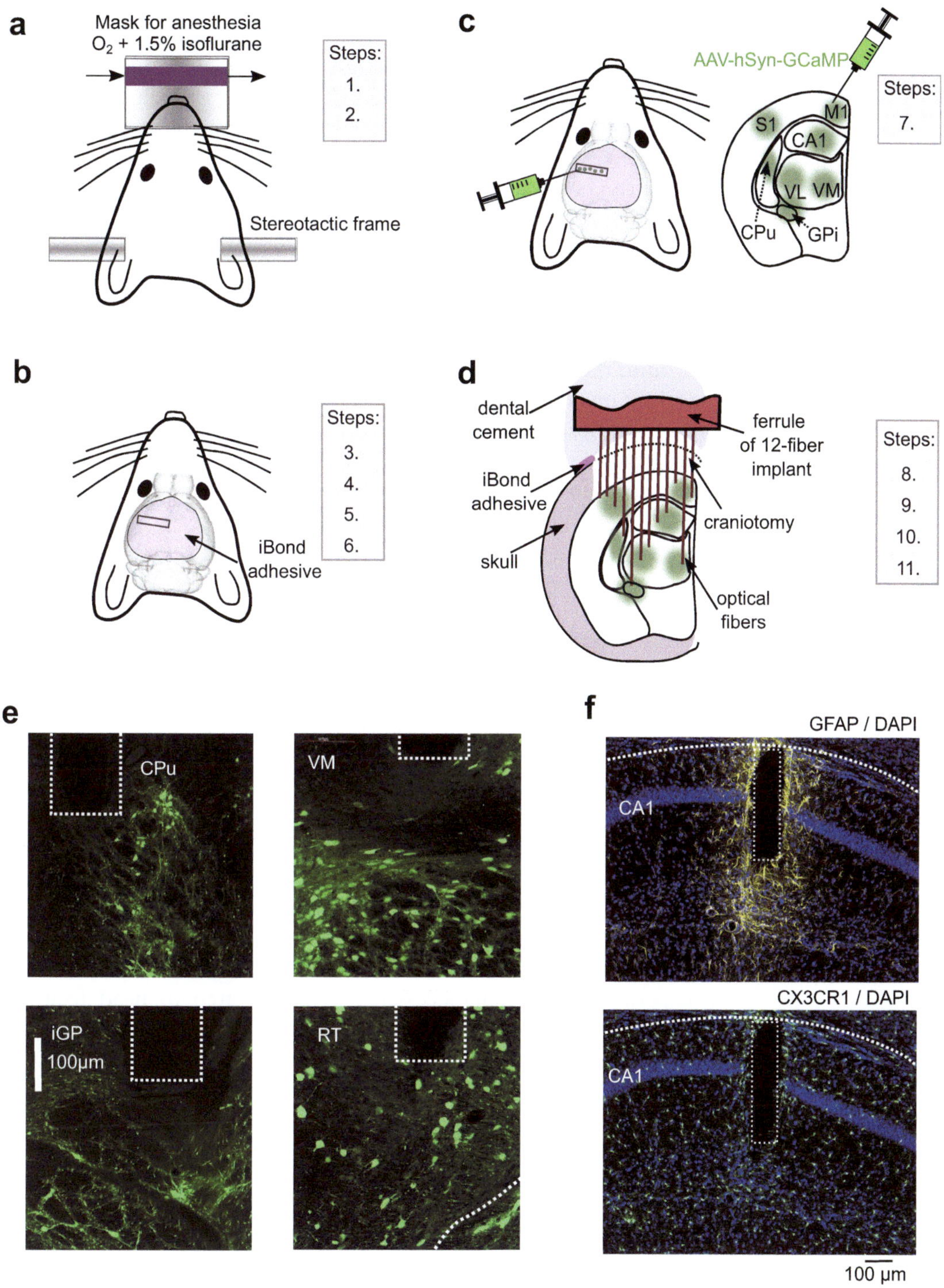

Fig. 3 Surgical procedures and implantation of the multi-fiber array. (**a**) Fixate the mouse in the stereotactic frame (1). Attach the mask (2) for the gas anesthesia (O_2 with 1.5% isoflurane). (**b**) Remove fur, skin, and expose the skull bone (3). Dry and remove connective tissue on the skull (4). Use i-Bond adhesive (5). Drill slit-shaped craniotomy (6). (**c**) Inject calcium sensor (e.g., GCaMP) into the target brain regions (7). (**d**) Align the multi-fiber implant to the brain (8) with the help of reference anatomical landmarks (e.g., bregma, midline,

places. This reduced the pressure on the cortex as we advanced the 12-fiber array into the brain. While moving fibers into the brain, we monitor the depth of the longest fiber but also note down the depth coordinate reading when shorter fibers touch the surface. Based on this reading the expected distances can be compared with the actual geometric distances so that the advancement into the brain can be slightly adjusted to achieve the best possible placement of the fiber tips.

Implant Fixation We seal the craniotomy with Vaseline, which melts at body temperature and completely covers the craniotomy. Next, we apply dental cement (Tetric EvoFlow A1) on the skull and around the implant followed by UV light curing (Fig. 3d). A lightweight metal head post is additionally cemented to the skull, allowing for head-fixation during the behavioral experiments. After 2 weeks of recovery, mice can enter the experimental workflow. In our experiments, mice were habituated to head-fixation and trained in a go/no-go texture discrimination task [94].

2.4 Experimental Timeline

In our first experiments we induced neuronal expression of GCaMP6m through multiple injections of AAV2.9-hSyn-GCaMP6m into 12 targeted brain regions [53, 57] (Table 1) of C57BL/6 mice (Fig. 3c, e). However, this approach is laborious and the difficulty and duration of the surgery scales with the number of injection sites. If available and appropriate for the research goal, the use of transgenic mice [72, 73] with neurons of interest already expressing GCaMP or another fluorescent activity sensor is an attractive alternative. Because the choice of indicator(s) and the method to express them in specific cells and regions depend on the specific aims of a project, no general recommendation can be made. We encourage the reader to consider these important points when designing their future experiments.

Three weeks after fiber implantation mice were handled by the experimenter and, if necessary, accommodated to head fixation. Please note that a 2- to 3-weeks post-surgery time period should be incorporated into the experimental timeline. First, virally induced expression of GECIs (e.g., GCaMP under the human

Fig. 3 (continued) lambda). Find the surface of the brain (9) by touching the exposed tissue with the longest fiber. Mark this as a zero-depth coordinate. Slowly insert the multi-fiber implant into the brain tissue (10) to the planned depth. Apply the cement around the ferrule to attach the implant to the skull (11). (**e**) Examples of histology showing GCaMP6 expression in neurons in CPu, iGP, VM, and RT. Positions of the 120-μm diameter fiber tips are indicated by dashed lines. (**f**) Top: Confocal image of astrocytes surrounding the fiber shaft in the hippocampal CA1 region (yellow). The CA1 pyramidal layer is marked by a dense band of DAPI-stained nuclei (blue). Bottom: Similar image with EGFP-stained microglia in the CX3CR1-EGFP mouse line. A moderate astroglia scar formation is detected near the optical fiber. Microglia density is not markedly changed, except for corpus callosum above CA1. (Modified with permission from Sych et al. [57])

Table 1
List of the brain regions

dCpu	Caudate putamen (striatum), dorsal-lateral part
vCPu	Caudate putamen (striatum), ventral-lateral part
iGP	Globus pallidus, internal part
VM	Ventromedial thalamic nucleus
VL	Ventrolateral thalamic nucleus
LD	Laterodorsal thalamic nucleus
RT	Reticular thalamic nucleus
CA1	CA1 field of the hippocampus, dorsal-lateral lamina
CA1Py	CA1 field of the hippocampus, pyramidal lamina
CA1Mol	CA1 field of the hippocampus, stratum lacunosum moleculare
DG	Dentate gyrus
M1	Primary motor cortex, posterior part
S1BF	Primary somatosensory cortex, barrel field

synapsin promoter) requires about 2 weeks to reach a sufficient expression level. In addition, homeostatic brain tissue regulation, i.e., post-surgery brain healing including recovery from initial astrocytic and microglia activation, often takes several weeks. When multiple optical fibers are implanted into the brain they push and displace tissue and may cause microlesions. Although we apply relatively thin fibers (with the integral fiber volume even for a 48-fiber implant being about five times less compared to typical chronic GRIN lens or cannula implants) tissue reactions may occur and sustain. To assess potential brain damage, we performed histological characterization of brain slices. One month post-surgery we did not find evidence for elevated microglia activity surrounding the implanted fibers in terms of cell counts and anatomical organization of microglia processes (Fig. 3f). Microglia around and below the optical fiber shafts was similar to the non-implanted contralateral hemisphere. Astrocytes formed a ~20-μm layer of scar tissue, not precluding the detection of GCaMP expressing neurons just below each fiber tip. We therefore conclude that experimental conditions 3–4 weeks after fiber implantation are suitable for in vivo photometric recordings.

2.5 Extraction of Multi-fiber Implants from the Brain

Multi-fiber implants have dense arrays of fibers positioned close to each other (standard pitch, i.e., center-to-center distance for the neighboring fibers, is 250 μm). When fibers are implanted into the brain, they displace the brain tissue. From our experience, we see that the brain tissue between fibers stays mostly intact, but sometimes superficial tissue at the cortical surface is attached to fibers. To register histological data from every subject to the reference brain atlas, the fiber tracks and all surrounding tissue should be maximally preserved, keeping all anatomical features for the brain regions of interest. One of the options to preserve the tissue, after perfusion of the animal, is to keep the head of the mouse in the fixative (1.5% PFA solution) for 1–2 weeks at +4 °C. Afterwards, remove the connective tissue around the head, the jaw, and the bottom of the skull with small scissors and keep the sample for one more week in the fixative solution. Next, access the brain from the ventral side and pull the brain gently out from the bottom. The brain should smoothly detach from the fiber array, leaving no connective tissue between neighboring fibers.

2.6 Histological Verification of Fiber Tip Positions Relative to the Target Brain Regions

To identify the position of each fiber tip relative to the targeted brain region, one needs to register histological images to the reference atlas. Initially we used a rather sophisticated approach consisting of multiple steps. Image processing was performed on grayscale confocal images taken from autofluorescence background in coronal histological sections. To obtain a sharp edge at the brain's midline in each coronal slice of the brain, all images were manually segmented from the background using the Lasso segmentation tool of the software Avizo 9.3.0 (FEI, Hillsboro, Oregon). The preprocessed images were then registered to the Allen Mouse Brain volumetric atlas [95] (downloaded from the Scalable Brain Atlas [96]) with the help of Avizo Software. The sharp midline edges of the brains and the atlas [97] reference data imposed an additional anatomical constrain and thus improved the alignment. The registration was implemented in two steps, first using only rigid-body transformations (translation and rotation), then allowing for isotropic scaling and, if needed, additional anisotropic scaling. All registered slices were validated by a human. Next, fiber tracks were manually segmented in each slice using Avizo's Lasso segmentation tool. Reconstructions of tracks allowed us to identify the fiber ends and obtain their coordinates in the atlas space.

Recently several open-source software packages were developed to track one- or multi-electrode shanks across the brain and to register them to the brain atlas [98, 99]. These software packages often include graphical user interface and, hence, greatly simplify the workflow. Above we described our initial workflow rather for educational purposes, to give an idea about the underlying logic of histological verification. However, we encourage the reader to use more recent packages, which integrate all the steps within one software. Tissue clearing [100], light-sheet imaging [101, 102], and subsequent 3D reconstruction is another newly developed alternative to track electrode and optical fiber shafts in the brain tissue [103].

3 Methods

3.1 Practical Tips for Photometric Recordings

To ensure high quality of the acquired photometry data, we recommend the following procedures before starting the recording. First, switch on the setup to warm up LEDs (or lasers) well in advance. Second, we also suggest to bleach optical fibers with the 470 nm or 405–425 nm excitation light for ~2 h prior to the start of recording (no animals connected). Third, before connecting the fiber bundle to the multi-fiber implant, the surface of the implant should be cleaned from dust and other debris to allow for tight contact between the fiber bundle to the implant(s). For this, we use fiber cleaning products: we hold a piece of the fiber cleaning cloth (FCC-7020, Thorlabs) with forceps, such that sharp forceps tips are covered in the cloth; next we spray the fiber cleaning fluid (FCS3, Thorlabs) on the cloth and gently brushed the surface of the implant and across the guiding pins. To connect the fiber bundle to the implant, mice are either head-fixed, or we hold the head still with the help of a small metal bar that was implanted on the side of the skull during the surgery. After connecting the mouse to the measurement setup, the baseline fluorescence should increase on all channels suggesting that this additional fluorescence comes from the cells expressing the fluorescent indicator (in our case GCaMP6m) and that fiber coupling is good. Fourth, we suggest to visually verify that no gap exists between the fiber bundle and the implant across the full width of the implant. Finally, examine the dynamics of the baseline fluorescence, which should be stable, not decaying less than 1% per minute. Otherwise, the excitation light intensity might be too high, causing substantial photobleaching of the fluorescent indicator, and therefore should be reduced. When both 470 nm and 405–425 nm control wavelengths are used, the photobleaching rate should be quantified independently for both wavelength ranges. On the longer time scale (minutes to hours) photobleaching may deplete the pool of available sensor molecules and therefore reduce the signal-to-noise ratio (SNR) of the recorded fluorescence signals. On the other hand, the excitation light intensity should also be set to a reasonably high level to provide sufficient SNR and enable $\Delta F/F$ measurement of indicator dynamics. In general, as photobleaching depends on the properties of the specific sensor as well as the experimental setup, optimization is required for each lab and project individually.

3.2 Example Multi-fiber Photometry Data

In this section we give examples of multi-fiber photometric recordings using a laser-based illumination setup as described in [57]. Multiple front pieces (two with 12 fibers and one with 24 fibers) were implanted on the left hemisphere of mice (C57BL/6, male, surgery done at 3 months age). We simultaneously recorded calcium signals from the targeted brain regions (Table 2) by exciting GCaMP fluorescence with 470-nm laser light

Table 2
List of the brain regions

12F anterior	lM1	Primary motor cortex, lateral part
	AI	Agranular insular cortex
	mM1	Primary motor cortex, medial part
	Pir	Piriform cortex
	LO	Lateral orbital cortex
	AOL	Anterior olfactory nucleus, lateral part
	AOD	Anterior olfactory nucleus, dorsal part
	M2	Secondary motor cortex
	VO	Ventral orbital cortex
	Cg1	Cingulate cortex, area 1
	PrL	Prelimbic cortex
	MO	Medial orbital cortex
24F medial, or 12F medial (selected subset)	S1BF	Primary somatosensory cortex, barrel field
	S2	Secondary somatosensory cortex, barrel field
	M1	Primary motor cortex, posterior part
	CPu	Caudate putamen (striatum), lateral part
	dCpu	Caudate putamen (striatum), dorsal-lateral part
	vCPu	Caudate putamen (striatum), ventral-lateral part
	BLA	Basolateral amygdaloid nucleus, anterior part
	BMA	Basomedial amygdaloid, anterior part
	LaDL	Lateral amygdaloid nucleus, dorsal part
	GP	Medial globus pallidus
	iGP	Globus pallidus, internal part
	eGP	Globus pallidus, external part
	LGP	Lateral globus pallidus
	VPM	Ventral posteromedial thalamic nucleus
	VPL	Ventral posterolateral thalamic nucleus
	VM	Ventromedial thalamic nucleus
	VL	Ventrolateral thalamic nucleus
	LD	Laterodorsal thalamic nucleus
	Po	Posterior thalamic nuclear group
	RT	Reticular thalamic nucleus
	CA1	CA1 field of the hippocampus, dorsal-lateral lamina
	CA1Py	CA1 field of the hippocampus, pyramidal lamina
	CA1Mol	CA1 field of the hippocampus, stratum lacunosum moleculare
	DG	Dentate gyrus
12F posterior	S	Subiculum
	vDG	Dentate gyrus, ventral part
	vCA1Mol	CA1, stratum lacunosum moleculare ventral part
	mV1	Primary visual cortex, medial part
	lV2	Secondary visual cortex, lateral part
	BIC	Nucleus of the brachium inferior colliculus
	SubB	Subbrachial nucleus
	RSG	Retrosplenial granular cortex
	InWh	Intermediate white layer of the superior colliculus
	DpMe	Deep mesencephalic nucleus
	SuG	Superficial gray, superior colliculus
	LPAG	Lateral periaqueductal gray

through 48 fibers in parallel. The exposure time for the sCMOS camera was set to 50 ms (20 Hz frame rate). We communicated with the Hamamatsu ORCA Flash4 camera via the drivers provided and with high-level functions written in LabView. Many types of cameras are supported by user-friendly libraries for Python, MATLAB, and LabView. Please note that if additional computational routines are included in the loop for acquiring frame-by-frame data, the camera will work slower as compared to the expected frame rate, with the effective rate given by the inverse of the sum of exposure time and read-out time. Useful computational routines, prior to saving data, may include integrating pixel intensities for the ROIs defined for each fiber facet image. If only preprocessed data are saved, the amount of information stored on the hard drive can be greatly reduced. Other useful computational routines may include parallel data acquisition of trigger signals for synchronization with the behavioral setup (e.g., behavioral training for goal-directed behaviors, behavioral video data, and any type of electrophysiological recording). The acquisition speed of the camera should relate to the properties of the indicator to minimally compromise its temporal dynamics. For example, typical impulse response profiles of GECIs show 10-20 ms rise time and 100 ms decay time. Therefore, a 20-Hz frame rate is adequate. However, if the camera is run in triggered mode to acquire interleaved frames with 470-ns and 415-nm LED illumination, the frame rate should be doubled to 40 fps to maintain an effective rate of 20 Hz for quasi-simultaneous recording at two excitation wavelengths. With advanced kinetic properties of new indicators [61], camera frame rate should be further increased.

In our experiment, mice were head-fixed and trained to perform a go/no-go tactile texture discrimination task [94]. Multi-fiber implants were placed in the left hemisphere targeting distributed regions across frontal cortex, the basal-ganglia thalamo-cortical network and the posterior forebrain (Fig. 4a; full list of regions in Table 2). When mice licked after a "go" texture, the trial was rewarded with a water drop (Hit trial), whereas licking after a "no-go" texture was punished with a short time-out and white sound noise (False alarm). Typically, it takes mice about 2 weeks to associate the "go" texture with reward and withhold licking for a "no-go" texture (Correct rejection, CR trial), thus becoming an expert in reliable discrimination between two types of tactile stimuli. We observed diverse time courses of $\Delta F/F$ signals across regions with amplitudes of up to 10–20%. Control experiments confirmed that the signals represented calcium-dependent fluorescence. Specifically, excitation at 425 nm, a wavelength close to the isosbestic point of GCaMP6, produced rather flat $\Delta F/F$ traces with significantly lower variance compared to 470-nm excitation, indicating minor contributions of intrinsic signal changes or motion artifacts (see Figure 1g in [57]). In each mouse, we

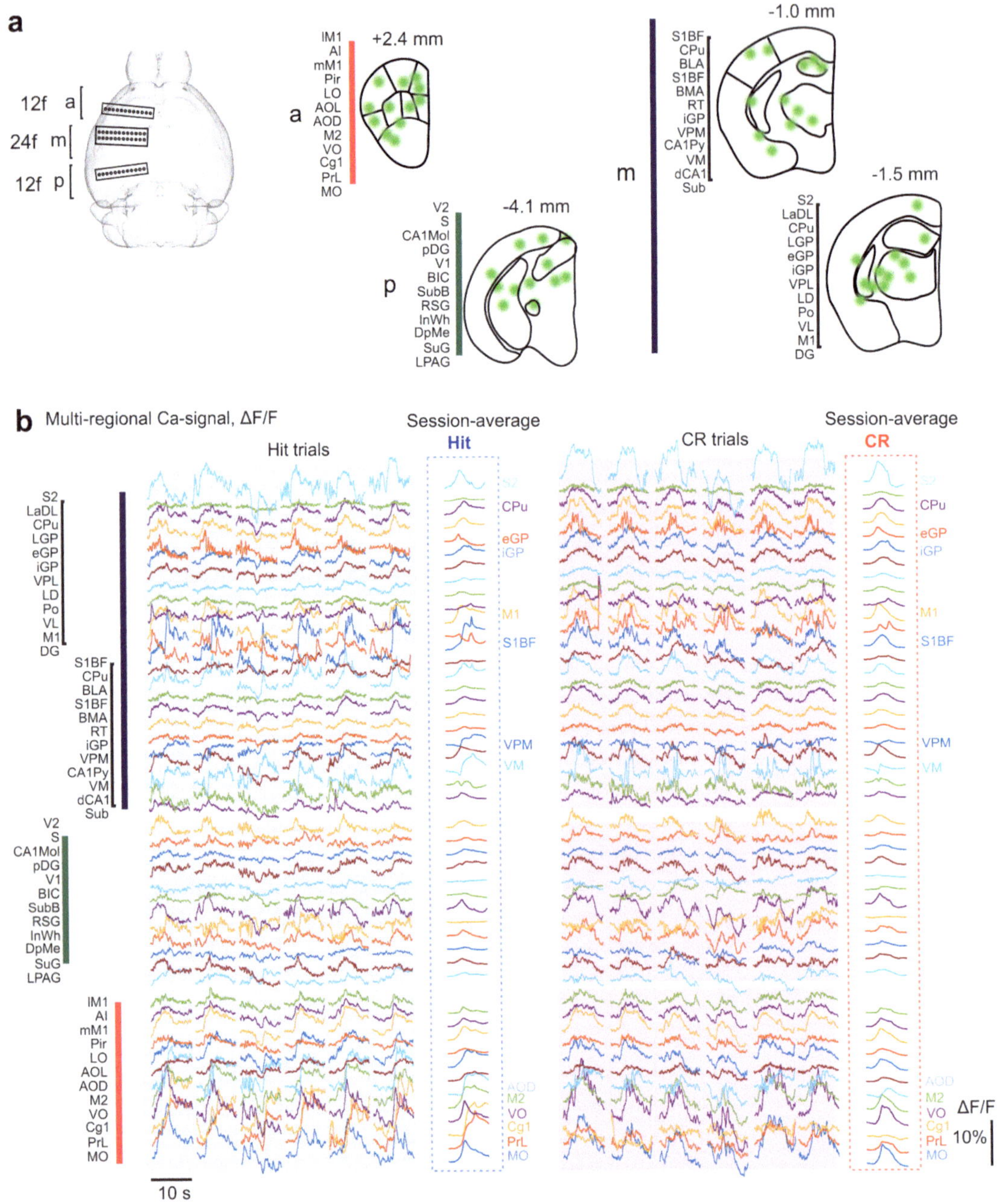

Fig. 4 Multi-fiber photometry of mesoscale brain activity in the behaving mouse. (**a**) Left: top view of three implanted multi-fiber arrays (anterior "a" network: 12-fiber arrays at +2.4 mm from bregma, posterior "p" network at −4.1 mm posterior from bregma, and medial "m" network targeted with 24-fiber array at about−1 to −1.4 mm posterior from bregma). Right: schematic representation of coronal sections, diffuse green dots are target sub-regions for each channel. Channels are arranged from lateral to medial position. (**b**) Measurement examples of mesoscale activity during a tactile texture discrimination task for head-fixed mice. Trial-related calcium signals in the "a," "p," and "m" networks during correct trials. Six Hit and CR trials each are shown from one example expert mouse. Session-averaged ΔF/F signals are displayed on the right for both Hit and CR trials. Labeling on the right indicates the subnetwork of task-specific regions differentially recruited during Hit and CR trials

recorded multiregional calcium dynamics and revealed trial-related ΔF/F signals above baseline in all regions. Notably, it was possible to perform multi-fiber photometric measurements chronically over several weeks, confirming the stability of the implants.

From the multi-fiber calcium recordings, we found distinct activity patterns across multiple regions [53] (Fig. 4b). Brain-wide networks were engaged both during correct trials, in which sensory stimuli predicted reward (Hit trials), and correct trials, in which no reward was available (CR trials). The temporal profile and amplitude of recorded ΔF/F signals were predictive of the trial type containing information about the task, stimulus, and choice (i.e., motor action). Session-averaged signals revealed a subset of regions with most prominent differences of the Hit vs. CR ΔF/F signals (labeled regions for one example mouse in Fig. 4b). Multi-fiber photometry recording thus facilitates the discovery of task-specific distributed functional networks, especially when it is combined with studying multiple behavioral tasks in the same subject.

3.3 Analytical Approaches for High-Density Multi-fiber Photometry

The following section briefly describes challenges for analysis and interpretation of mesoscale brain activity signals as measured by multi-fiber photometry. We do not provide a comprehensive overview of all available analytical approaches and refer to other reviews for more information on inference of multiregional interactions [104–108], population neural codes [109–112] and behavioral analysis [113–118]. One approach to account for behavioral contributions is to fit a generalized linear model (GLM) [119] and estimate how much variance it can explain. The explained variance is a measure of how well a particular model or hypothesis can account for the variability in the data. It is commonly used in neuroimaging studies to assess the degree to which a particular brain region or neural activity pattern is associated with a specific cognitive or behavioral measure. The GLM can contain a set of normalized behavioral variables weighted by the model coefficients and fitted by linear regression to the training dataset (Fig. 5a). The explained variance and residual errors are then computed on testing dataset not used for training. Usually k-fold cross-validation is applied to train the model and prevent over-fitting (e.g., in case of fivefold cross validation, 80% of the data are used for training, and testing is performed on the remaining 20%). The average performance of the model across all k iterations is used as an estimate of its performance on unseen data. The advantage of the GLM is that it represents the explained variance for each trial (Fig. 5b) allowing researchers to judge how much of their data is explained and during which trial periods.

An alternative approach to account for task- and motor-related contributions is to define particular types of trials based on behavioral outcomes or movement patterns and to train a statistical model to discriminate these categories based on neural data. A

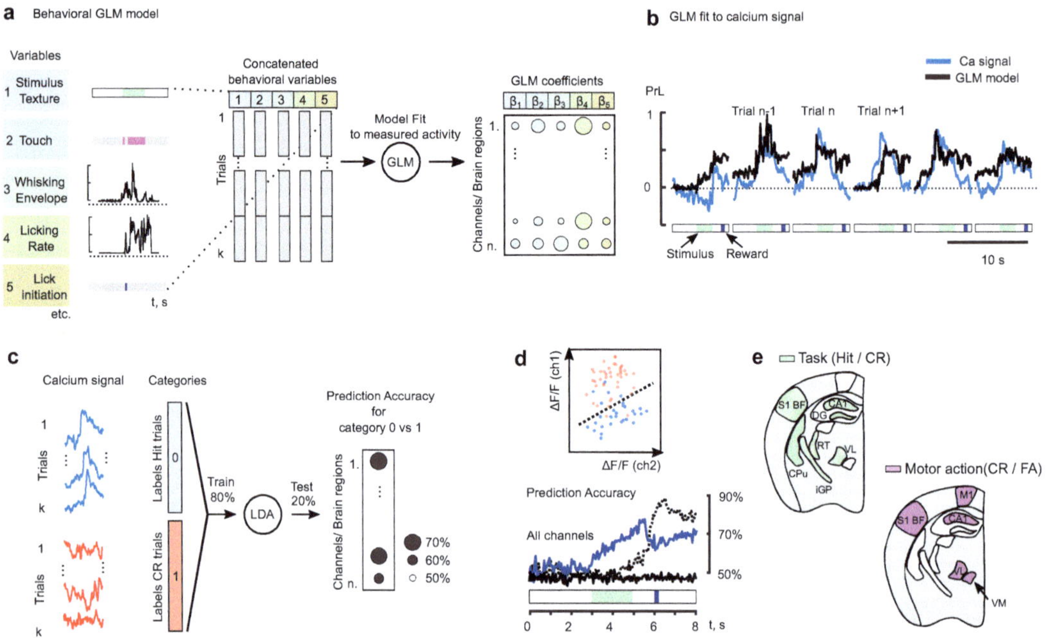

Fig. 5 Analytical approaches to explain mesoscale brain activity during tactile discrimination with behavior- and task-related components. (**a**) Schematic description of the GLM analysis. All behavioral variables are normalized, concatenated across trials, and fed to train the GLM model. The model infers β_n weights for n behavioral variables by fitting behavioral data-points with weights β_n to the given normalized calcium signal. By repeating the same GLM analysis across all brain regions (channels) one can describe how various behavioral variables are represented in each given region. (**b**) Example of the calcium signal recorded with multi-fiber photometry in prelimbic cortex (blue) and the fitted GLM model (black) across several correct and rewarded trials (Hit). The period of sensory texture presentation is indicated by the 2-s green bar below, the reward period with the blue bar. In some trials, the GLM model described the calcium signal well. (**c**) Schematic description of the Linear Discriminant Analysis (LDA). Calcium signals along with the trial labels, defining the category, are fed to train the LDA. The model learns to predict the type of the trial based on the input brain activity (usually for a defined time window Δt in the trial). Prediction accuracy across channels informs us about how strongly the respective signals are related to the given task. (**d**) LDA may enable dimensionality reduction once the model is trained on signals across multiple channels that form the multidimensional space. Top: schematic representation of two-dimensional plane deduced from $\Delta F/F$ signals recorded in two channels and for two labels (blue and red). LDA can separate into two categories (e.g., two types of trials) by finding the borderline between two distributions. Bottom: Hit vs. CR prediction accuracy across trial time. LDA trained on the activity across all channels in naïve mice (dashed line) and in expert mice (blue line). Black line represents control by shuffling labels. Green and blue bars indicate stimulus and reward windows. (**e**) Map of brain regions with task-related and motor-related prediction accuracy

classifier model (e.g., Linear Discriminant Analysis (LDA), Support Vector Machines (SVM), or Random Forest based on Decision Trees) is applied to the labeled neural data (Fig. 5c). k-fold cross-validation is used to test how well the model can classify across given categories. If the classifier is able to predict the trial identity in at least a fraction of trials, we may conclude that neural activity represents task-relevant processing of the distinct information.

Compared to GLM models, classifiers do not give information about the explained variance within each trial but rather evaluate a session-based classification accuracy. For example, in our study [53] we classified Hit vs. CR, i.e., task-relevant information, by applying LDA to $\Delta F/F$ signals in each region individually and for all regions simultaneously (time window before the first lick occurred; randomly selected 80% training data; 20% test data; fivefold cross-validated). Prediction accuracy for LDA trained on all channels increased across the trial (Fig. 5d) and generally was higher compared to LDA trained on single channels, consistent with distributed task-related activity. Next, to verify whether Hit vs. CR classification reflects motor-related behavior, we applied LDA during the same early stimulus window to classify CR vs. FA trials (i.e., same stimulus, different motor command). The accuracy of task- and motor-related classification depended on the brain region (Fig. 5e). For example, task-related classification accuracy was high in the somatosensory barrel cortex (and other subcortical regions), while motor-related classification was higher in the primary motor cortex and ventral medial and lateral nuclei of thalamus. The two types of comparisons (Hit vs. CR and CR vs. FA) also had distinct dynamics throughout task learning. While motor-related classification peaked during the pre-learning phase and the first learning sessions, it gradually decreased back to the chance level for late expert sessions. In contrast, task-related classification showed gradually increasing accuracy towards expert sessions.

Note that both computational approaches have certain limitations. The results of such analyses should be carefully considered, and the underlying assumptions recognized. For example, it is rather difficult to have a complete description of motor behavior, making the model incomplete [120]. In the case of GLM-based analysis, this may lead to residual activity being misinterpreted as decision-related. The separation between sensory and motor aspects is already difficult for single neurons due to their mixed selectivity. Mixed selectivity refers to the ability of a neuron to respond to multiple types of stimuli or inputs. As an example, neurons have been found in the primary visual cortex (V1) that respond to complex patterns [121, 122] (combinations of edges and color, etc.) but also motor behaviors [123], such as paw movement. The problem becomes even more nuanced when considering that photometry typically averages across ensembles of neurons. Averaging multiple complex neuronal responses into one signal clearly poses a challenge. One of the available solutions could be to temporally separate decision- and motor-related activity by the task design (e.g., tasks with delayed motor response [87, 124]) or by selecting the brief period during the trial when the sensory stimulus is presented but a motor response has not yet detected [53]. Another suggestion is to carefully measure mouse behavior with a high-speed camera and at different view angles, allowing to account for and model as many motor variables as possible. Note, however, that both strategies cannot exclude that the observed activity can be interpreted as a preparatory motor activity (pre-motor activity).

3.4 Multi-fiber Photometry in Freely Moving Animals

The use of lightweight fiber bundles and miniaturized fiber arrays also allows for experiments where the subject can move freely. We showcased this opportunity by conducting proof-of-concept measurements of brain activity in mice as they explore a new environment and a novel object [57]. We compared activity changes in various brain regions during different behaviors such as approaching a novel object and rearing, to baseline activity during free exploration of the cage. Results showed increased activity in specific brain regions during object approach, and different activity patterns during rearing compared to baseline behavior. Upon object approach $\Delta F/F$ significantly increased in S1BF, CPu, GP, in basolateral/basomedial amygdala (BLA/BMA), and in ventral posterolateral/posteromedial thalamic nuclei (VPL/VPM). Rearing, on the other hand, was correlated with increased $\Delta F/F$ values in DG and interestingly also with decreased $\Delta F/F$ values in S1BF and CPu as compared to baseline behavior. Further experiments and a more detailed analysis of behavioral variables are needed to better understand the link between mesoscale brain activity and naturalistic behaviors.

We also conducted multi-fiber photometry in two interacting mice by connecting separate branches of the fiber bundle to each animal and measuring the fluorescence signals with one photometry setup [57]. Two male mice (one with a 12-fiber implant, the other with a 36-fiber implant) were placed in the same cage and they approached each other multiple times during a session. Several regions in the amygdala were active during these social interactions (BLA, BMA, and a dorsal part of lateral amygdaloid nucleus LaDL, $p < 0.01$; one-way ANOVA test for $\Delta F/F$ distribution of samples for a relative distance between two mice "far," i.e., >20 cm, or "close," <10 cm). In addition, a brain-wide but specific network consisting of frontal cortical regions (anterior lateral motor cortex M1, insular cortex Ai, secondary motor cortex M2, lateral and medial orbital cortices LO and MO), sensory cortices (somatosensory barrel S1BF, anterior olfactory nucleus AOL, piriform cortex Pir), subcortical basal ganglia (dorsal striatum CPu, pallidum GP), and thalamic regions (ventral thalamus VL and VPL) was recruited upon the close encounter with another animal (significantly increased $\Delta F/F$; $p < 0.01$, one-way ANOVA test). This finding is consistent with recent studies [125], suggesting that neuronal ensembles in amygdala encode context- and brain-state-dependent information relevant for social interaction, while other areas such as S1BF and CPu may reflect multiple aspects of social touch and olfactory processing initiated by social interaction.

A recent study explored the activity in multiple regions of the limbic system during social interactions of mice [126]. During sexual and aggressive behaviors, calcium signals of estrogen receptor alpha (Esr1)-expressing cells were recorded in 13 brain regions using multi-fiber photometry. The authors found that conspecific sensory information, as well as social action initiation signals, are widely distributed in the limbic system and can be decoded from network activity. They also observed striking increases in functional connectivity in the network during the action initiation phase of social behaviors, whereas other behaviors were accompanied by a more dissociated network state.

Overall, the application of high-density fiber arrays in freely moving animals can help us better understand the long-range circuits underlying complex naturalistic behaviors, including social interactions. The use of multi-fiber photometry allows for the simultaneous measurement of multiple subjects using only one relatively simple fluorescence setup that connects to the fiber bundle, which can handle multiple animals. Simultaneous measurements in multiple animals in a single experiment may provide valuable insights into the neural basis of social interactions and other behaviors that involve multiple individuals. Additionally, the use of high-density fiber arrays in freely moving animals allows for the measurement of brain activity in a more naturalistic setting compared to behaviors under head-fixation conditions.

3.5 Tracking Longitudinal Changes of Brain Activity during Task Learning

How large-scale brain activity reorganizes during task learning, linking stimulus-related sensory representations to task-relevant motor representations and thereby instantiating behavioral changes, remains largely unknown. Neuroimaging studies in humans can reveal changes in functional brain networks during task learning [127, 128]. However, such methods, mostly based on fMRI or EEG, have either limited temporal or spatial resolution. New methods to interrogate the large-scale brain dynamics for previously challenging longitudinal studies have emerged in animals. One promising approach is to repeatedly image neuronal activity during the entire training process. This chronic imaging can be done with high resolution in specific regions [129] or across multiple regions [130–132] at lower resolution. So far, these methods were limited to more easily accessible areas of the dorsal cortex using wide-field calcium imaging [30, 78–80, 87, 130, 131, 133]. Using multi-fiber photometry, activity in multiple subcortical regions can be tracked longitudinally. For example, we have applied high-density multi-fiber photometry to track large-scale brain activity changes when mice learn a texture discrimination task [53]. We focused on the cortico-basal ganglia-thalamo-cortical network (CBGTC network), which is believed to be crucial in forming stimulus–response associations. Mice were trained until they reached an expert performance level of 70% correct trials, which occurred within 1–15 days. The data were aligned to the first session after reaching expert performance and divided into a "naïve" phase (sessions before first expert session) and an "expert" phase. In one group of mice, we used 12-fiber implants which targeted regions of the CBGTC network and the hippocampus, while in another group of mice, 48 fibers targeted the same set of regions plus additional regions. Using viral expression of a GECI, we simultaneously recorded bulk calcium signals in all targeted brain regions across many trials from the naive to expert phase. By analyzing the correlation of calcium signals with behavioral variables separately in the group of naïve and expert mice, we found that the representation of the reward-predicting stimulus, the texture touch, became stronger during learning, while the representation of licking action and signals related to reward itself decreased in parallel in several brain regions. That is, nearly all regional $\Delta F/F$ signals showed increased correlation with the texture touch in expert mice, most prominently in cortical S1BF and M1, thalamic VM, and basal ganglia CPu and iGP ($p \leq 0.01$ for S1BF, M1, VM, and CPu; $p = 0.03$ for iGP). Correlation with whisking also tended to increase in most regions, though not reaching significance. Correlation to licking decreased in basal ganglia CPu and iGP, cortical S1BF, thalamic VM, LD, RT, and hippocampal DG, CA1Py regions ($p \leq 0.01$ for CPu, LD, DG, CA1Py, $p \leq 0.05$ for iGP, VM, RT, S1BF; list of regions provided in Tables 1 and 2).

We reported learning-dependent changes for four major behavioral parameters: touch, whisking envelope, licking rate, and reaction time. While whisking envelope, licking rate, and reaction time experienced significant changes over the course of learning, touch probability stayed largely constant. Under the hypothesis that the observed calcium signals exclusively reflect motor behavior, one would expect their correlation to stay relatively constant with learning. However, this is not what we observed. Instead, when evaluating these correlations as a function of learning, we found that they show large changes, specific to individual brain regions. This indicates that behavioral changes cannot fully explain the learning-related changes of brain activity. Our results suggest that the mesoscale network spanning cortex, basal ganglia, thalamus, and hippocampus underwent a significant functional reorganization during task learning [53].

Recently multi-fiber photometry was used together with a virtual-reality based behavioral task to observe neural activity patterns associated with memory reorganization and stabilization on the time scale of several weeks [134]. By simultaneous measurements in multiple brain regions (anterior cingulate cortex, entorhinal cortex, basolateral amygdala, anterior thalamus, and hippocampus) researchers discovered a unique and prominent neural correlate of memory in the anteromedial thalamus during memory retrieval. Activity in anterior thalamus also stabilized task-relevant cortical representations, further strengthening the hypothesis that anterior thalamus plays a key role for long-term memory formation [135].

4 Conclusions

Multi-fiber arrays complement existing neural interfaces for read-out and optical control of large-scale network dynamics in the mammalian brain. Above we have highlighted the versatility of multi-fiber arrays and the advances they can bring to the field of circuit neuroscience. We discuss further prospects in terms of technology and applications. As with single-fiber approaches [31, 32], multi-fiber approaches will gain their full strength through combination with genetic tools for cell-type or pathway-specific expression of fluorescent indicators (for neuronal calcium levels [61], voltage [136–140], neuromodulator release such as dopamine [141], noradrenaline [142], glial dynamics [33, 143] etc.). In addition to pathway-specific labeling, soma-targeted or nuclear expression of calcium indicators [65, 66] could circumvent ambiguities in signal assignment and dissect mixed components in fiber-optic bulk signals.

High-density multi-fiber photometry can be further extended and improved. For example, fibers can be made less invasive by modifying fiber tips through fiber tapering [144, 145]. Alternatively, other strategies to functionalize the implanted optical fibers can be used, for example, one fiber tip can accommodate multiple tens-of-micron-sized apertures along the fiber facet or embed multiple plasmonic nanoparticles to locally enhance fluorescence [146, 147]. However, such approaches are hardly scalable, limiting their applicability for in vivo applications. Additionally, collection of fluorescence photons across widely distributed fiber apertures also represents a considerable challenge, which can be solved with the state-of-the art design of integrated photonic waveguides [148]. Combining optical fiber probes with electrophysiological recordings is another direction where a lot of recent research efforts has been directed to, such probes allow simultaneous recording of spiking neural activity and its optogenetic interrogation [149–152]. Recently, a new set of multichannel flexible probes based on integrated circuits was developed. Such probes combine optical fibers and conducting sites for electrophysiological recording, integrated pre-amplifiers [153] and μ-LEDs for optogenetic manipulations [154, 155].

In summary, high-density multi-fiber photometry complements the multitude of techniques for optical interrogation of neural circuits by filling a niche and providing a relatively simple and low-cost method to study brain circuitry on the mesoscale. Its application is highly versatile, reaching from targeted investigations of specific communicating neuronal subtypes in head-fixed mice, to studies of distributed cross-regional signal flow in freely behaving animals, to coupled simultaneous recordings of brain activity in naturally interacting mice. With increasing recognition of the strengths, versatility, and the further potential of multi-fiber approaches, their application is likely to expand in the near future.

5 Notes

1. Fiber photometry signals report changes of average (bulk) fluorescence intensity in a specific region of interest, representing, for example, the population-average activity from all labeled neurons and neuropil located in the recording volume. The effective recording volume is defined by the properties of optical fibers [50, 51] (i.e., numerical aperture (NA) and fiber diameter), by the extent of labeling, and by the tissue properties (i.e., scattering and absorption of excitation and emission light as well as patterns of blood vessels in the vicinity of the fiber tip). Researchers may change the recording volume by using optical fibers with different properties, for example, fibers with higher numerical aperture (NA) and larger diameter sample fluorescence from larger recording volumes.
2. Because action potentials cause changes in intracellular calcium concentration, GECI fluorescence changes can be used to infer underlying neuronal spiking. In addition, more localized calcium signals occur in axonal and dendritic processes that are linked to synaptic transmission and dendritic integration. For glial cells such as astrocytes, intracellular calcium signals also play essential roles and are a prominent readout of glial activity. Because fiber photometry records a bulk fluorescence signal, it may represent a mixture of components depending on specificity and locality of labeling. For example, if only neuronal somata are labeled with a calcium indicator (using a soma-targeted indicator version), the photometric signal mainly will represent spiking activity in the local population. Recent progress in the development of molecular sensors allow such precise targeting to somata [65, 66], or axons [67]. However, more typically the recorded fluorescence contains additionally dendritic, synaptic, and axonal calcium signal components (i.e., neuropil signals), which may reflect input signals to the selected brain region, originating in distant regions. In this situation, when the fluorescence signals recorded from somata and neuropil are distinct, the volume-average of the fiber photometric signal poses a further interpretational challenge.

One approach to experimentally characterize and validate photometric calcium signals is to simultaneously record neuronal spiking activity with electrodes. In an earlier study we implanted single 100-μm diameter optical fibers glued together with four tetrodes for extracellular recording of multiunit activity (Supplementary Figure 3 in [57]). Tetrodes were positioned around the fiber and cut down to the fiber tip. The probe was implanted in the motor cortex of mice. Mice were trained in a go/no-go tactile discrimination task allowing us to compare optical and electrophysiological recordings across multiple trials of repeated goal-directed behavior. We did not observe large differences between the optical and electrical recordings [57], indicating that under our conditions the fluorescence signal is a useful proxy of local spiking activity. However, note that in our analysis we averaged both signals across trials, representing their activity in terms of probability distributions relative to the trial-related events of sensory stimulation and reward delivery. In another study [156], calcium signals recorded with fiber photometry in striatum primarily reflected non-somatic changes of calcium concentration possibly reflecting dendritic and synaptic signals.

3. The optical properties of the tissue and the dynamics of blood vessels might be changing in vivo, contributing to intrinsic optical signal changes. In brain tissue, the hemodynamic response, reflecting changes in hemoglobin oxygenation/deoxygenation ratio in the blood and in flavoprotein concentrations, can lead to intrinsic optical signal changes [91, 92, 157]. These changes may contaminate activity indicator signals, especially since hemodynamic responses themselves are tightly coupled to endogenous brain activity [93]. Thus, hemodynamic responses must be assessed in control experiments (e.g., by expressing GFP) or may need to be corrected with a dual-color fluorescence measurement. For GCaMP recordings, one approach is to add a violet LED illumination path, with an excitation wavelength in the 405–420 nm range, where GCaMP fluorescence is approximately calcium-independent. The violet-excited reference signal can then be used to correct the signals recorded with other excitation wavelengths (e.g., 470–490 nm for GCaMP) to uncover the "true" functional signal related to cellular activities.

References

1. Bassett DS, Wymbs NF, Porter MA, Mucha PJ, Carlson JM, Grafton ST (2011) Dynamic reconfiguration of human brain networks during learning. PNAS 108:7641–7646. https://doi.org/10.1073/pnas.1018985108

2. Cohen MX, van Gaal S (2013) Dynamic interactions between large-scale brain networks predict behavioral adaptation after perceptual errors. Cereb Cortex 23:1061–1072. https://doi.org/10.1093/cercor/bhs069

3. Cole MW, Reynolds JR, Power JD, Repovs G, Anticevic A, Braver TS (2013) Multi-task connectivity reveals flexible hubs for adaptive task control. Nat Neurosci 16:1348–1355. https://doi.org/10.1038/nn.3470
4. Ferster D, Miller KD (2000) Neural mechanisms of orientation selectivity in the visual cortex. Annu Rev Neurosci 23:441–471. https://doi.org/10.1146/annurev.neuro.23.1.441
5. Petreanu L, Gutnisky DA, Huber D, Xu N, O'Connor DH, Tian L, Looger L, Svoboda K (2012) Activity in motor–sensory projections reveals distributed coding in somatosensation. Nature 489:299–303. https://doi.org/10.1038/nature11321
6. Leong ATL, Chan RW, Gao PP, Chan Y-S, Tsia KK, Yung W-H, Wu EX (2016) Long-range projections coordinate distributed brain-wide neural activity with a specific spatiotemporal profile. PNAS 113:E8306–E8315. https://doi.org/10.1073/pnas.1616361113
7. Steinmetz NA, Zatka-Haas P, Carandini M, Harris KD (2019) Distributed coding of choice, action and engagement across the mouse brain. Nature 576:266–273. https://doi.org/10.1038/s41586-019-1787-x
8. Clancy KB, Orsolic I, Mrsic-Flogel TD (2019) Locomotion-dependent remapping of distributed cortical networks. Nat Neurosci 22:778–786. https://doi.org/10.1038/s41593-019-0357-8
9. Lee JH, Durand R, Gradinaru V, Zhang F, Goshen I, Kim D-S, Fenno LE, Ramakrishnan C, Deisseroth K (2010) Global and local fMRI signals driven by neurons defined optogenetically by type and wiring. Nature 465:788–792. https://doi.org/10.1038/nature09108
10. Han Z, Chen W, Chen X, Zhang K, Tong C, Zhang X, Li CT, Liang Z (2019) Awake and behaving mouse fMRI during Go/No-Go task. NeuroImage 188:733–742. https://doi.org/10.1016/j.neuroimage.2019.01.002
11. Jung WB, Im GH, Jiang H, Kim S-G (2021) Early fMRI responses to somatosensory and optogenetic stimulation reflect neural information flow. Proc Natl Acad Sci 118: e2023265118. https://doi.org/10.1073/pnas.2023265118
12. Macé E, Montaldo G, Cohen I, Baulac M, Fink M, Tanter M (2011) Functional ultrasound imaging of the brain. Nat Methods 8: 662–664. https://doi.org/10.1038/nmeth.1641
13. Steinmetz NA, Koch C, Harris KD, Carandini M (2018) Challenges and opportunities for large-scale electrophysiology with Neuropixels probes. Curr Opin Neurobiol 50:92–100. https://doi.org/10.1016/j.conb.2018.01.009
14. Kuhlman SJ, Huang ZJ (2008) High-resolution labeling and functional manipulation of specific neuron types in mouse brain by Cre-activated viral gene expression. PLoS One 3:e2005. https://doi.org/10.1371/journal.pone.0002005
15. Betley JN, Sternson SM (2011) Adeno-associated viral vectors for mapping, monitoring, and manipulating neural circuits. Hum Gene Ther 22:669–677. https://doi.org/10.1089/hum.2010.204
16. Haggerty DL, Grecco GG, Reeves KC, Atwood B (2020) Adeno-associated viral vectors in neuroscience research. Mol Ther Methods Clin Dev 17:69–82. https://doi.org/10.1016/j.omtm.2019.11.012
17. Fenno LE, Ramakrishnan C, Kim YS, Evans KE, Lo M, Vesuna S, Inoue M, Cheung KYM, Yuen E, Pichamoorthy N, Hong ASO, Deisseroth K (2020) Comprehensive dual- and triple-feature intersectional single-vector delivery of diverse functional payloads to cells of behaving mammals. Neuron 107: 836–853.e11. https://doi.org/10.1016/j.neuron.2020.06.003
18. Sheintuch L, Rubin A, Brande-Eilat N, Geva N, Sadeh N, Pinchasof O, Ziv Y (2017) Tracking the same neurons across multiple days in Ca2+ imaging data. Cell Rep 21:1102–1115. https://doi.org/10.1016/j.celrep.2017.10.013
19. Chen JL, Voigt FF, Javadzadeh M, Krueppel R, Helmchen F (2016) Long-range population dynamics of anatomically defined neocortical networks. elife 5. https://doi.org/10.7554/eLife.14679
20. Sofroniew NJ, Flickinger D, King J, Svoboda K (2016) A large field of view two-photon mesoscope with subcellular resolution for in vivo imaging. elife 5. https://doi.org/10.7554/eLife.14472
21. Ji N, Freeman J, Smith SL (2016) Technologies for imaging neural activity in large volumes. Nat Neurosci 19:1154–1164. https://doi.org/10.1038/nn.4358
22. Stirman JN, Smith IT, Kudenov MW, Smith SL (2016) Wide field-of-view, multi-region, two-photon imaging of neuronal activity in the mammalian brain. Nat Biotechnol 34: 857–862. https://doi.org/10.1038/nbt.3594

23. Ota K, Oisi Y, Suzuki T, Ikeda M, Ito Y, Ito T, Uwamori H, Kobayashi K, Kobayashi M, Odagawa M, Matsubara C, Kuroiwa Y, Horikoshi M, Matsushita J, Hioki H, Ohkura M, Nakai J, Oizumi M, Miyawaki A, Aonishi T, Ode T, Murayama M (2021) Fast, cell-resolution, contiguous-wide two-photon imaging to reveal functional network architectures across multi-modal cortical areas. Neuron 109:1810–1824.e9. https://doi.org/10.1016/j.neuron.2021.03.032
24. Yu C-H, Stirman JN, Yu Y, Hira R, Smith SL (2021) Diesel2p mesoscope with dual independent scan engines for flexible capture of dynamics in distributed neural circuitry. Nat Commun 12:6639. https://doi.org/10.1038/s41467-021-26736-4
25. Clough M, Chen IA, Park S-W, Ahrens AM, Stirman JN, Smith SL, Chen JL (2021) Flexible simultaneous mesoscale two-photon imaging of neural activity at high speeds. Nat Commun 12:6638. https://doi.org/10.1038/s41467-021-26737-3
26. Cardin JA, Crair MC, Higley MJ (2020) Mesoscopic imaging: shining a wide light on large-scale neural dynamics. Neuron 108:33–43. https://doi.org/10.1016/j.neuron.2020.09.031
27. Lim DH, Mohajerani MH, LeDue J, Boyd J, Chen S, Murphy TH (2012) In vivo large-scale cortical mapping using Channelrhodopsin-2 stimulation in transgenic mice reveals asymmetric and reciprocal relationships between cortical areas. Front Neural Circuits 6. https://doi.org/10.3389/fncir.2012.00011
28. Minderer M, Liu W, Sumanovski LT, Kügler S, Helmchen F, Margolis DJ (2012) Chronic imaging of cortical sensory map dynamics using a genetically encoded calcium indicator. J Physiol 590:99–107. https://doi.org/10.1113/jphysiol.2011.219014
29. Vanni MP, Murphy TH (2014) Mesoscale transcranial spontaneous activity mapping in GCaMP3 transgenic mice reveals extensive reciprocal connections between areas of somatomotor cortex. J Neurosci 34:15931–15946. https://doi.org/10.1523/JNEUROSCI.1818-14.2014
30. Gilad A, Gallero-Salas Y, Groos D, Helmchen F (2018) Behavioral strategy determines frontal or posterior location of short-term memory in neocortex. Neuron 99:814–828.e7. https://doi.org/10.1016/j.neuron.2018.07.029
31. Adelsberger H, Garaschuk O, Konnerth A (2005) Cortical calcium waves in resting newborn mice. Nat Neurosci 8:988–990. https://doi.org/10.1038/nn1502
32. Gunaydin LA, Grosenick L, Finkelstein JC, Kauvar IV, Fenno LE, Adhikari A, Lammel S, Mirzabekov JJ, Airan RD, Zalocusky KA, Tye KM, Anikeeva P, Malenka RC, Deisseroth K (2014) Natural neural projection dynamics underlying social behavior. Cell 157:1535–1551. https://doi.org/10.1016/j.cell.2014.05.017
33. Schlegel F, Sych Y, Schroeter A, Stobart J, Weber B, Helmchen F, Rudin M (2018) Fiber-optic implant for simultaneous fluorescence-based calcium recordings and BOLD fMRI in mice. Nat Protoc 13:840–855. https://doi.org/10.1038/nprot.2018.003
34. Barretto RPJ, Messerschmidt B, Schnitzer MJ (2009) In vivo fluorescence imaging with high-resolution microlenses. Nat Methods 6:511–512. https://doi.org/10.1038/nmeth.1339
35. Flusberg BA, Nimmerjahn A, Cocker ED, Mukamel EA, Barretto RPJ, Ko TH, Burns LD, Jung JC, Schnitzer MJ (2008) High-speed, miniaturized fluorescence microscopy in freely moving mice. Nat Methods 5:935–938. https://doi.org/10.1038/nmeth.1256
36. Ghosh KK, Burns LD, Cocker ED, Nimmerjahn A, Ziv Y, Gamal AE, Schnitzer MJ (2011) Miniaturized integration of a fluorescence microscope. Nat Methods 8:871–878. https://doi.org/10.1038/nmeth.1694
37. Ziv Y, Burns LD, Cocker ED, Hamel EO, Ghosh KK, Kitch LJ, Gamal AE, Schnitzer MJ (2013) Long-term dynamics of CA1 hippocampal place codes. Nat Neurosci 16:264–266. https://doi.org/10.1038/nn.3329
38. Aharoni D, Khakh BS, Silva AJ, Golshani P (2019) All the light that we can see: a new era in miniaturized microscopy. Nat Methods 16:11–13. https://doi.org/10.1038/s41592-018-0266-x
39. Skocek O, Nöbauer T, Weilguny L, Martínez Traub F, Xia CN, Molodtsov MI, Grama A, Yamagata M, Aharoni D, Cox DD, Golshani P, Vaziri A (2018) High-speed volumetric imaging of neuronal activity in freely moving rodents. Nat Methods 15:429–432. https://doi.org/10.1038/s41592-018-0008-0
40. Meng G, Liang Y, Sarsfield S, Jiang W, Lu R, Dudman JT, Aponte Y, Ji N (2019) High-throughput synapse-resolving two-photon fluorescence microendoscopy for deep-brain volumetric imaging in vivo. eLife 8:e40805. https://doi.org/10.7554/eLife.40805

41. Dombeck DA, Harvey CD, Tian L, Looger LL, Tank DW (2010) Functional imaging of hippocampal place cells at cellular resolution during virtual navigation. Nat Neurosci 13: 1433–1440. https://doi.org/10.1038/nn.2648
42. Pilz G-A, Carta S, Stäuble A, Ayaz A, Jessberger S, Helmchen F (2016) Functional imaging of dentate granule cells in the adult mouse hippocampus. J Neurosci 36:7407–7414. https://doi.org/10.1523/JNEUROSCI.3065-15.2016
43. Schoenfeld G, Carta S, Rupprecht P, Ayaz A, Helmchen F (2021) Vivo calcium imaging of CA3 pyramidal neuron populations in adult mouse hippocampus. eNeuro 8. https://doi.org/10.1523/ENEURO.0023-21.2021
44. Papadopoulos IN, Farahi S, Moser C, Psaltis D (2012) Focusing and scanning light through a multimode optical fiber using digital phase conjugation. Opt Express 20: 10583–10590. https://doi.org/10.1364/OE.20.010583
45. Turtaev S, Leite IT, Altwegg-Boussac T, Pakan JMP, Rochefort NL, Čižmár T (2018) High-fidelity multimode fibre-based endoscopy for deep brain in vivo imaging. Light Sci Appl 7:92. https://doi.org/10.1038/s41377-018-0094-x
46. Leite IT, Turtaev S, Boonzajer Flaes DE, Čižmár T (2021) Observing distant objects with a multimode fiber-based holographic endoscope. APL Photonics 6:036112. https://doi.org/10.1063/5.0038367
47. Du Y, Turtaev S, Leite IT, Lorenz A, Kobelke J, Wondraczek K, Čižmár T (2022) Hybrid multimode – multicore fibre based holographic endoscope for deep-tissue neurophotonics. Light Adv Manuf 3:408–416. https://doi.org/10.37188/lam.2022.029
48. Papadopoulos IN, Farahi S, Moser C, Psaltis D (2013) High-resolution, lensless endoscope based on digital scanning through a multimodeoptical fiber. Biomed Opt Express 4:260–270. https://doi.org/10.1364/BOE.4.000260
49. Farahi S, Ziegler D, Papadopoulos IN, Psaltis D, Moser C (2013) Dynamic bending compensation while focusing through a multimode fiber. Opt Express 21:22504. https://doi.org/10.1364/OE.21.022504
50. Pfefer TJ, Schomacker KT, Nishioka NS (2001) Effect of fiber optic probe design on fluorescent light propagation in tissue, pp 410–416
51. Pfefer TJ, Schomacker KT, Ediger MN, Nishioka NS (2002) Multiple-fiber probe design for fluorescence spectroscopy in tissue. Appl Opt 41:4712–4721. https://doi.org/10.1364/AO.41.004712
52. Osorio-Forero A, Cardis R, Vantomme G, Guillaume-Gentil A, Katsioudi G, Devenoges C, Fernandez LMJ, Lüthi A (2021) Noradrenergic circuit control of non-REM sleep substates. Curr Biol 31: 5009–5023.e7. https://doi.org/10.1016/j.cub.2021.09.041
53. Sych Y, Fomins A, Novelli L, Helmchen F (2022) Dynamic reorganization of the cortico-basal ganglia-thalamo-cortical network during task learning. Cell Rep 40: 111394. https://doi.org/10.1016/j.celrep.2022.111394
54. Muir J, Lorsch ZS, Ramakrishnan C, Deisseroth K, Nestler EJ, Calipari ES, Bagot RC (2018) In vivo fiber photometry reveals signature of future stress susceptibility in nucleus accumbens. Neuropsychopharmacology 43:255–263. https://doi.org/10.1038/npp.2017.122
55. Ineichen C, Greter A, Baer M, Sigrist H, Sautter E, Sych Y, Helmchen F, Pryce CR (2022) Basomedial amygdala activity in mice reflects specific and general aversion uncontrollability. Eur J Neurosci 55:2435–2454. https://doi.org/10.1111/ejn.15090
56. Kim CK, Yang SJ, Pichamoorthy N, Young NP, Kauvar I, Jennings JH, Lerner TN, Berndt A, Lee SY, Ramakrishnan C, Davidson TJ, Inoue M, Bito H, Deisseroth K (2016) Simultaneous fast measurement of circuit dynamics at multiple sites across the mammalian brain. Nat Methods 13:325–328. https://doi.org/10.1038/nmeth.3770
57. Sych Y, Chernysheva M, Sumanovski LT, Helmchen F (2019) High-density multi-fiber photometry for studying large-scale brain circuit dynamics. Nat Methods 16(6): 553–560. https://doi.org/10.1038/s41592-019-0400-4
58. Miyawaki A, Llopis J, Heim R, McCaffery JM, Adams JA, Ikura M, Tsien RY (1997) Fluorescent indicators for Ca2+based on green fluorescent proteins and calmodulin. Nature 388:882–887. https://doi.org/10.1038/42264
59. Akerboom J, Chen T-W, Wardill TJ, Tian L, Marvin JS, Mutlu S, Calderón NC, Esposti F, Borghuis BG, Sun XR, Gordus A, Orger MB, Portugues R, Engert F, Macklin JJ, Filosa A, Aggarwal A, Kerr RA, Takagi R, Kracun S, Shigetomi E, Khakh BS, Baier H, Lagnado L, Wang SS-H, Bargmann CI, Kimmel BE, Jayaraman V, Svoboda K, Kim DS, Schreiter ER, Looger LL (2012)

Optimization of a GCaMP calcium indicator for neural activity imaging. J Neurosci 32: 13819–13840. https://doi.org/10.1523/JNEUROSCI.2601-12.2012

60. Chen T-W, Wardill TJ, Sun Y, Pulver SR, Renninger SL, Baohan A, Schreiter ER, Kerr RA, Orger MB, Jayaraman V, Looger LL, Svoboda K, Kim DS (2013) Ultrasensitive fluorescent proteins for imaging neuronal activity. Nature 499:295–300. https://doi.org/10.1038/nature12354
61. Zhang Y, Rózsa M, Liang Y, Bushey D, Wei Z, Zheng J, Reep D, Broussard GJ, Tsang A, Tsegaye G, Narayan S, Obara CJ, Lim J-X, Patel R, Zhang R, Ahrens MB, Turner GC, Wang SS-H, Korff WL, Schreiter ER, Svoboda K, Hasseman JP, Kolb I, Looger LL (2023) Fast and sensitive GCaMP calcium indicators for imaging neural populations. Nature 615:884–891. https://doi.org/10.1038/s41586-023-05828-9
62. Hires SA, Tian L, Looger LL (2008) Reporting neural activity with genetically encoded calcium indicators. Brain Cell Bio 36:69–86. https://doi.org/10.1007/s11068-008-9029-4
63. Akerboom J, Carreras Calderón N, Tian L, Wabnig S, Prigge M, Tolö J, Gordus A, Orger MB, Severi KE, Macklin JJ, Patel R, Pulver SR, Wardill TJ, Fischer E, Schüler C, Chen T-W, Sarkisyan KS, Marvin JS, Bargmann CI, Kim DS, Kügler S, Lagnado L, Hegemann P, Gottschalk A, Schreiter ER, Looger LL (2013) Genetically encoded calcium indicators for multi-color neural activity imaging and combination with optogenetics. Front Mol Neurosci 6. https://doi.org/10.3389/fnmol.2013.00002
64. Bethge P, Carta S, Lorenzo DA, Egolf L, Goniotaki D, Madisen L, Voigt FF, Chen JL, Schneider B, Ohkura M, Nakai J, Zeng H, Aguzzi A, Helmchen F (2017) An R-CaMP1.07 reporter mouse for cell-type-specific expression of a sensitive red fluorescent calcium indicator. PLoS One 12: e0179460. https://doi.org/10.1371/journal.pone.0179460
65. Chen Y, Jang H, Spratt PWE, Kosar S, Taylor DE, Essner RA, Bai L, Leib DE, Kuo T-W, Lin Y-C, Patel M, Subkhangulova A, Kato S, Feinberg EH, Bender KJ, Knight ZA, Garrison JL (2020) Soma-targeted imaging of neural circuits by ribosome tethering. Neuron 107:454–469.e6. https://doi.org/10.1016/j.neuron.2020.05.005
66. Shemesh OA, Linghu C, Piatkevich KD, Goodwin D, Celiker OT, Gritton HJ, Romano MF, Gao R, Yu C-C (Jay), Tseng H-A, Bensussen S, Narayan S, Yang C-T, Freifeld L, Siciliano CA, Gupta I, Wang J, Pak N, Yoon Y-G, JFP U, Guner-Ataman B, Noamany H, Sheinkopf ZR, Park WM, Asano S, Keating AE, Trimmer JS, Reimer J, Tolias AS, Bear MF, Tye KM, Han X, Ahrens MB, Boyden ES (2020) Precision calcium imaging of dense neural populations via a cell-body-targeted calcium indicator. Neuron 107:470–486.e11. https://doi.org/10.1016/j.neuron.2020.05.029
67. Broussard GJ, Liang Y, Fridman M, Unger EK, Meng G, Xiao X, Ji N, Petreanu L, Tian L (2018) In vivo measurement of afferent activity with axon-specific calcium imaging. Nat Neurosci 21:1272–1280. https://doi.org/10.1038/s41593-018-0211-4
68. Barnett LM, Hughes TE, Drobizhev M (2017) Deciphering the molecular mechanism responsible for GCaMP6m's Ca2+-dependent change in fluorescence. PLoS One 12:e0170934. https://doi.org/10.1371/journal.pone.0170934
69. Huang L, Ledochowitsch P, Knoblich U, Lecoq J, Murphy GJ, Reid RC, de Vries SE, Koch C, Zeng H, Buice MA, Waters J, Li L (2021) Relationship between simultaneously recorded spiking activity and fluorescence signal in GCaMP6 transgenic mice. eLife 10: e51675. https://doi.org/10.7554/eLife.51675
70. Shemetov AA, Monakhov MV, Zhang Q, Canton-Josh JE, Kumar M, Chen M, Matlashov ME, Li X, Yang W, Nie L, Shcherbakova DM, Kozorovitskiy Y, Yao J, Ji N, Verkhusha VV (2021) A near-infrared genetically encoded calcium indicator for in vivo imaging. Nat Biotechnol 39:368–377. https://doi.org/10.1038/s41587-020-0710-1
71. Shcherbakova DM (2021) Near-infrared and far-red genetically encoded indicators of neuronal activity. J Neurosci Methods 362: 109314. https://doi.org/10.1016/j.jneumeth.2021.109314
72. Madisen L, Garner AR, Shimaoka D, Chuong AS, Klapoetke NC, Li L, van der Bourg A, Niino Y, Egolf L, Monetti C, Gu H, Mills M, Cheng A, Tasic B, Nguyen TN, Sunkin SM, Benucci A, Nagy A, Miyawaki A, Helmchen F, Empson RM, Knopfel T, Boyden ES, Reid RC, Carandini M, Zeng H (2015) Transgenic mice for intersectional targeting of neural sensors and effectors with high specificity and performance. Neuron 85:942–958. https://doi.org/10.1016/j.neuron.2015.02.022
73. Daigle TL, Madisen L, Hage TA, Valley MT, Knoblich U, Larsen RS, Takeno MM, Huang L, Gu H, Larsen R, Mills M, Bosma-

Moody A, Siverts L, Walker M, Graybuck LT, Yao Z, Fong O, Nguyen TN, Garren E, Lenz G, Chavarha M, Pendergraft J, Harrington J, Hirokawa KE, Harris JA, Nicovich PR, McGraw M, Ollerenshaw DR, Smith K, Baker CA, Ting JT, Sunkin SM, Lecoq J, Lin MZ, Boyden ES, Murphy GJ, da Costa N, Waters J, Li L, Tasic B, Zeng H (2018) A suite of transgenic driver and reporter mouse lines with enhanced brain cell type targeting and functionality. Cell 174:465–480.e22. https://doi.org/10.1016/j.cell.2018.06.035

74. Kasatkina LA, Verkhusha VV (2022) Transgenic mice encoding modern imaging probes: properties and applications. Cell Rep 39: 110845. https://doi.org/10.1016/j.celrep.2022.110845

75. MacDowell CJ, Buschman TJ (2020) Low-dimensional spatiotemporal dynamics underlie cortex-wide neural activity. Curr Biol 30:2665–2680.e8. https://doi.org/10.1016/j.cub.2020.04.090

76. Ferezou I, Bolea S, Petersen CC (2006) Visualizing the cortical representation of whisker touch: voltage-sensitive dye imaging in freely moving mice. Neuron 50:617–629

77. Ferezou I, Haiss F, Gentet LJ, Aronoff R, Weber B, Petersen CCH (2007) Spatiotemporal dynamics of cortical sensorimotor integration in behaving mice. Neuron 56:907–923. https://doi.org/10.1016/j.neuron.2007.10.007

78. Musall S, Kaufman MT, Juavinett AL, Gluf S, Churchland AK (2019) Single-trial neural dynamics are dominated by richly varied movements. Nat Neurosci 22:1677–1686. https://doi.org/10.1038/s41593-019-0502-4

79. Mohajerani MH, Chan AW, Mohsenvand M, LeDue J, Liu R, McVea DA, Boyd JD, Wang YT, Reimers M, Murphy TH (2013) Spontaneous cortical activity alternates between motifs defined by regional axonal projections. Nat Neurosci 16:1426–1435. https://doi.org/10.1038/nn.3499

80. Allen WE, Kauvar IV, Chen MZ, Richman EB, Yang SJ, Chan K, Gradinaru V, Deverman BE, Luo L, Deisseroth K (2017) Global representations of goal-directed behavior in distinct cell types of mouse neocortex. Neuron 94:891–907.e6. https://doi.org/10.1016/j.neuron.2017.04.017

81. Salkoff DB, Zagha E, McCarthy E, McCormick DA (2020) Movement and performance explain widespread cortical activity in a visual detection task. Cereb Cortex 30:421–437. https://doi.org/10.1093/cercor/bhz206

82. Wagner MJ, Kim TH, Kadmon J, Nguyen ND, Ganguli S, Schnitzer MJ, Luo L (2019) Shared cortex-cerebellum dynamics in the execution and learning of a motor task. Cell 177:669–682.e24. https://doi.org/10.1016/j.cell.2019.02.019

83. Roth MM, Dahmen JC, Muir DR, Imhof F, Martini FJ, Hofer SB (2016) Thalamic nuclei convey diverse contextual information to layer 1 of visual cortex. Nat Neurosci 19:299–307. https://doi.org/10.1038/nn.4197

84. Sauerbrei BA, Guo J-Z, Cohen JD, Mischiati M, Guo W, Kabra M, Verma N, Mensh B, Branson K, Hantman AW (2020) Cortical pattern generation during dexterous movement is input-driven. Nature 577:386–391. https://doi.org/10.1038/s41586-019-1869-9

85. Nakajima M, Schmitt LI, Halassa MM (2019) Prefrontal cortex regulates sensory filtering through a basal ganglia-to-thalamus pathway. Neuron 103:445–458.e10. https://doi.org/10.1016/j.neuron.2019.05.026

86. Pinto L, Rajan K, DePasquale B, Thiberge SY, Tank DW, Brody CD (2019) Task-dependent changes in the large-scale dynamics and necessity of cortical regions. Neuron 104:810–824.e9. https://doi.org/10.1016/j.neuron.2019.08.025

87. Gallero-Salas Y, Han S, Sych Y, Voigt FF, Laurenczy B, Gilad A, Helmchen F (2021) Sensory and behavioral components of neocortical signal flow in discrimination tasks with short-term memory. Neuron 109:135–148.e6. https://doi.org/10.1016/j.neuron.2020.10.017

88. Mitelut C, Zhang Y, Sekino Y, Boyd JD, Bollanos F, Swindale NV, Silasi G, Saxena S, Murphy TH (2022) Mesoscale cortex-wide neural dynamics predict self-initiated actions in mice several seconds prior to movement. eLife 11:e76506. https://doi.org/10.7554/eLife.76506

89. Vesuna S, Kauvar IV, Richman E, Gore F, Oskotsky T, Sava-Segal C, Luo L, Malenka RC, Henderson JM, Nuyujukian P, Parvizi J, Deisseroth K (2020) Deep posteromedial cortical rhythm in dissociation. Nature 586:87–94. https://doi.org/10.1038/s41586-020-2731-9

90. Wang Z, Fei X, Liu X, Wang Y, Hu Y, Peng W, Wang Y, Zhang S, Xu M (2022) REM sleep is associated with distinct global cortical dynamics and controlled by occipital cortex. Nat Commun 13:6896. https://doi.org/10.1038/s41467-022-34720-9

91. Ma Y, Shaik MA, Kim SH, Kozberg MG, Thibodeaux DN, Zhao HT, Yu H, Hillman

EMC (2016) Wide-field optical mapping of neural activity and brain haemodynamics: considerations and novel approaches. Philos Trans R Soc B 371:20150360. https://doi.org/10.1098/rstb.2015.0360
92. Sirotin Y, Das A (2010) Spatial relationship between flavoprotein fluorescence and the hemodynamic response in the primary visual cortex of alert macaque monkeys. Front Neuroenerg 2
93. Valley MT, Moore MG, Zhuang J, Mesa N, Castelli D, Sullivan D, Reimers M, Waters J (2020) Separation of hemodynamic signals from GCaMP fluorescence measured with wide-field imaging. J Neurophysiol 123: 356–366. https://doi.org/10.1152/jn.00304.2019
94. Chen JL, Carta S, Soldado-Magraner J, Schneider BL, Helmchen F (2013) Behaviour-dependent recruitment of long-range projection neurons in somatosensory cortex. Nature 499:336–340. https://doi.org/10.1038/nature12236
95. Oh SW, Harris JA, Ng L, Winslow B, Cain N, Mihalas S, Wang Q, Lau C, Kuan L, Henry AM, Mortrud MT, Ouellette B, Nguyen TN, Sorensen SA, Slaughterbeck CR, Wakeman W, Li Y, Feng D, Ho A, Nicholas E, Hirokawa KE, Bohn P, Joines KM, Peng H, Hawrylycz MJ, Phillips JW, Hohmann JG, Wohnoutka P, Gerfen CR, Koch C, Bernard A, Dang C, Jones AR, Zeng H (2014) A mesoscale connectome of the mouse brain. Nature 508:207–214. https://doi.org/10.1038/nature13186
96. Bakker R, Tiesinga P, Kötter R (2015) The scalable brain atlas: instant web-based access to public brain atlases and related content. Neuroinformatics 13:353–366. https://doi.org/10.1007/s12021-014-9258-x
97. Paxinos and Franklin's the mouse brain in stereotaxic coordinates – 4th edition. https://www.elsevier.com/books/paxinos-and-franklins-the-mouse-brain-in-stereotaxic-coordinates/paxinos/978-0-12-391057-8. Accessed 14 Aug 2018
98. Shamash P, Carandini M, Harris K, Steinmetz N (2018) A tool for analyzing electrode tracks from slice histology 447995
99. Fuglstad JG, Saldanha P, Paglia J, Whitlock JR (2022) HERBS: Histological E-data Registration in rodent Brain Spaces 2021.10.01.462770
100. Tomer R, Ye L, Hsueh B, Deisseroth K (2014) Advanced CLARITY for rapid and high-resolution imaging of intact tissues. Nat Protoc 9:1682–1697. https://doi.org/10.1038/nprot.2014.123
101. Voigt FF, Kirschenbaum D, Platonova E, Pagès S, Campbell RAA, Kastli R, Schaettin M, Egolf L, van der Bourg A, Bethge P, Haenraets K, Frézel N, Topilko T, Perin P, Hillier D, Hildebrand S, Schueth A, Roebroeck A, Roska B, Stoeckli ET, Pizzala R, Renier N, Zeilhofer HU, Karayannis T, Ziegler U, Batti L, Holtmaat A, Lüscher C, Aguzzi A, Helmchen F (2019) The mesoSPIM initiative: open-source light-sheet microscopes for imaging cleared tissue. Nat Methods 16:1105–1108. https://doi.org/10.1038/s41592-019-0554-0
102. Tomer R, Lovett-Barron M, Kauvar I, Andalman A, Burns VM, Sankaran S, Grosenick L, Broxton M, Yang S, Deisseroth K (2015) SPED light sheet microscopy: fast mapping of biological system structure and function. Cell 163:1796–1806. https://doi.org/10.1016/j.cell.2015.11.061
103. Kahan A, Greenbaum A, Jang MJ, Robinson JE, Cho JR, Chen X, Kassraian P, Wagenaar DA, Gradinaru V (2021) Light-guided sectioning for precise in situ localization and tissue interface analysis for brain-implanted optical fibers and GRIN lenses. Cell Rep 36: 109744. https://doi.org/10.1016/j.celrep.2021.109744
104. McIntosh AR, Jirsa VK (2019) The hidden repertoire of brain dynamics and dysfunction. Netw Neurosci 3:994–1008. https://doi.org/10.1162/netn_a_00107
105. Shine JM, Bissett PG, Bell PT, Koyejo O, Balsters JH, Gorgolewski KJ, Moodie CA, Poldrack RA (2016) The dynamics of functional brain networks: integrated network states during cognitive task performance. Neuron 92:544–554. https://doi.org/10.1016/j.neuron.2016.09.018
106. Severino FPU, Ban J, Song Q, Tang M, Bianconi G, Cheng G, Torre V (2016) The role of dimensionality in neuronal network dynamics. Sci Rep 6:29640. https://doi.org/10.1038/srep29640
107. Foti NJ, Fox EB (2019) Statistical model-based approaches for functional connectivity analysis of neuroimaging data. Curr Opin Neurobiol 55:48–54. https://doi.org/10.1016/j.conb.2019.01.009
108. Williamson RC, Doiron B, Smith MA, Yu BM (2019) Bridging large-scale neuronal recordings and large-scale network models using dimensionality reduction. Curr Opin Neurobiol 55:40–47. https://doi.org/10.1016/j.conb.2018.12.009
109. Cunningham JP, Yu BM (2014) Dimensionality reduction for large-scale neural

recordings. Nat Neurosci 17:1500–1509. https://doi.org/10.1038/nn.3776

110. Elsayed GF, Cunningham JP (2017) Structure in neural population recordings: an expected byproduct of simpler phenomena? Nat Neurosci 20:1310–1318. https://doi.org/10.1038/nn.4617

111. Pandarinath C, O'Shea DJ, Collins J, Jozefowicz R, Stavisky SD, Kao JC, Trautmann EM, Kaufman MT, Ryu SI, Hochberg LR, Henderson JM, Shenoy KV, Abbott LF, Sussillo D (2018) Inferring single-trial neural population dynamics using sequential auto-encoders. Nat Methods 15:805–815. https://doi.org/10.1038/s41592-018-0109-9

112. Whiteway MR, Butts DA (2019) The quest for interpretable models of neural population activity. Curr Opin Neurobiol 58:86–93. https://doi.org/10.1016/j.conb.2019.07.004

113. Wiltschko AB, Johnson MJ, Iurilli G, Peterson RE, Katon JM, Pashkovski SL, Abraira VE, Adams RP, Datta SR (2015) Mapping sub-second structure in mouse behavior. Neuron 88:1121–1135. https://doi.org/10.1016/j.neuron.2015.11.031

114. Berman GJ (2018) Measuring behavior across scales. BMC Biol 16:23. https://doi.org/10.1186/s12915-018-0494-7

115. Mathis A, Mamidanna P, Cury KM, Abe T, Murthy VN, Mathis MW, Bethge M (2018) DeepLabCut: markerless pose estimation of user-defined body parts with deep learning. Nat Neurosci 21:1281–1289. https://doi.org/10.1038/s41593-018-0209-y

116. Lauer J, Zhou M, Ye S, Menegas W, Schneider S, Nath T, Rahman MM, Di Santo V, Soberanes D, Feng G, Murthy VN, Lauder G, Dulac C, Mathis MW, Mathis A (2022) Multi-animal pose estimation, identification and tracking with DeepLabCut. Nat Methods 19:496–504. https://doi.org/10.1038/s41592-022-01443-0

117. Hsu AI, Yttri EA (2021) B-SOiD, an open-source unsupervised algorithm for identification and fast prediction of behaviors. Nat Commun 12:5188. https://doi.org/10.1038/s41467-021-25420-x

118. Luxem K, Sun JJ, Bradley SP, Krishnan K, Yttri E, Zimmermann J, Pereira TD, Laubach M (2023) Open-source tools for behavioral video analysis: setup, methods, and best practices. eLife 12:e79305. https://doi.org/10.7554/eLife.79305

119. Weber AI, Pillow JW (2017) Capturing the dynamical repertoire of single neurons with generalized linear models. Neural Comput 29:3260–3289. https://doi.org/10.1162/neco_a_01021

120. Fomins A, Sych Y, Helmchen F (2022) Conservative significance testing of tripartite statistical relations in multivariate neural data. Netw Neurosci 6:1243–1274. https://doi.org/10.1162/netn_a_00259

121. Tang S, Lee TS, Li M, Zhang Y, Xu Y, Liu F, Teo B, Jiang H (2018) Complex pattern selectivity in macaque primary visual cortex revealed by large-scale two-photon imaging. Curr Biol 28:38–48.e3. https://doi.org/10.1016/j.cub.2017.11.039

122. Johnston WJ, Palmer SE, Freedman DJ (2020) Nonlinear mixed selectivity supports reliable neural computation. PLoS Comput Biol 16:e1007544. https://doi.org/10.1371/journal.pcbi.1007544

123. Ayaz A, Saleem AB, Schölvinck ML, Carandini M (2013) Locomotion controls spatial integration in mouse visual cortex. Curr Biol 23:890–894. https://doi.org/10.1016/j.cub.2013.04.012

124. Guo ZV, Li N, Huber D, Ophir E, Gutnisky D, Ting JT, Feng G, Svoboda K (2014) Flow of cortical activity underlying a tactile decision in mice. Neuron 81:179–194. https://doi.org/10.1016/j.neuron.2013.10.020

125. Fustiñana MS, Eichlisberger T, Bouwmeester T, Bitterman Y, Lüthi A (2021) State-dependent encoding of exploratory behaviour in the amygdala. Nature 592:267–271. https://doi.org/10.1038/s41586-021-03301-z

126. Guo Z, Yin L, Osakada T, Lischinsky J, Chien J, Dai B, Urtecho A, Tong X, Chen ZS, Lin D (2023) Neural dynamics in the limbic system during male social behaviors 2023.03.12.532199

127. Finc K, Bonna K, He X, Lydon-Staley DM, Kühn S, Duch W, Bassett DS (2020) Dynamic reconfiguration of functional brain networks during working memory training. Nat Commun 11:2435. https://doi.org/10.1038/s41467-020-15631-z

128. Mohr H, Wolfensteller U, Betzel RF, Mišić B, Sporns O, Richiardi J, Ruge H (2016) Integration and segregation of large-scale brain networks during short-term task automatization. Nat Commun 7:13217. https://doi.org/10.1038/ncomms13217

129. Poort J, Khan AG, Pachitariu M, Nemri A, Orsolic I, Krupic J, Bauza M, Sahani M, Keller GB, Mrsic-Flogel TD, Hofer SB (2015) Learning enhances sensory and multiple

non-sensory representations in primary visual cortex. Neuron 86:1478–1490. https://doi.org/10.1016/j.neuron.2015.05.037
130. Makino H, Ren C, Liu H, Kim AN, Kondapaneni N, Liu X, Kuzum D, Komiyama T (2017) Transformation of cortex-wide emergent properties during motor learning. Neuron 94:880–890.e8. https://doi.org/10.1016/j.neuron.2017.04.015
131. Peters AJ, Chen SX, Komiyama T (2014) Emergence of reproducible spatiotemporal activity during motor learning. Nature 510: 263–267. https://doi.org/10.1038/nature13235
132. Gilad A, Helmchen F (2020) Spatiotemporal refinement of signal flow through association cortex during learning. Nat Commun 11: 1744. https://doi.org/10.1038/s41467-020-15534-z
133. Kauvar IV, Machado TA, Yuen E, Kochalka J, Choi M, Allen WE, Wetzstein G, Deisseroth K (2020) Cortical observation by synchronous multifocal optical sampling reveals widespread population encoding of actions. Neuron 107:351–367.e19. https://doi.org/10.1016/j.neuron.2020.04.023
134. Toader AC, Regalado JM, Li YR, Terceros A, Yadav N, Kumar S, Satow S, Hollunder F, Bonito-Oliva A, Rajasethupathy P (2023) Anteromedial thalamus gates the selection and stabilization of long-term memories. Cell 186:1369–1381.e17. https://doi.org/10.1016/j.cell.2023.02.024
135. Aggleton JP, O'Mara SM (2022) The anterior thalamic nuclei: core components of a tripartite episodic memory system. Nat Rev Neurosci 23:505–516. https://doi.org/10.1038/s41583-022-00591-8
136. Peterka DS, Takahashi H, Yuste R (2011) Imaging voltage in neurons. Neuron 69:9–21. https://doi.org/10.1016/j.neuron.2010.12.010
137. Marshall JD, Li JZ, Zhang YP, Gong YY, St-Pierre F, Lin MZ, Schnitzer MJ (2016) Cell-type-specific optical recording of membrane voltage dynamics in freely moving mice. Cell 167:1650. https://doi.org/10.1016/j.cell.2016.11.021
138. Chien M-P, Brinks D, Testa-Silva G, Tian H, Phil Brooks F, Adam Y, Bloxham B, Gmeiner B, Kheifets S, Cohen AE (2021) Photoactivated voltage imaging in tissue with an archaerhodopsin-derived reporter. Sci Adv 7:eabe3216. https://doi.org/10.1126/sciadv.abe3216
139. Xiao S, Lowet E, Gritton HJ, Fabris P, Wang Y, Sherman J, Mount RA, Tseng H, Man H-Y, Straub C, Piatkevich KD, Boyden ES, Mertz J, Han X (2021) Large-scale voltage imaging in behaving mice using targeted illumination. iScience 24:103263. https://doi.org/10.1016/j.isci.2021.103263
140. Abdelfattah AS, Zheng J, Singh A, Huang Y-C, Reep D, Tsegaye G, Tsang A, Arthur BJ, Rehorova M, Olson CVL, Shuai Y, Zhang L, Fu T-M, Milkie DE, Moya MV, Weber TD, Lemire AL, Baker CA, Falco N, Zheng Q, Grimm JB, Yip MC, Walpita D, Chase M, Campagnola L, Murphy GJ, Wong AM, Forest CR, Mertz J, Economo MN, Turner GC, Koyama M, Lin B-J, Betzig E, Novak O, Lavis LD, Svoboda K, Korff W, Chen T-W, Schreiter ER, Hasseman JP, Kolb I (2023) Sensitivity optimization of a rhodopsin-based fluorescent voltage indicator. Neuron 111:1547–1563.e9. https://doi.org/10.1016/j.neuron.2023.03.009
141. Patriarchi T, Cho JR, Merten K, Howe MW, Marley A, Xiong W-H, Folk RW, Broussard GJ, Liang R, Jang MJ, Zhong H, Dombeck D, von Zastrow M, Nimmerjahn A, Gradinaru V, Williams JT, Tian L (2018) Ultrafast neuronal imaging of dopamine dynamics with designed genetically encoded sensors. Science 360:eaat4422. https://doi.org/10.1126/science.aat4422
142. Feng J, Zhang C, Lischinsky JE, Jing M, Zhou J, Wang H, Zhang Y, Dong A, Wu Z, Wu H, Chen W, Zhang P, Zou J, Hires SA, Zhu JJ, Cui G, Lin D, Du J, Li Y (2019) A genetically encoded fluorescent sensor for rapid and specific in vivo detection of norepinephrine. Neuron 102:745–761.e8. https://doi.org/10.1016/j.neuron.2019.02.037
143. Schulz K, Sydekum E, Krueppel R, Engelbrecht CJ, Schlegel F, Schroter A, Rudin M, Helmchen F (2012) Simultaneous BOLD fMRI and fiber-optic calcium recording in rat neocortex. Nat Methods 9:597–602. https://doi.org/10.1038/Nmeth.2013
144. Pisanello F, Mandelbaum G, Pisanello M, Oldenburg IA, Sileo L, Markowitz JE, Peterson RE, Patria AD, Haynes TM, Emara MS, Spagnolo B, Datta SR, Vittorio MD, Sabatini BL (2017) Dynamic illumination of spatially restricted or large brain volumes via a single tapered optical fiber. Nat Neurosci 20:1180–1188. https://doi.org/10.1038/nn.4591
145. Pisano F, Pisanello M, Lee SJ, Lee J, Maglie E, Balena A, Sileo L, Spagnolo B, Bianco M, Hyun M, De Vittorio M, Sabatini BL, Pisanello F (2019) Depth-resolved fiber photometry with a single tapered optical fiber implant. Nat Methods 16:1185–1192.

https://doi.org/10.1038/s41592-019-0581-x

146. Ahn H, Kim S, Kim Y, Kim S, Choi J, Kim K (2021) Plasmonic sensing, imaging, and stimulation techniques for neuron studies. Biosens Bioelectron 182:113150. https://doi.org/10.1016/j.bios.2021.113150

147. Bauch M, Toma K, Toma M, Zhang Q, Dostalek J (2014) Plasmon-enhanced fluorescence biosensors: a review. Plasmonics 9:781–799. https://doi.org/10.1007/s11468-013-9660-5

148. Zorzos AN, Boyden ES, Fonstad CG (2010) Multiwaveguide implantable probe for light delivery to sets of distributed brain targets. Opt Lett 35:4133–4135. https://doi.org/10.1364/OL.35.004133

149. Wu F, Stark E, Im M, Cho I-J, Yoon E-S, Buzsáki G, Wise KD, Yoon E (2013) An implantable neural probe with monolithically integrated dielectric waveguide and recording electrodes for optogenetics applications. J Neural Eng 10:056012. https://doi.org/10.1088/1741-2560/10/5/056012

150. Anikeeva P, Andalman AS, Witten I, Warden M, Goshen I, Grosenick L, Gunaydin LA, Frank LM, Deisseroth K (2012) Optetrode: a multichannel readout for optogenetic control in freely moving mice. Nat Neurosci 15:163–170. https://doi.org/10.1038/nn.2992

151. Kim EGR, Tu H, Luo H, Liu B, Bao S, Zhang J, Xu Y (2015) 3D silicon neural probe with integrated optical fibers for optogenetic modulation. Lab Chip 15:2939–2949. https://doi.org/10.1039/C4LC01472C

152. Dufour S, De Koninck Y (2015) Optrodes for combined optogenetics and electrophysiology in live animals. Neurophoton 2:031205–031205. https://doi.org/10.1117/1.NPh.2.3.031205

153. Chou N, Shin H, Kim K, Chae U, Jang M, Jeong U-J, Hwang K-S, Yi B, Lee SE, Woo J, Cho Y, Lee C, Baker BJ, Oh S-J, Nam M-H, Choi N, Cho I-J (2022) A multimodal multishank fluorescence neural probe for cell-type-specific electrophysiology in multiple regions across a neural circuit. Adv Sci 9:2103564. https://doi.org/10.1002/advs.202103564

154. Yu J, Ling W, Li Y, Ma N, Wu Z, Liang R, Pan H, Liu W, Fu B, Wang K, Li C, Wang H, Peng H, Ning B, Yang J, Huang X (2021) A multichannel flexible optoelectronic fiber device for distributed implantable neurological stimulation and monitoring. Small 17:2005925. https://doi.org/10.1002/smll.202005925

155. Eriksson D, Schneider A, Thirumalai A, Alyahyay M, de la Crompe B, Sharma K, Ruther P, Diester I (2022) Multichannel optogenetics combined with laminar recordings for ultra-controlled neuronal interrogation. Nat Commun 13:985. https://doi.org/10.1038/s41467-022-28629-6

156. Legaria AA, Matikainen-Ankney BA, Yang B, Ahanonu B, Licholai JA, Parker JG, Kravitz AV (2022) Fiber photometry in striatum reflects primarily nonsomatic changes in calcium. Nat Neurosci 25:1124–1128. https://doi.org/10.1038/s41593-022-01152-z

157. Takahashi M, Urushihata T, Takuwa H, Sakata K, Takado Y, Shimizu E, Suhara T, Higuchi M, Ito H (2018) Imaging of neuronal activity in awake mice by measurements of flavoprotein autofluorescence corrected for cerebral blood flow. Front Neurosci 11

Chapter 6

Widefield Imaging Combined with a Transparent Electrographic Probe

Chi Ren, Xin Liu, Duygu Kuzum, and Takaki Komiyama

Abstract

Simultaneous recording from multiple brain regions is essential for understanding large-scale interactions among distributed neural networks. Recent advances in recording techniques have brought about increased knowledge of how different brain regions cooperate in various cognitive processes. Here, we describe a protocol combining widefield calcium imaging with electrical recordings using a flexible, insertable, and transparent microelectrode probe (Neuro-FITM). This multimodal approach allows simultaneous recording of neural activity from the dorsal cortex and subcortical brain regions in awake, head-fixed mice. This protocol provides guidelines on how to implement this technique, including surgical preparation, recording setup, and data processing.

Key words Widefield imaging, Electrical recording, Transparent electrographic probe, Flexible microelectrode, Multimodal recording

1 Introduction

The brain functions through communication across multiple brain regions to drive behavior and cognition. The properties of such large-scale interactions cannot be deduced from the characterization of individual regions separately. Instead, it requires simultaneous recording from multiple regions. Recent advances in recording techniques have enabled researchers to monitor large-scale brain activity at unprecedented spatiotemporal scales and deepened our understanding of how different regions cooperate to generate perception and behavior [1–4].

Widefield imaging is well suited for recording neural activity with extensive spatial coverage through one-photon excitation [2, 3, 5, 6]. Although this technique normally does not resolve individual cells, it enables simultaneous capturing of neural dynamics across brain areas with a sufficient spatial and temporal resolution to resolve behaviorally relevant information. Meanwhile,

James E. Niemeyer et al. (eds.), *Awake Behaving Mesoscopic Brain Imaging*, Neuromethods, vol. 214,
https://doi.org/10.1007/978-1-0716-4120-0_6,

recent improvements in genetically encoded fluorescent indicators [7, 8], especially the calcium indicator GCaMP family, have increased the signal-to-noise ratio in widefield imaging. GCaMP fluorescence is sensitive to changes in intracellular calcium levels, which are dominated by action potentials, and therefore reports neuronal spiking activity with bright fluorescent transients. The use of transgenic mice allows broad and stable endogenous expression of GCaMP in specific neuronal populations [9, 10]. These advantages make widefield calcium imaging with GCaMP an attractive approach to characterizing large-scale neural dynamics in behaving animals.

On the other hand, widefield calcium imaging only has access to the cortex because of the strong scattering of the excitation and emission light of GCaMP (~485 and ~520 nm, respectively) in brain tissue. Attempts have been made to compensate for this limited recording depth of widefield calcium imaging by combining it with other recording modalities, such as fMRI and electrical recordings [11–14]. These multimodal recording techniques have been effective in a systematic investigation of the functional relationships between cortical or subcortical regions [13, 14].

To achieve simultaneous widefield imaging and electrical recordings, we developed a transparent, insertable, and flexible microelectrode probe (Neuro-FITM) [13]. Transparency minimizes the formation of optical shadows and thus prevents blocking of the field of view (FOV) in widefield imaging. The high flexibility allows the bending of the probe shank and the connected amplifier board away to fit into the working distance of the microscope objective. These two features enable vertical implantation of the Neuro-FITM probe to the target brain region, minimizing implantation damage compared to other opaque and rigid probes (e.g., Neuropixels [14], Neuronexus [12]) that have to be inserted obliquely when combined with widefield imaging. With Neuro-FITM, we have performed simultaneous widefield calcium imaging of the dorsal cortex and electrical recordings of the hippocampus in awake, head-fixed mice [13]. We found a selective and diverse interaction between the hippocampus and large-scale cortical activity during awake hippocampal sharp-wave ripples (SWRs), which might be unlikely to be revealed by classical recording techniques. In this protocol, we use our study as an example to provide guidelines on implementing this multimodal recording technique and hope to facilitate a broader application in investigating the interactions between the cortex and other subcortical structures.

2 Materials

2.1 Transparent Electrographic Probe for Electrical Recording

The Neuro-FITM probe was designed as a three-layered structure (Fig. 1). The transparent and flexible Parylene-C material was used as the substrate (14-μm thick) and top encapsulation layer (2-μm thick), with Au wires (2-μm wide, 100-nm thick) in between. The probe contains 32 or 64 circular openings as the electrode recording sites, each with a diameter of 10 μm connected to the Au wire. To lower the impedance for single-unit recording, platinum nanoparticles (PtNPs) were deposited onto electrodes and the impedance was reduced to ~ 150 kΩ at 1 kHz. For the probe designed to record in the mouse hippocampus, the electrodes are aligned in two rows with 50- or 20-μm spacing between adjacent sites, and the distance between the top and bottom sites is 750 μm, which is long enough to record from the hippocampal CA1 region in the dorsal-ventral axis. To achieve a shuttle-free insertion into the brain, the length of the probe was carefully engineered to 1.8 mm, which is long enough to reach the CA1 region of the mouse hippocampus, yet short enough to prevent buckling during insertion. Note that the Neuro-FITM probe can be fabricated in various configurations depending on the needs of specific experiments (*see* **Note 1**).

2.2 Transgenic Mouse Lines for Widefield Imaging

To image the cortex-wide activity of excitatory neurons, we crossed tetO-GCaMP6s transgenic mice (JAX 024742) [10] to CaMKIIa-tTA transgenic mice (JAX 003010) [15] to generate double-transgenic offspring. In these double-transgenic mice, the expression of GCaMP6s depends on the presence of the tTA protein, and the latter is under the regulatory control of the excitatory neuron-specific CaMKIIa promoter. Therefore, the expression of GCaMP6s in the cortex is restricted to excitatory neurons. Both homozygotes and heterozygotes can be used as parents for

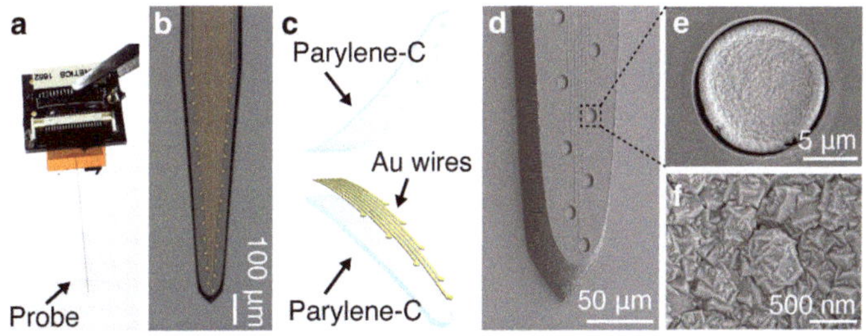

Fig. 1 Design of the Neuro-FITM probe. (**a**) A Neuro-FITM probe connected to the customized printed circuit board. (**b**) Microscope image showing the layout of the 32 electrodes at the probe tip. (**c**) Schematic showing an exploded view of the three-layered structure of Neuro-FITM. (**d**) SEM image of the probe showing 10-μm-diameter electrodes and 2-μm-wide Au wires. (**e**) Magnified view of a single recording site. (**f**) SEM image showing PtNPs deposited onto the Au surface at each electrode. (Adapted from [13])

crossing. When heterozygous parents are used, PCR genotyping is required to identify offspring carrying both transgenic alleles.

Although other transgenic mouse lines (e.g., Thy1-GCaMP6s (GP4.3) [9]) also express GCaMP6s specifically in excitatory neurons, we favor this crossing setup as the double-transgenic offspring exhibit a strong and uniform expression of GCaMP6s in most cortical excitatory neurons and therefore yield bright optical signals that can be directly observed through the intact skull [10]. This is particularly beneficial to chronic imaging experiments as lower excitation light power is required to achieve a good imaging quality, reducing the risk of phototoxicity during repeated imaging. In addition, because of the bright fluorescent signals in these double-transgenic mice, the expression of GCaMP6s can also be confirmed by fluorescence imaging through the intact skull in an early step of the surgical procedure. In addition, we did not observe aberrant cortical activity in these double-transgenic mice, which has been reported in other GCaMP6-expressing transgenic mouse lines [16]. Note that researchers should choose transgenic mouse lines based on the neurons of interest and the needs of experiments (*see* **Note 2**).

We used both adult (between 6 weeks and 4 months old) male and female mice in our experiments. Mice were group housed in disposable plastic cages with standard bedding in a room with a reversed light cycle (12 h:12 h). Temperatures and humidity ranged from 18 to 23 °C and 40% to 60%, respectively. Experiments were performed during the dark period. All procedures were performed in accordance with protocols approved by the University of California San Diego (UCSD) Institutional Animal Care and Use Committee and guidelines of the National Institute of Health (NIH).

2.3 Surgery Setup and Key Reagents for the Skull-Intact Surgery and Electrode Implantation

Please see Table 1 for a detailed list of key equipment and reagents for surgical preparation. The headbar and headbar grips were custom manufactured with stainless steel (Fig. 2a). Other designs can also be used if the head fixation can be achieved without blocking the FOV over the dorsal cortex. The Neuro-FITM probe holder was custom designed and 3D-printed with Vero white (Fig. 2b).

To avoid blocking the FOV of widefield calcium imaging, we used thin stainless steel wires as ground and reference electrodes and assembled them and the headbar together (Fig. 2c). One end of the wire was soldered to the male pin of a jumper wire for easy connection during recording. Stainless steel wires were isolated from the headbar with heat shrink tubes and fixed to the headbar with cyanoacrylate glue.

Table 1
Key equipment and reagents for surgical preparation

Item	Source	Identifier
Stereomicroscope and anesthesia system		
Leica M125 stereomicroscope	Leica Microsystems	M125
Fiber optic light source with gooseneck	Leica Microsystems	KL 1500 LED
Motorized micromanipulator	Sutter instrument	Cat#MP-285
Mouse gas anesthesia system	Vetequip	Cat#901806, Cat#941444
Mouse gas anesthesia head holder (nose cone and bite bar)	KOPF	Cat#1923-B
Surgical tools		
Dental drill handpiece	Midwest Tradition	N/A
Drill bit (FG 1/4)	Henry Schein	Cat#1007205
Scalpel (#3)	Fine Science Tools	Cat#10003-12
Fine scissors	Fine Science Tools	Cat#14060-11
Forceps	Fine Science Tools	Cat#11223-20
Double-end carver	Henry Schein	Cat#101-0333
Reagents		
Vetbond	3M™	Cat#70200742529
Cyanoacrylate glue	Krazy Glue	All Purpose, Precision Tip
Dental acrylic cement	Lang Dental	Contemporary Lang Dental Ortho-Jet™ Powder and Jet Liquid
Customized items		
Stainless steel headbar	Custom-made	N/A
Neuro-FITM probe holder	Custom-made	N/A
Headbar grips	Custom-made	N/A
Stage for head-fixation	Custom-made with pieces from Thorlabs	N/A
Other		
Stainless steel wires	A-M Systems	Cat#791900
Gelfoam	Ethicon	1975

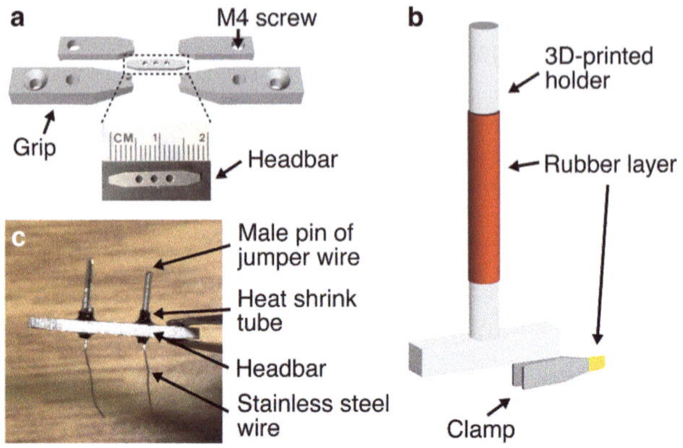

Fig. 2 Design of the head-fixation strategy (**a**), the Neuro-FITM probe holder (**b**), and the assembly of the ground/reference wires and the headbar (**c**). We added rubber layers to the rod and the clamp tips to increase the grip force between the micromanipulator and the holder, and the clamp and the probe, respectively

2.4 Recording Systems

For in vivo electrical recordings, we used the Intan RHD recording system (Intan Technologies). The Neuro-FITM probe was attached to a customized adaptor board that routed the electrical signals to the Intan RHD2132 amplifier boards (headstages) connected to the RHD USB interface board (Part #C3100). Other recording systems, such as Open Ephys, can also be used.

For widefield calcium imaging, we used a commercial fluorescence microscope (Axio Zoom.V16, Zeiss, objective lens (1×, 0.25 numerical aperture)) and a CMOS camera (ORCA-Flash4.0 v.2, Hamamatsu) to image cortex-wide activity through the intact skull. The software for image acquisition was HCImage Live (Hamamatsu). There are a number of vendors that produce widefield microscopes. Widefield microscopes can also be custom-built [5]. In general, scientific cameras with a large FOV, high quantum efficiency, and fast frame rates (~15–60 Hz for imaging GCaMP6) are suitable for widefield imaging.

We used TTL pulses to align the widefield imaging and electrical recordings offline. The TTL pulse triggers the start of widefield image acquisition and is also recorded by the electrical recording system.

3 Methods

3.1 Surgical Procedure

1. Prepare the surgical tools and the animal.

Sterilize the surface of the surgery table, surgical tools, and the assembly of the ground/reference electrodes and the headbar.

Anesthetize the adult mouse with isoflurane (3% in 100% O_2 for induction, 1–2% in 100% O_2 for maintenance). Transfer the anesthetized mouse to an isoflurane ventilator with a nose cone and

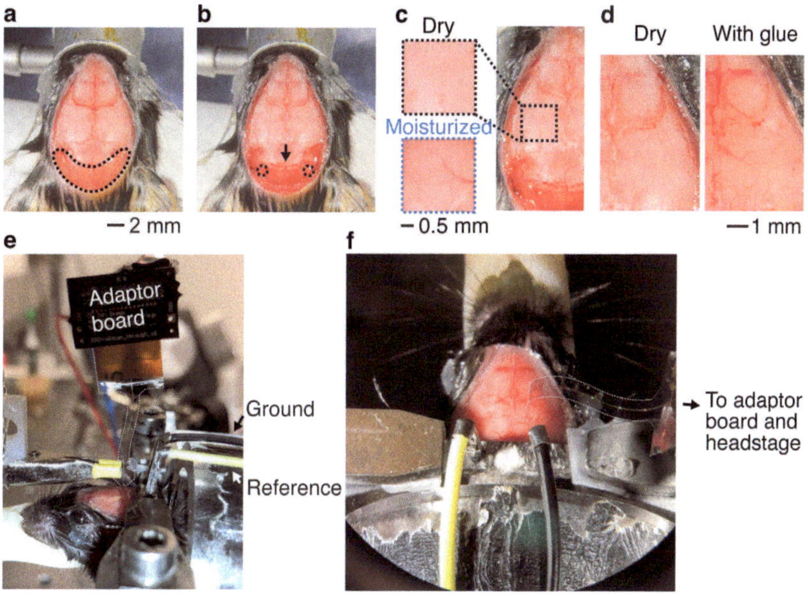

Fig. 3 Surgical procedures of the skull-intact preparation for widefield imaging and Neuro-FITM probe implantation. (**a–d**) Critical steps for skull-intact preparation (**a, c, d**) and implantation of the assembly of the ground/reference wires and the headbar (**b**). (**e**) The Neuro-FITM probe was held vertically by the custom-built probe holder during insertion. (**f**) The probe was bent to the right side of the mouse after insertion and fixed to the skull. Note that in (**e**, **f**), the probe shank is largely invisible. The edge of the shank is marked by white dashed lines for visualization. (Adapted from [6, 13])

a bite bar. The mouse is kept on a servo-controlled heating pad to maintain body temperature, and the breathing rate is kept at ~1 breath/s during the entire surgery. Cover the eyes with Vaseline (Fig. 3a) to protect against drying and potential exposure to irritant chemicals (e.g., cyanoacrylate glue and dental acrylic cement) in the following surgical steps.

2. Expose the skull.

Clean and antisepticise the scalp with 70% ethanol and betadine. Remove a circular piece of the scalp to expose the skull. The expression of GCaMP6s can also be confirmed by fluorescence imaging through the intact skull at this point if necessary. Detach the muscles from the occipital bone (Fig. 3a, muscles within the black dashed line) with a scalpel and push the muscles away. This is to expose enough skull surface over the cerebellum to implant the assembly of the ground/reference electrodes and the headbar. It also extends the bone interface for cyanoacrylate glue and dental acrylic cement, which is critical for the steady implantation of the headbar. If bleeding happens, insert small pieces of gel foam into the gaps between the occipital bone and muscles to stop bleeding.

Carefully remove the soft tissues on the skull surface using a scalpel (Fig. 3a). Be particularly gentle when scratching along the bone sutures to avoid bleeding under the skull. A clean and smooth skull surface is critical to achieving even optical access throughout the dorsal cortex for widefield imaging. Therefore, it is important to avoid making deep scratches or notches on the skull when cleaning the soft tissues (*see* **Note 3**). Wipe the skull to remove any remaining soft tissues with cotton swabs soaked in 3% hydrogen peroxide.

3. Implant the assembly of the ground/reference electrodes and the headbar.

Remove the gel foam. Open two small craniotomies (~0.5 mm in diameter) on the skull over the cerebellum for implanting the reference and ground wires (Fig. 3b, black circles). The distance between the two small craniotomies should be close to that between the stainless steel wires. Adjust the angle of the wires to ensure they can reach the cerebellum through the craniotomies. Trim the wires to the proper length. It should be slightly longer than the distance between the cerebellum and the occipital bone surface to ensure a secure touch between the wires and the cerebellum. Place the assembly on the occipital bone (Fig. 3b, black arrow). Make sure the reference and ground wires touch the cerebellum, and the headbar is parallel to the transverse plane of the skull. Secure the assembly and fill the gaps between the occipital bone and muscles with cyanoacrylate glue and dental acrylic.

4. Seal the skull with cyanoacrylate glue.

Before applying cyanoacrylate glue, make sure the skull surface is dry. This is a deterministic step in creating a clear FOV for widefield imaging as the cyanoacrylate glue becomes opaque when encountering a wet surface. The skull should become white and less transparent as it dries (see Fig. 3c for a comparison between dry and moisturized skulls). If drying is too slow, gently blow the skull with a compressed air duster to facilitate drying and ensure dryness.

After the skull is dry, apply a small amount of cyanoacrylate glue to the skull over the dorsal cortex. Evenly distribute the glue over the entire skull using a double-end carver to form a thin layer. Wait for the glue to cure to form a solid, smooth, and transparent layer. Curing usually takes ~5 min. Repeat this procedure three to four times to build a solid layer of cyanoacrylate glue (~0.5-mm thick) to protect the skull. The skull should gradually become more transparent within 20 min and the brain blood vessels should become clearly visible (Fig. 3d).

5. Transfer the animal to the portable head-fixation stage.

To avoid damaging the implanted probe during head fixation, we prefer transferring the mouse to a portable head-fixation stage used in widefield imaging before probe implantation. Starting from this step, the mouse will be head-fixed by the custom-built headbar.

6. Implant the Neuro-FITM probe.

Here we use implanting the Neuro-FITM probe into the mouse hippocampal CA1 pyramidal layer as an example to introduce the implantation procedure. Open a craniotomy (~0.5 mm in diameter, ~1.5–1.7 mm lateral, and ~2.1–2.3 mm posterior to bregma) above the target location with a dental drill in the right hemisphere for probe insertion. Carefully remove the dura over the exposed brain surface with a hooked insulin syringe needle (30 G). This is to facilitate the insertion as breaking the dura may generate extra resistance force that can bend the probe.

As the probe cannot afford many twists after being fixed onto the skull, we first connect the probe, the adaptor board, and the Intan recording headstage together to avoid damaging the probe after implantation. The Neuro-FITM probe is held vertically above the insertion site by a custom-built probe holder attached to a micromanipulator (Fig. 3e), and the Intan recording headstage is held by a helping hands tool. It is important to place the headstage close to the probe so that the probe tip can be lowered smoothly during insertion. Insert the probe into the target location at ~45 μm/s by manual control of the micromanipulator. Secure the probe to the skull with a small drop of Vetbond once it reaches the target. After the Vetbond becomes solid, carefully release the probe from the holder and bend the exposed part of the probe to the right side of the animal. Finally, fix the adaptor board and the headstage onto the right headbar grip arm on the stage (Fig. 3f).

7. Intraoperative electrical recording.

Intraoperative electrical recording is an effective method of determining whether electrodes have been delivered to the target brain region, especially when the target region exhibits some featured electrophysiological events. We also performed intraoperative electrical recording to ensure that the probe was inserted successfully into the hippocampal CA1 pyramidal layer. SWRs are typical biomarkers of neural activity in the hippocampal CA1 pyramidal layer [17]. SWRs arise from synchronous neural activity and exhibit fast oscillations with large amplitudes (Fig. 4b), which can be easily spotted in real time during electrical recordings. As SWRs mostly occur during sleep and quiet awake status, we turned off the isoflurane before probe implantation to bring the mouse awake. We first inserted the probe tip until one-third of the electrodes reached the target depth (~1 mm for CA1 pyramidal layer) and then monitored the electrical signals for ~30 min. If no SWRs were detected, we lowered the probe by ~100 μm and sought SWRs in intraoperative electrical signals again. We repeated this procedure until SWRs appeared and then fixed the probe or terminated the experiments if no SWRs were detected after 2-h intraoperative electrical recordings.

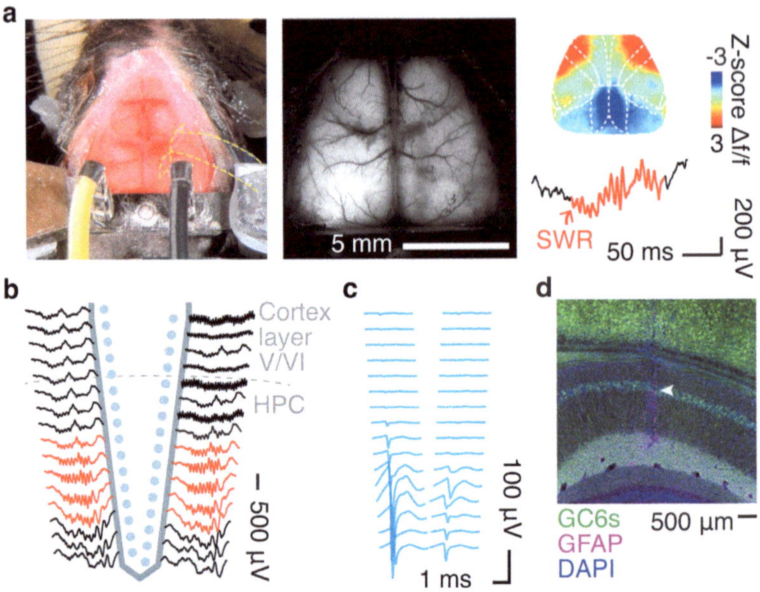

Fig. 4 Simultaneous multimodal recordings from the cortex and the hippocampus with the Neuro-FITM probe. (**a**) Left: surgical preparation for widefield imaging and Neuro-FITM probe implantation, showing the probe shank bent to the right side of the animal. The edge of the shank is marked by yellow dashed lines for visualization. Middle: example FOV of widefield calcium imaging. Note that the probe shank was largely invisible and generated minimal shadows on the overlaying cortex. Right: example of simultaneously recorded cortical activity and hippocampal SWR. (**b**) Representative LFP signals recorded by the Neuro-FITM probe. Multiple channels (red) close to the CA1 pyramidal layer detected SWRs. HPC: hippocampus. (**c**) Spike waveforms of one example neuron recorded across 32 channels. (**d**) Penetrating trajectory of the Neuro-FITM probe visualized by immunostaining against GFAP. The arrowhead indicates the trajectory in the CA1 pyramidal layer. GC6s: GCaMP6s. (Adapted from [13])

3.2 Electrical Recordings and Widefield Imaging of Spontaneous Activity in Awake Mice

1. Electrical recordings of hippocampal activity.

Electrical recording is performed using the Intan RHD recording system (Intan Technologies) at a sampling rate of 30 kHz. Electrical noises are minimized by shielding the recording setup and the mouse with a Faraday cage connected to the utility power ground. Likewise, the ground of the Intan USB interface board is also connected to the utility power ground. The ground and reference pins on the mouse head are wired to the corresponding pins of the Intan headstage. Electrical signals are acquired with Intan RHX software. Besides channels for recording neural activity, a digital-in channel is used to audit the TTL signal for synchronization with widefield imaging.

2. Widefield imaging of cortex-wide activity.

The commercial system we used for widefield calcium imaging consists of a microscope (AxioZoom.V16 equipped with an objective lens (1 ×, 0.25 NA, FWD 56 mm), Zeiss), a camera (ORCA-Flash4.0, Hamamatsu), and an imaging light source (HXP200C, Zeiss). The filter set for imaging GCaMP signals includes a bandpass filter for the excitation light (485 ± 17 nm), a beamsplitter (500 nm), and a tunable bandpass filter centered at 520 nm for the emission light. As our imaging light source is a mercury short-arc lamp, we usually turn it on ~5–10 min ahead to warm it up to achieve a steady illumination. Widefield images are acquired with HCImage Live software (Hamamatsu). The frame size is set to 512 × 512 pixels and the frame rate is set to 30 Hz (*see* **Note 4**). To ensure synchronization with electrical recordings, the start of image acquisition is triggered by a TTL pulse, which is a synchronous signal for offline alignment between multiple recording systems.

Once the mouse has fully woken up after surgery, place the portable head-fixation stage under the microscope and adjust its position until the cortex is in the center of the FOV (~11 mm × 11 mm). Then, adjust the height of the objective until the blood vessels are in focus. Due to the curvature of the cortex, the lateral blood vessels around the edges will be out of focus and appear slightly blurry (Fig. 4a, middle). This adjusting step may take ~5–10 min or even longer for beginners; therefore, it is important to minimize the power of excitation light during this step to avoid photobleaching of GCaMP signals.

After the FOV has been settled, adjust the power of the excitation light. To avoid saturation, the intensity of the brightest pixels (usually located in the retrosplenial cortex) should be ~50–60% of the full dynamic range of the camera sensor. For imaging GCaMP6s in cortical excitatory neurons, the light power is ~10–20 mW to obtain a good signal quality without inducing significant photobleaching during ~3-h imaging. It should be noted that around this power level, the imaging depth is down to ~200 μm for ~485-nm light, covering cortical layer 1 and shallow layer 2/3. Due to the rapid decay of light intensity in brain tissue (decreasing to 50% at ~100 μm and 10% at ~200 μm) [18], the majority of widefield signals likely come from layer 1. Cortical layer 1 mainly consists of dense neuropils, including dendrites from local neurons in layers 2/3 and 5, and axons innervating these layer 1 dendrites. Although the majority of widefield calcium signals reflect local activity [19], the contributions of long-range axonal projections are not negligible, which raises extra caution in interpreting imaging results.

Due to the large amount of data generated by electrical recordings (~0.5 GB/min for 32 channels sampled at 30 kHz) and widefield imaging (~1 GB/min for image size of 512 × 512 pixels sampled at 30 Hz), we limit the duration of each recording session to be ~1 h to accommodate the computational capability for data processing.

3.3 Processing of Electrical Recording Data

In general, we first apply a 60-Hz notch filter to the raw electrical signals to remove power-frequency noises and then apply bandpass filters to extract signals in the desired frequency range. For example, to identify SWR events in our experiments, the electrical signals from the channels near the CA1 pyramidal layer were bandpass filtered at 120–250 Hz (Fig. 4b). An eighth-order Butterworth filter was used in both forward and reverse directions to prevent phase distortion.

The activity of individual neurons is isolated by spike sorting from electrical signals (Fig. 4c). In brief, spike sorting is a process of grouping spikes into clusters based on the similarity of their shapes. Because each neuron tends to fire spikes of a particular shape, the clusters correspond to the activity of different putative neurons. Many spike-sorting algorithms (e.g., Kilosort [20], MountainSort [21], and SpyKING CIRCUS [22]) have been well established, and readers can find a comprehensive comparison of their performance on SpikeForest (https://spikeforest.flatironinstitute.org). For spike sorting, we first high-pass filter the electrical signals at 250 Hz with a third-order Butterworth filter and whiten the data to remove the correlation between nearby channels. Then the Kilosort algorithm identifies the best templates and the putative clusters of neurons, along with their spike timing and amplitudes. These preliminary results are further manually curated by splitting different neurons, merging the same neurons, and labeling low-amplitude inseparable spikes as multiunit activities. Furthermore, the hippocampus pyramidal cells and interneurons can also be classified based on the firing rates and the asymmetry of spike waveforms [23].

3.4 Processing of Widefield Imaging Data

We usually first bin widefield images from 512 × 512 pixels (~21.5 × 21.5 mm^2/pixel) to 128 × 128 pixels (~86 × 86 mm^2/pixel). Binning sacrifices spatial resolution, but also reduces noise and compresses data size for processing. After binning, the spatial resolution (~86 × 86 mm^2/pixel) is still sufficient to resolve activity within individual cortical regions.

1. Obtain Δf/f time series for each pixel in widefield images.

To account for different fluorescence intensities across pixels, we calculate Δf/f time series for each pixel to measure the relative changes in fluorescence intensity caused by activity. Here f is the time-varying baseline fluorescence, and Δf is the difference between the fluorescence intensity and the baseline. Given that each imaging session is ~1 h, the f for a given time point is estimated as the tenth percentile value over 180 s around it. For the start and end of each imaging session, the following and preceding 90-s windows are used to determine the baseline, respectively. The length of the sliding window for f estimation can be adjusted based on the duration of the imaging session. For example, for a 5-min imaging session, we often choose a 30-s sliding window for f estimation. The goal is to estimate an f close to the fluorescence intensity where no apparent activation occurs. An ideal f should filter out fast signal transients but capture slow drifts in the fluorescence trace (e.g., the gradual decay of signal intensity during long-time imaging or accumulated intensity drifts due to the slow decay time of GCaMP6). After normalization, the Δf/f time series should show a more stable baseline level while fast transients are retained (Fig. 5a).

2. Registration across multiple imaging sessions of the same mouse.

When there are multiple imaging sessions from the same animal, images across different sessions can be aligned first before further processing, such as dimensionality reduction and ROI identification (*see* **Note 5**). For image registration, we first calculate the average frame of each imaging session, and then take the average frame of the first imaging session as the reference image. Registration can be performed by manually selecting landmarks on the average frame and the reference image or using automated algorithms (such as "imregconfig" and "imregtform" in MATLAB) to generate a geometric transformation matrix, which aligns the average frame to the reference image. Then, the geometric transformation matrix is applied to its corresponding imaging session to complete registration.

3. Remove hemodynamic artifacts from widefield signals.

The raw fluorescence signals of widefield calcium imaging are contaminated by hemodynamic changes, as the excitation and emission wavelengths of GCaMP (~485 and ~520 nm, respectively) reside in the absorption spectrum of oxyhemoglobin and deoxyhemoglobin. Several methods are available to correct hemodynamic contamination. One prevalent method is using a secondary wavelength of light (e.g., violet light at ~405 nm) to estimate calcium-invariant reflectance changes caused by hemodynamics, which can then be used for a regression-based subtraction of hemodynamic signals [5, 10, 24, 25]. Another analytical correction method that we often use is to extract hemodynamic components using principal component analysis (PCA) followed by independent component analysis (ICA), and then reconstruct the corrected widefield signals from the remaining components that reflect neural activity [13, 19, 26] (Fig. 5b). This analytical method may require large computing power when imaging data size increases. Different ICA algorithms can be applied based on the available computing power. See reference [27] for a detailed comparison between different algorithms.

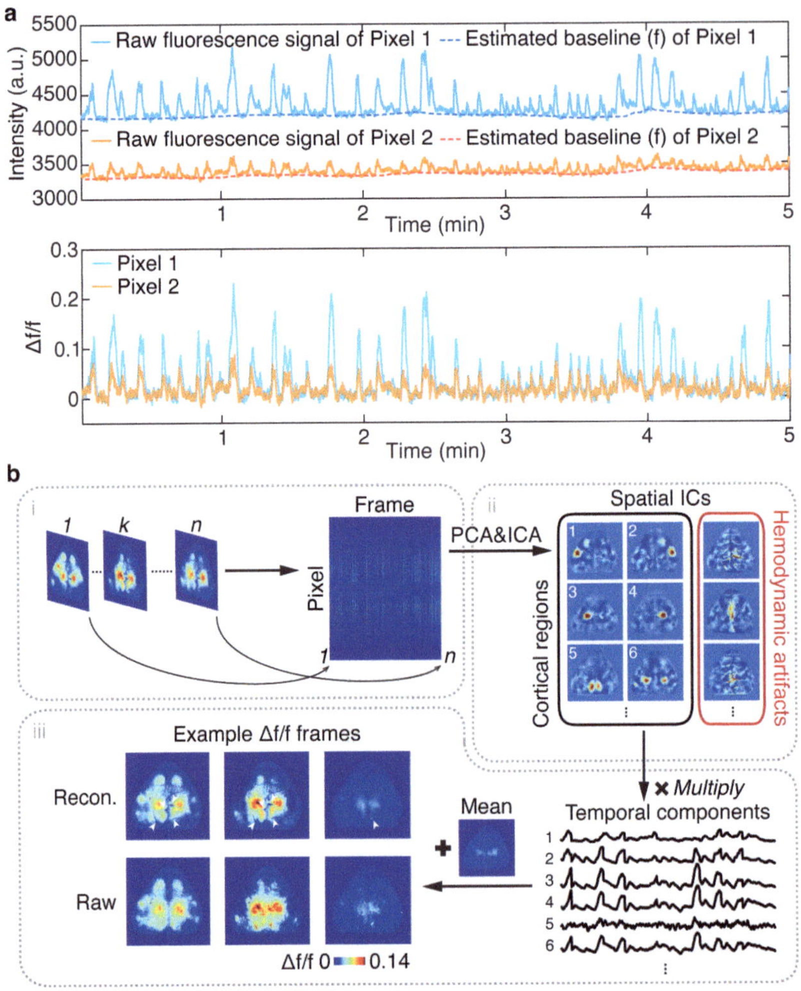

Fig. 5 Processing of the widefield imaging data. (**a**) Raw fluorescence signals, estimated baselines (f), and Δf/f traces of two example pixels with different brightness are shown here. Here we chose a 30-s sliding window to estimate f for a 5-min imaging session. Note that the f captures the differences in fluorescence intensity between two pixels and the slow drifts in raw fluorescence traces (potentially due to the slow decay time of GCaMP6s). After normalization, the Δf/f time series shows a more stable baseline level while fast transients are retained. (**b**) Procedures of using PCA-and-ICA analysis to remove hemodynamic artifacts from widefield signals. The 3D Δf/f image stack is first transformed to a 2D matrix for PCA where pixels are treated as variables and frames are treated as observations (i). Then, spatial components are extracted by PCA-and-ICA analysis. Here, most spatial ICs correspond to known cortical regions or hemodynamic artifacts. ICs attributable to artifacts (e.g., hemodynamic signals) are visually identified and excluded (ii). Last, Δf/f images are reconstructed by multiplying the remaining ICs corresponding to cortical regions with their time series. As the initial PCA subtracts the mean of each pixel, the mean value from the original Δf/f image stack is added. The resulting reconstructed images retain the activity of cortical regions while effectively reducing artifacts. White arrowheads indicate the elimination of hemodynamic signals in reconstructed images compared to the raw images (iii). (Adapted from [6])

Alternatively, repeating the same experiments in animals expressing activity-insensitive GFP can be used as a control to test whether the observed widefield fluorescence changes mainly reflect neural activity rather than hemodynamic artifacts [10, 28].

3.5 Examine the Implantation Accuracy by Post-hoc Immunostaining

As the insertion trajectory is difficult to detect due to the small dimension of the Neuro-FITM probe itself and the shuttle-free implantation strategy, we usually leave the probe in the brain for 4–5 weeks before perfusion to allow glial scar formation, which is a good indication of the probe location. We stain for glial fibrillary acidic protein (GFAP), a marker for astrocytes activated during glial scar formation around the probe (Fig. 4d). To quantify the implantation accuracy, we measure the distance between the target location and the actual location of the probe tip based on the immunostaining results. Across 6 mice in our past experiments, the distance was 100 ± 33 μm in the medial–lateral direction, 113 ± 18 μm in the anteroposterior direction, and 87 ± 24 μm in the vertical direction, suggesting a good accuracy of probe implantation.

4 Notes

1. Neuro-FITM probes can also be fabricated with smaller site spacing (20 μm) for potential use in a tetrode configuration, and with longer shanks to reach deeper mouse brain structures or to use in rats and primates. To implant Neuro-FITM probes with very long shanks, using a rigid shuttle or adding a bioresorbable stiffening layer will be needed to provide extra support during the insertion step. Electrodes that fit into the working distance of the widefield microscope and generate minimized shadows can also be combined with widefield imaging [12, 14, 29, 30].
2. The choice of transgenic mouse lines should be based on the needs of the study. For example, specific expression of reporters in cortical layer 2/3 excitatory neurons can be achieved by using Cux2-Cre transgenic mice [31]. If a particular transgenic line is not available, expression of the indicator throughout the cortex can also be achieved using AAV-PHP.eB, an AAV variant that crosses the blood-brain barrier and delivers genes systemically to the brain [32]. See reference [33] for a detailed comparison between transgenic and AAV-PHP.eB-mediated expression of GCaMP6s in widefield calcium imaging.
3. If bone bleeding happens, press the bleeding spot with a dry sterile cotton swab until bleeding stops (usually within 5 min). Rinse the skull with saline several times to remove any blood stains.

4. The pixel number and frame rate vary depending on the experimental design and the kinetics of the indicator. For imaging GCaMP signals, the frame size is often set between 512 × 512 to 1024 × 1024 pixels, and the frame rate is in a range of 15–60 Hz.
5. We recommend doing image registration on imaging data combined across sessions before running dimensionality reduction analyses, such as PCA and ICA. Registration can minimize the effects of variability in head position across sessions so that the top components from PCA and ICA will be focused on variances caused by brain activity.

References

1. Steinmetz NA, Zatka-Haas P, Carandini M, Harris KD (2019) Distributed coding of choice, action and engagement across the mouse brain. Nature 576:266–273
2. Cardin JA, Crair MC, Higley MJ (2020) Mesoscopic imaging: shining a wide light on large-scale neural dynamics. Neuron 108:33–43
3. Ren C, Komiyama T (2021) Characterizing cortex-wide dynamics with wide-field calcium imaging. J Neurosci 41:4160–4168
4. Urai AE, Doiron B, Leifer AM, Churchland AK (2022) Large-scale neural recordings call for new insights to link brain and behavior. Nat Neurosci 25:11–19
5. Couto J, Musall S, Sun XR et al (2021) Chronic, cortex-wide imaging of specific cell populations during behavior. Nat Protoc 16: 3241–3263
6. Ren C, Komiyama T (2021) Wide-field calcium imaging of cortex-wide activity in awake, head-fixed mice. STAR Protoc 2:100973
7. Sabatini BL, Tian L (2020) Imaging neurotransmitter and neuromodulator dynamics in vivo with genetically encoded indicators. Neuron 108:17–32
8. Lin MZ, Schnitzer MJ (2016) Genetically encoded indicators of neuronal activity. Nat Neurosci 19:1142–1153
9. Dana H, Chen T-W, Hu A et al (2014) Thy1-GCaMP6 transgenic mice for neuronal population imaging in vivo. PLoS One 9:e108697
10. Wekselblatt JB, Flister ED, Piscopo DM, Niell CM (2016) Large-scale imaging of cortical dynamics during sensory perception and behavior. J Neurophysiol 115:2852–2866
11. Lake EMR, Ge X, Shen X et al (2020) Simultaneous cortex-wide fluorescence Ca^{2+} imaging and whole-brain fMRI. Nat Methods 17: 1262–1271
12. Clancy KB, Orsolic I, Mrsic-Flogel TD (2019) Locomotion-dependent remapping of distributed cortical networks. Nat Neurosci 22:778–786
13. Liu X, Ren C, Lu Y et al (2021) Multimodal neural recordings with Neuro-FITM uncover diverse patterns of cortical–hippocampal interactions. Nat Neurosci 24:886–896
14. Peters AJ, Fabre JMJ, Steinmetz NA et al (2021) Striatal activity topographically reflects cortical activity. Nature 591:420–425
15. Mayford M, Bach ME, Huang YY et al (1996) Control of memory formation through regulated expression of a CaMKII transgene. Science 274:1678–1683
16. Steinmetz NA, Buetfering C, Lecoq J et al (2017) Aberrant cortical activity in multiple GCaMP6-expressing transgenic mouse lines. eNeuro 4:ENEURO.0207-17.2017
17. Buzsaki G (2015) Hippocampal sharp wave-ripple: a cognitive biomarker for episodic memory and planning. Hippocampus 1188:1073–1188
18. Yizhar O, Fenno LE, Davidson TJ et al (2011) Optogenetics in neural systems. Neuron 71:9–34
19. Makino H, Ren C, Liu H et al (2017) Transformation of cortex-wide emergent properties during motor learning. Neuron 94:880–890.e8
20. Pachitariu M, Steinmetz N, Kadir S et al (2016) Kilosort: realtime spike-sorting for extracellular electrophysiology with hundreds of channels. bioRxiv:061481. https://doi.org/10.1101/061481
21. Chung JE, Magland JF, Barnett AH et al (2017) A fully automated approach to spike sorting. Neuron 95:1381–1394.e6

22. Yger P, Spampinato GLB, Esposito E et al (2018) A spike sorting toolbox for up to thousands of electrodes validated with ground truth recordings in vitro and in vivo. elife 7: e34518
23. Zutshi I, Fu ML, Lilascharoen V et al (2018) Recurrent circuits within medial entorhinal cortex superficial layers support grid cell firing. Nat Commun 9:3701
24. Ma Y, Shaik M, Kim S et al (2016) Wide-field optical mapping of neural activity and brain haemodynamics: considerations and novel approaches. Philos Trans R Soc Lond Ser B Biol Sci 371:20150360
25. Valley MT, Moore MG, Zhuang J et al (2020) Separation of hemodynamic signals from GCaMP fluorescence measured with wide-field imaging. J Neurophysiol 123:356–366
26. Ren C, Peng K, Yang R et al (2022) Global and subtype-specific modulation of cortical inhibitory neurons regulated by acetylcholine during motor learning. Neuron 110:2334–2350.e8
27. Sahonero-Alvarez G, Calderon H (2017) A comparison of SOBI, FastICA, JADE and infomax algorithms. Paper presented at the 8th international multi-conference on complexity, informatics and cybernetics proceedings, 2017
28. Vanni MP, Murphy TH (2014) Mesoscale transcranial spontaneous activity mapping in GCaMP3 transgenic mice reveals extensive reciprocal connections between areas of somatomotor cortex. J Neurosci 34:15931–15946
29. Xiao D, Vanni MP, Mitelut CC et al (2017) Mapping cortical mesoscopic networks of single spiking cortical or sub-cortical neurons. elife 6:e19976
30. Karimi Abadchi J, Nazari-Ahangarkolaee M, Gattas S et al (2020) Spatiotemporal patterns of neocortical activity around hippocampal sharp-wave ripples. elife 9:e51972
31. Franco SJ, Gil-Sanz C, Martinez-Garay I et al (2012) Fate-restricted neural progenitors in the mammalian cerebral cortex. Science 337: 746–749
32. Chan KY, Jang MJ, Yoo BB et al (2017) Engineered adeno-associated viruses for efficient and noninvasive gene delivery throughout the central and peripheral nervous systems. Nat Neurosci 20:1172–1179
33. Michelson NJ, Vanni MP, Murphy TH (2019) Comparison between transgenic and AAV-PHP.eB-mediated expression of GCaMP6s using in vivo wide-field functional imaging of brain activity. Neurophotonics 6: 025014

Part III

Imaging Behavior and Motor Systems

Chapter 7

Wide-Field Calcium Imaging of Mesoscale Networks Underlying the Encoding of Skilled Voluntary Movement

Jessica Lucchesi, Alessandro Scaglione, Eros Quarta, Anna Letizia Allegra Mascaro, and Francesco Saverio Pavone

Abstract

The cortex-wide mechanisms coordinating skilled voluntary movements in rodents are still largely unexplored. The combination of optical imaging techniques, such as wide-field microscopy, with the use of transgenic animals made it possible to monitor mesoscale cortical activity during the execution of these goal-directed behaviors. Here we describe the development of a setup for large-scale functional imaging integrated within a customized behavioral platform for training mice to perform a goal-directed movement of reach-to-grasp (RtG). We report a protocol to perform the surgery for the realization of the optical window, the methods to motivate the animal to learn and perform the RtG task in the behavioral platform, and the image processing and data analysis of the behavioral data and the fluorescent calcium image will be described.

Key words Reach-to-grasp, Wide-field fluorescence microscopy, Calcium imaging, Goal-directed movement, Behavioral platform, Cortical activity, In vivo

1 Introduction

In mammals, the planning and execution of a goal-directed movement are driven by the activity of neural circuits of the cerebral cortex and subcortical regions reaching the spinal cord, where thanks to motor neurons, the contraction of the different muscles of the body is orchestrated to achieve fine control of limb movement [1, 2].

Reaching and grasping movements are extensively performed and finely controlled in primates [3] and in rodents to achieve several objectives, one of which is the collection and consumption of food [4]. When rodents explore in search of food, they proceed with locating the target through olfactory stimuli and with the help of the whiskers, after which they grab the object with their mouth and sit on their hind paws to consume the food with the help of

James E. Niemeyer et al. (eds.), *Awake Behaving Mesoscopic Brain Imaging*, Neuromethods, vol. 214,
https://doi.org/10.1007/978-1-0716-4120-0_7,

their front paws. Although the use of the forelimbs is not the primary action of choice for gathering food, rodents have sufficiently developed abilities to perform goal-directed motor tasks using the paws. Despite the substantial phylogenetic difference between rodents and primates, similarities between the two orders occur at neural, skeletal, muscular, and behavioral levels [4].

The orchestrated activity from multiple cortical regions with different roles and functions and a hierarchical organization subserves the planning, execution, and control of the movement.

A large contribution comes from the primary and secondary motor cortex along with the participation of somatosensory regions [5]. However, given the complexity of goal-directed movements such as reaching and grasping, it is plausible that there is recruitment of multiple areas of the cortex so extending the study over the entire cortex could allow us to expand the repertoire of neuronal information underlying these movements. In the past, the study of neuronal activity and the involvement of the cerebral cortex related to movement was conducted by performing studies on surgical lesions or by electrical stimulation of the various brain regions [5, 6].

In recent years, thanks to the development of transgenic mice expressing genetically encoded calcium indicators (GECI) [7–13] combined with wide-field optical imaging techniques [14–18], it became possible to monitor the activity of neurons and the role of different cortical areas simultaneously enabling investigation on how the activity of the whole cortex is organized during the execution of motor tasks.

Thanks to wide-field microscopy techniques, it has been possible to overcome the traditional view of neuroscience according to which the cortical contribution during the execution of voluntary movements emerges exclusively from somato-motor regions of the hemisphere contralateral to the limb that performed the movement. Recent studies have shown that there is extensive involvement of various cortical regions during the execution of behavioral tasks [15, 19, 20] and it is confirmed that a global network distributed over the entire cortex correlates with unilateral limb movement during the execution of goal-directed movements, particularly during the reach-to-grasp task in mice [17]. Here we detail the procedure and materials to perform an experiment of wide-field imaging in transgenic mice expressing the calcium indicator GCaMP6f (Thy1-GCaMP6f line) to study the distributed cortical activation during a reach-to-grasp behavior, as described in the work of Quarta and colleagues [17].

2 Materials

2.1 Materials for the Realization of the Optical Window on the Dorsal Cortex and Food Pellet for Reach-to-Grasp

1. Isoflurane.
2. Ophthalmic gel: 42.5% Liquid Paraffin, 53.7% White Vaseline (excipients: lanolin alcohol).
3. Dexamethasone: 0.3 mL dexamethasone, 10 mL saline solution.
4. Carprofen: 0.1 mL carprofen, 10 mL saline solution.
5. Stereotaxic apparatus: surgery apparatus equipped with ear bars and centering scope 40X.
6. Cotton stick: Absorbent cotton swab.
7. Betadine: povidone-iodine (10% iodine), glycerol, macrogol lauryl ether, dihydrated dibasic sodium phosphate, citric acid monohydrate, sodium hydroxide, purified water.
8. Lidocaine: lidocaine hydrocloride 2 g (excipients: chlorobutane).
9. Scalpel: single-use sterile steel blade.
10. Saline solution: 0.9% NaCl.
11. Dental drill: precise DC Micro-Drill, high-speed side switch adjustable, conceived for brain microsurgery operations. Drill bits 0.5 mm diameter.
12. Permanent marker: ultra fine (~0.5 mm line width): permanent black ink.
13. Cover glass: optically transparent 5 mm diameter round cover glass 1 mm thick, *see* **Note 1**.
14. Acrylic glue: cold-curing acrylic denture preparation material.
15. Dental cement: self-curing adhesive resin cement: 4-methacryloyloxyethyl, methyl methacrylate, tri-n-butyl borane, polymethyl methacrylate.
16. Heating pad: temperature probe and heating pad.
17. Head-post: aluminum bar (30 × 2.5 × 1 mm).
18. Insulin syringe: total volume 1 mL × 27G.
19. Chocolate-flavored pellets: 10 mg, 5TUL Purified 10 mg pellets: sucrose, casein, maltodextrin, corn starch, corn oil, cellulose, minerals, silicon dioxide, vitamins, magnesium stearate, DL-methionine, color and/or flavor added where applicable.

3 Methods

3.1 Animal Model: Breeding and Husbandry of Transgenic Mice

The choice of the animal model for these experiments is crucial both for the kind of tasks that the animal can be trained to perform and for the technique that allows for monitoring the neural activity cortex-wide. To image cortical activity during the reach-to-grasp (RtG) experiments we use transgenic mice, without preference for sex, that express the GCaMP6f neural indicator (C57BL/6J-Tg (Thy1-GCaMP6f)GP5.17Dkim/J) to monitor neural activity over almost the entire cortex. Thy1-GCaMP6f-WPRE transgenic mice express the GCaMP6f indicator under the control of the Thy1 promoter to drive the expression at the neuronal level. WPRE is the posttranscriptional regulatory element of the marmot hepatitis virus that together with the poly-adenylation (pA) tail increases mRNA stability and consequently GCaMP protein expression [9, 21]. In the brain, GCaMP is uniformly expressed throughout the cortex with the highest expression in layers 5 and 6 (L5 and L6) compared to layer 2/3 (L2/3) while no expression was recorded in layer 4 (L4) cells. The transgenic line was acquired from Jackson Laboratory (Stock #025393, www.jax.org/strain/025393) and we bred progenitors as heterozygous (Male [Hemizygous for Tg (Thy1-GCaMP6f)GP5.17Dkim] x Female [Non Carrier for Tg (Thy1-GCaMP6f)GP5.17Dkim]) in our local facility. In addition to GCaMP6f transgenic animals, GFPM mice (Tg(Thy1-EGFP) MJrs) expressing enhanced green fluorescent protein (EGFP) under the control of a modified Thy1 promoter region were also used to evaluate the effect of the hemodynamic contribution. These transgenic mice express EGFP in sparse subsets of neurons providing a bright Golgi-like stain. We bought the transgenic line from Jackson Laboratory (Stock #007788, www.jax.org/strain/00 7788) and bred progenitors as heterozygous (Male [Hemizygous for Tg(Thy1-EGFP)MJrs] x Female [Non Carrier for Tg(Thy1-EGFP)MJrs]) in our local facility.

Since gestation and lactation of pups require a rather prolonged period of female inactivity, it is preferable to use males expressing the GCaMP6f and GFPM indicator and wild-type females when choosing the pair of progenitors. In this way, fewer male progenitors can be used on a larger number of females. For this reason, it is important to execute PCR testing to separate animals with and without the expression of GCaMP6f and GFPM. In general, we have a 50% yield of positive mice over the total number of newborns; however, depending on the breeder we noted significant departures from this ratio. To avoid the appearance of congenital defects we try to replenish our breeders every year. It is important to check the pups for congenital diseases and perform a careful observation of their behavior to avoid training animals that have congenital defects.

3.2 Food and Water Restriction

Training rodents to perform a motor task can sometimes be difficult, especially if the training is performed under head-fixed conditions. These conditions arise when, in addition to the behavioral analysis, a cortical activity monitoring study is performed with optical imaging methods, in which high stability of the subject is required to avoid introducing movement artifacts. To study learning while performing motor tasks and motivating the animal, it is preferable to set the task with a reward system; this can be done by providing a positive or negative reward.

Starting with the pioneering work of B.F. Skinner [22], positive reinforcement via rewards is being used to induce wanted behaviors. While the role of both reward and punishment appear to contribute to learning, at least in humans [23, 24], the reward-based approaches are grounded on physiological and ethical reasons [25]. For this reason, the paradigm developed to induce the motor task included a reward that was provided whenever the task was performed correctly. The type of reward may be liquid or solid.

The choice of the type of reward mainly depends on the behavioral paradigm used for learning the motor task. If the caloric restriction does not sufficiently motivate the animals or leads to a small number of trials, it is possible (by making some changes to the behavioral apparatus) to switch to a liquid reward. *See* **Note 2**.

The most commonly used solid rewards are sucrose-based pellets with different flavors. Since the individual pellets have a fixed weight, they allow you to provide an exact amount of reward with each attempt; however, weighing between 10 and 20 mg (for mice) and between 40 and 50 mg (for rats), the amount of reward provided per attempt tends to be high compared to the daily food intake of mice (4.4 ± 0.1 g/mouse) [26]. *See* **Note 3**.

Independently of the reward type, to motivate animals to perform a task such as the reach-to-grasp, mice are put on caloric or water restriction. Considering the fixed weight of the pellets, and the reproducibility of the reward delivery system, a behavioral task was developed inspired by the work of Guo and colleagues [27] in which a solid reward is provided.

For this reason, animals are placed on calorie restriction to 80–90% of their original body weight by limiting daily food intake. The weight of the mice was monitored daily, and the amount of food provided was calculated and adjusted according to the animal's weight changes. *See* **Note 4**. The animals were subjected to caloric restriction. Considering that the mice consume a daily food intake of between 12 and 18 g/100 g body weight/day [28], a theoretical daily intake of 12 g/100 g body weight/day was considered to calculate the amount of food to be supplied daily, from which we calculated 80% to determine the actual amount of food to be provided to the animals so that they would have a weight loss within physiological parameters. This amount of food is subject to slight daily changes according to weight changes. Given the amount of pellets provided is small compared to the standard daily quantity, the piece of food was placed directly into the home cage in contact with the bedding to facilitate consumption. The daily ratio of food was provided a few hours after the conclusion of the training, thus ruling out the possibility of animals not performing the RtG motor task since they will still receive a reward at the end of the training. On the following day, any leftover food not consumed from the previous day was removed from the cage to prevent the animals from consuming it while waiting for the training to begin.

3.3 Animal Preparation: Surgical Procedures to Prepare the Optical Window

To visualize neuronal dynamics during the execution and learning of the reaching and grasping motor tasks, it is necessary to create an optical window on the dorsal surface of the mouse skull. Depending on the biological question and the brain area to be investigated, it is possible to customize the size and characteristics of the optical window during surgery.

Since we are interested in recording the cortical activity distributed over both hemispheres, it is preferable to employ surgery that is easy to perform, to increase reproducibility, and that the procedure is long-lasting since a rather long time is required to train the animals to learn the motor task.

To record neural dynamics using optical imaging techniques, one can proceed either by removing a portion of the skull via a craniotomy to gain direct access to the region of interest or by skull thinning. Both procedures are rather invasive and difficult to perform, since performing the craniotomy of very large areas (such as both hemispheres) or thinning such a fine structure as that of the skull of mice requires a lot of dexterity and experience, increasing the chances of postoperative complications. In addition, the animal's inflammatory processes and recovery time are greatly increased following such invasive surgical procedures. To reduce the invasiveness of surgery and monitor cortical activity over very large fields of view, recently it is also possible to conduct wide-field optical imaging through the intact skull [16–18, 29, 30]. This technique requires rapid execution times resulting in simple execution. By reducing the invasiveness, the inflammatory processes, postoperative complications, and recovery times will also be reduced, allowing one to train the animal and monitor the neuronal activity for a very long time (even a few months) [30].

To protect the exposed area following the removal of the skin flap, a thin and homogeneous layer of transparent dental cement is applied over the entire surface. Depending on the level of detail required during imaging, it is possible to apply only the dental cement or add two semicircular slides on both hemispheres. To conduct wide-field neural activity studies, the application of dental cement alone is a sufficient condition to perform functional connectivity studies of large-scale cortical areas parceled according to anatomical atlases such as that of the Allen Institute Mouse Brain

Atlas. The addition of the coverslip glasses, having a different refractive index, allows a clearer image of the brain with a more detailed view of the whole network of blood vessels of the superficial cerebral layers. The application of one or more very thin coverslip glasses complicates the procedural step. The bonding step is extremely delicate, and it is possible that microbubbles form below the surface, compromising the quality of the imaging, or that the slides break during application. Considering the fast catalysis times of the cement, in case of glass fracture, its replacement during surgery is difficult to perform. Furthermore, during the months of animal training, the coverslips could be chipped or fractured following impacts of the animal's head inside the home cage.

The detailed protocol with all the surgical steps for the execution of the optical window will be described below, exposing both cortical hemispheres, to perform optical imaging on the intact skull with and without the application of the two coverslip glasses.

3.3.1 Optical Window on the Mouse Dorsal Cortex

1. Administer Enrofloxacin (10 mg/kg) subcutaneously to mice the day before surgery and then daily for the next 3 days for antibiotic prophylaxis.
2. Sterilize all instruments with 70% ethanol and subsequently with high temperatures to ensure an excellent microbiological standard.
3. Weigh the animal.
4. Anesthetize the animal with oxygen and isoflurane (5% for induction, 1–2% for maintenance). *See* **Note 5**.
5. Fix the animal's head (using ear bars) in the stereotaxic apparatus (Model 1900, Kopf Instruments, Tujunga, California, USA) to avoid movements that could affect the course of the surgery (Fig. 1a). *See* **Notes 6** and 7.
6. Apply a heating pad under the mouse's ventral area to reduce the risk of hypothermia. Body temperature is monitored continuously by a rectal probe and maintained in the range of 36.5–37 °C (ThermoStar, RWD, San Diego, California, USA). NOTE: Mice were monitored by the operator (by observation of chest movements) to assess that the respiratory rate was in the optimal range of ~55–65 respirations per minute.
7. Apply ophthalmic gel (Lacrilube 42.5% liquid paraffin +53.7% white vaseline, AbbVie S.r.l., Latina, Italy) on the eyeball to create a protective film to prevent ocular dehydration.
8. Remove the hair using first the scissors and then the depilatory cream exposing the dorsal surface of the animal's head and limiting the infective processes.

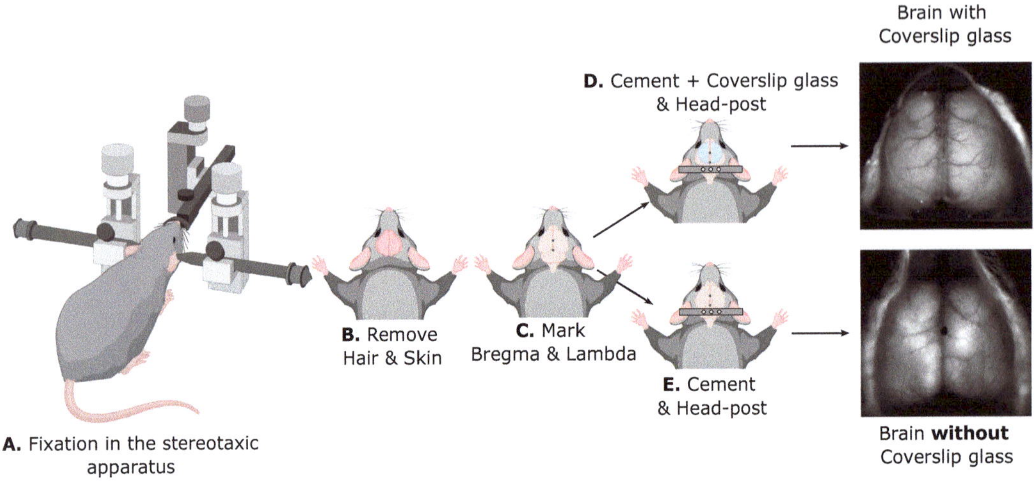

Fig. 1 Surgical steps for the realization of the optical window: (**a**) Place and fix the mouse in the previously aligned stereotaxic apparatus. (**b**) Remove the hairs on the dorsal surface of the skull, sterilize the exposed surface, and remove a flap of skin to create a window to visualize both cortical hemispheres. (**c**) Mark bregma and lambda on the skull. Now you can proceed in two ways: (**d**) apply the dental cement, the two coverslip glasses, and the head-post posterior to lambda, or (**e**) cover the entire optical window with only the dental cement and add the head-post posterior to lambda. (Image created with Biorender.com)

9. Disinfect the shaved area with 70% ethanol and 10% Betadine (Betadine 10%, Meda Pharma S.p.A, Milan, Italy) three times.
10. Subcutaneous administration of corticosteroid anti-inflammatory (Dexamethasone sodium phosphate, 5 mg/Kg) to reduce phlogosis processes and the local application of a 2% lidocaine hydrochloride solution applied topically as a local anesthetic.
11. With the help of tweezers and scissors, cut a flap of skin to open a window on the skull of approximately 2–2.5 cm^2 and expose both hemispheres (Fig. 1b).
12. Keep the exposed surface hydrated with sodium chloride 0.9% (Sodium chloride 0.9%, Eurospital S.p.A., Trieste, Italy), and clean from any hairs.
13. Apply lidocaine hydrochloride solution again on the surface of the skull before scraping it with a scalpel. *See* **Note 8**.
14. Position with the stereotaxic apparatus and the monocle on Bregma (the midline bony landmark where the coronal and sagittal sutures meet, between the frontal and two parietal bones) and then on Lambda (point of conjunction between the two parietals and the occipital bone), which is generally – 4.2 mm from Bregma.
15. Mark the two anatomical landmarks with a permanent marker (Fig. 1c). *See* **Note 9**.

16. Apply a layer of transparent dental cement (Super-Bond Universal Starter Kit, Sun Medical, Furutaka, Japan) to the entire cortical surface making sure that the edges were carefully covered and not to have left any flap of tissue exposed to air to prevent any infection (Fig. 1e).
17. Optional: Apply two coverslips to improve the resolution of the field of view (Fig. 1d). In this case, two semicircular coverslips, one for each hemisphere, are placed over the transparent dental cement before it dries.
18. Glue the custom-made aluminum bar (head-post), previously disinfected with ethanol 70%, and place it posterior to the Lambda on the animal's head. *See* **Note 10**.
19. Subcutaneous administration of analgesics and nonsteroidal anti-inflammatory drugs (Carprofen, 5 mg/Kg) and 0.2 mL of lactated Ringer's solution (Lactated Ringers, Eurospital S.p. A., Trieste, Italy) at the end of surgery will facilitate recovery.
20. Move the animal to the induction chamber and administer pure oxygen until the first indications of awakening.
21. Place the animal in its home cage inside the animal facility on a heating pad for 1 h until fully awakened. In the postoperative hours, the mouse should be scrupulously followed to evaluate both the recovery of motor activity and the response to stimuli.
22. Administer multimodal analgesic therapy for 3 days (Carprofen (5 mg/Kg) subcutaneously) to limit postoperative pain.
23. The animals received 1 week of recovery from surgery where they were given ad libitum water and food.

After the surgery and the application of the aluminum head-post (and possibly some coverslip glasses), the animals are housed individually in the cages to prevent interactions with other subjects from causing injury to the other animals or from damaging the window optics compromising the training process. *See* **Notes 11** and **12**.

3.4 Behavioral Apparatus to Perform Reach-to-Grasp Task

To train the animals to learn the RtG task, a customized behavioral setup inspired by the work of Guo and colleagues [27] was developed. The experimental apparatus (Fig. 2a) consists of a Thorlabs base in which various components and accessories have been mounted to facilitate learning of the motor task and carrying out the training sessions.

A custom-made attachment structure (Fig. 2b) is mounted in the apparatus to insert the aluminum head-post (Fig. 2c) and perform the training under head-fixed conditions. The structure has an inverted L shape, in which the terminal part is formed by two metal supports, with a rectangular opening, in which it is possible to fix the head-post by tightening the two metal plates with a

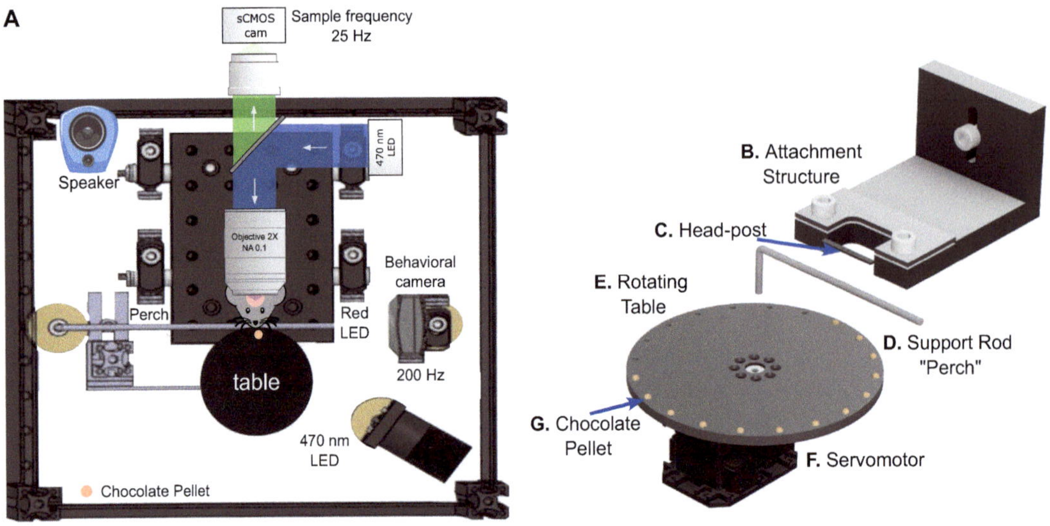

Fig. 2 Rendering of the experimental setup and the behavioral apparatus
(**a**) The experimental setup includes a wide-field microscope to monitor cortical activity with a sampling frequency of 25 Hz. The behavioral apparatus consists of a rotating table, a behavioral camera (sampling frequency of 200 Hz), a red LED, a speaker, and an LED to illuminate the stage and the front of the animal. Image modified from Quarta et al., 2022 [17]. (**b**) Representation of the attachment structure in which the head-post (**c**) is inserted to fix the animal's head and perform optical imaging in head-fixed conditions. (**d**) Support rod (perch) for positioning the front paws of the animal. (**e**) Rotating table, connected to the servomotor (**f**), for distributing chocolate pellets (**g**). (Image created with Biorender.com)

hexagonal key (Fig. 2b). The rectangular aperture must be large enough to allow the head-post to enter and view the entire optical window. The presence of a slot and screw on the vertical portion of the structure allows the anchoring system to be adjusted in height (Fig. 2b).

A support rod (perch) is added to position the animal's front paws (Fig. 2d). A small red LED is inserted near the perch (visible in the field of view of the behavioral camera but not from the animal), which is used as a visual reference for synchronization of the behavioral task. A rotating table (Fig. 2e) with concavity on the circumference, is placed in front of the animal to accommodate the target of the RtG motor task. The rotation of the table, to advance to the next pellet, could be controlled either by the operator, via a button, or automatically by the servo motor (Fig. 2f) equipped with a contactless magnetic rotary encoder to ensure consistent pellet positioning (Servomotor Dynamixel MX-28AT, Robotis, Lake Forest, California, USA) and controlled by a controller board (ArbotiX-M. Robocontroller, Trousser Robotics, Downers Grove, Illinois, USA). The rotation of the table is preceded by the emission of an acoustic tone from a speaker near the animal to associate the sound with the arrival of the reward. Since our video cameras do not record audio, the speaker sound is synchronized with the illumination of a red LED within the field of view of the video camera. An ad hoc program written in Arduino programming language is used to control time interval, angle, and rotation speed. Chocolate-flavored pellets (10 mg, 5TUL Purified 10 mg pellets, catalog 1811529, TestDiet, Richmond, Indiana, USA) (Fig. 2g) are placed in the concavities of the rotating table and used to induce the movement of RtG. A high-speed camera (Fig. 2a) is used to monitor the animal's behavior (CM3-U3-13YC-CS, Chameleon3, FLIR, Wilsonville, Oregon, USA) with 2.8–8 mm varifocal objective lenses (LENS-30F2-V80CS, Fujinon, Tokyo) positioned on the side of the paw of the animal that will perform the movement. A sampling rate of 200 frames/sec is used to perform a detailed analysis of the movement, and a 470 nm LED illuminated both the stage and the front of the animal to create a high-contrast image that allows one to view the details of the paws and mouth (Fig. 2a). Video recordings are checked using FlyCapture software. The behavioral apparatus is mounted on a motorized stage that allowed the platform to be moved vertically to adjust the focus of the wide-field microscope objective. The setup is enclosed and isolated by a wooden panel with acoustic foam inside to conduct the experiments in complete darkness without adding sound artifacts.

3.5 Wide-Field Microscopy to Monitor Bilateral Cortical Activity

The wide-field epifluorescence microscope used is a custom-made system developed in our laboratories consisting of a 470 nm LED (M470L5, 470 nm, 650 mW; Thorlabs Inc., Newton, New Jersey, USA) to excite the GCaMP6f indicator. Lighting and dimming of this LED are set by a USB Controller Hub and power supply (KCH301; Thorlabs Inc., Newton, New Jersey, USA) and a T-Cube LED Driver (LEDD1B; Thorlabs Inc., Newton, New Jersey, USA). The illumination emitted by the LED was collimated by an aspherical condenser (ACL2520U-A; Thorlabs Inc., Newton, New Jersey, USA) and reflected by an elliptical mirror (BBE1-E02, 1" Broadband Dielectric Elliptical Mirror; Thorlabs Inc., Newton, New Jersey, USA) placed in a right-angle mirror mount (KCB1EC/M, Kinematic Elliptical Mirror Mount; Thorlabs Inc., Newton, New Jersey, USA) to deflect light to a 100 mm focal lens (f = 100 mm) (AC254–100-A-ML; Thorlabs Inc., Newton, New Jersey, USA) placed before the cube (DFM1/M) containing the filter and mirror kit for GFP (MDF-GFP; Thorlabs Inc., Newton, New Jersey, USA). Inside are: the 482/18 bandpass excitation filter (MDF-GFP; Thorlabs Inc., Newton, New Jersey, USA) to select only the wavelengths to excite GCaMP and the 495 nm dichroic mirror (DC FF 495-DI02 Semrock, Rochester, New York, USA) that is responsible for reflecting the excitation wavelength and directing it to the rear aperture of the objective (TL2X-SAP 2X Super Apochromatic Microscope Objective, 0.1 NA, 56.3 mm WD; Thorlabs Inc., Newton, New Jersey, USA). Once the sample is illuminated, the emitted fluorescence is collected again by the objective lens and directed into the cube. The fluorescence with a longer wavelength than the excitation light will pass through the dichroic mirror and be filtered by a 525/50 bandpass excitation filter for GFP (MDF-GFP; Thorlabs Inc., Newton, New Jersey, USA). At the top end of the cube, an achromatic Doublet lens with a 100 mm focal length was added before focusing the image on the CMOS sensor of the high-speed camera (Orca Flash 4.0 V2; Hamamatsu Digital camera; Bridgewater, New Jersey, USA). The illumination provided by the optical pathway made it possible to illuminate the entire dorsal surface of the cortex and perform imaging of both hemispheres. An iris of 1–2 mm was put above the animal's head to prevent the collimated light beam from striking the animal's eyes, causing them to close or creating a visual artifact. The fluorescence signal related to cortical activity was acquired at a sampling rate of 25 Hz, with a resolution of 512 × 512 pixels with a Field of View (FOV) of 12 × 12 mm (16-bit depth) using HCImage Live software. The synchronization of the components of the behavioral apparatus and wide-field microscopy was controlled by a National Instrument via a TTL signal.

3.6 Experimental Design to Train Animals to Perform the Reach-to-Grasp Task

One week after surgery to create the optical window on the dorsal surface of the skull, the animals are placed on food restriction and began acclimatization to head-fixed conditions. In the first days, animals are manually locked for a few seconds using the head-post inside the home cage.

Approximately 2 weeks after surgery, the mice are placed in the behavioral platform and locked via the head-post to the anchor structure by habituating (for 1 week) them at increasing times until a period of at least 30 minutes per day per mouse. At this stage, although no recordings are made, it is advisable to simulate real conditions by turning on both the LED of the wide-field microscope to illuminate the optical window and the LED to illuminate the behavioral apparatus. *See* **Note 13**.

Subsequently the acclimatization phase, both in the home cage and in the behavioral apparatus, some chocolate pellets are provided in the animal's cage. This serves the dual purpose of providing a reward for the task and familiarizing the mouse with the target recognition of RtG movement.

Each day, the cages of the animals that will perform the training are brought into the experimental room for at least 30–45 min before the start of the recordings for habituation. After fixing them in the anchor structure, the rodent head is placed under the microscope to align both cortical hemispheres. To maintain good head alignment throughout the training days, it is advisable to use a reference image acquired on the first day and use the two anatomical landmarks, Bregma and Lambda, as a reference. The correct positioning of the animal's head, in addition to ensuring the acquisition of images of both hemispheres in the same orientation and facilitating the alignment phase during data analysis, allows the position of the animal's head from the rotating table not to vary excessively, thus complicating the learning of the motor task. The perch is adjusted both in height and distance from the table to position both the front paws of the animal; finally, the table loaded with all the chocolate pellets is positioned. In the first training sessions, the table with the pellets is kept close to the mouth and nose of the animal to facilitate the licking process. When this task is performed easily, the table is moved away and gradually lowered to make licking more complicated and to stimulate the animal to perform the reach-to-grasp task using the paw. To induce movement with only one of the two paws, it is preferable to place the chocolate pellet not in a position central to the animal's mouth but slightly lateral (to the left or right of the nose). Over the training sessions, the pellet position is gradually moved away from the animal's snout, thus making licking more difficult to perform. The animals will begin to use the forelimb to reach the pellet and make attempts to bring it to the mouth.

To ensure consistent comparison of neural activity across behavioral states, the execution of an RtG movement requires the animal to be in a state of rest and keep both paws on the perch. Following the acoustic signal (compulsory with the lighting of the red LED) the table rotates to advance to the next pellet, the animal lifts its paw from the perch to start the reaching movement towards the pellet to grasp it and bring it towards the mouth. When the RtG task was learned, the table is gradually moved away to encourage performing the movement with maximum limb extension. When the RtG movement is fully and correctly executed (generally 3 or 4 weeks of daily training are required for the animal to perform the movement of RtG), the animal has completed the motor task learning process, and it is possible to move on to a new training phase in which the subject improves task performance in terms of execution speed and movement accuracy. *See* **Note 14**.

The training sessions of the behavioral task included 5 s of total silence, in which the LEDs of the wide-field microscope and behavioral apparatus are turned on. Next we have the emission of an acoustic stimulus emitted by the speaker simultaneous with the lighting of the small red LED (near the animal's paws and visible from the side behavioral chamber) followed (200 milliseconds later) by the rotation of the table, which will bring the first pellet in front of the animal's mouth. Subsequent rotations will occur at regular 20-s intervals *see* **Note 15**.

During the sessions, both the recordings of the behavior through the high-speed camera in the behavioral apparatus and the images of the cortical activity reporting the fluorescence signal of the calcium activity acquired with the wide-field microscope were acquired. The animals are trained 5 days a week, initially for a duration of about 20 min. In the following weeks, the daily sessions will increase to a total duration of about 30–45 min. At the end of the training, the animals are placed in the home cage inside the animal facility.

3.7 Data Recording and Synchronization of the Wide-Field Microscope and Behavioral Apparatus

The behavior is acquired at 200 Hz using a high-speed camera (CM3-U3-13YC-CS, Chameleon3, Teledyne FLIR, Wilsonville, Oregon, USA) with 2.8–8 mm varifocal objective lenses (LENS-30F2-V80CS, Fujinon, Tokyo) and managed with FlyCapture Software (Teledyne FLIR, Wilsonville, Oregon, USA).

The emission of the acoustic tone, lighting of the red LED, and the rotation of the stage controlled by the controller board (ArbotiX-M Robocontroller, Trousser Robotics, Downers Grove, Illinois, USA) is regulated by a custom Arduino program.

The fluorescence signal related to calcium activity is acquired by a CMOS camera, which acquired images with a frame rate of 25 Hz, a resolution of 512 × 512 pixels controlled via the HCImage Live Software provided by Hamamatsu.

The synchronization of all behavioral and fluorescence signal components is controlled by National Instruments using a common TTL trigger.

3.8 Behavioral Analysis of the Reach-to-Grasp Movement

To analyze the behavioral data and conduct the kinematic analysis of the limb that performed the task, the recordings acquired with the lateral camera are analyzed and the movement of RtG is examined in all its components (Fig. 3a).

The following criteria are used to define successful movements:

1. The subject in the seconds before the onset of movement has to be in a state of rest.
2. Both front paws should be placed on the perch.
3. The limb performing the motor task must start from the perch and head towards the rotating table to reach the target (chocolate pellet).
4. Once the target has been reached (reaching), the paw grasps the pellet (grasping) and through the rotation of the paw, brings the target to the mouth (pellet at the mouth).

Once the correct movements are identified, the video file is cut to select the individual events and perform a detailed analysis of the motor task. The analysis of the movement for the generation of the relative trajectories is conducted using the ImageJ program (https://imagej.net/) by manually tracing the position of the metacarpal (Fig. 3B) from the beginning of the movement (MoveON) to its conclusion, the moment in which the pellet is brought to the mouth. The RtG MoveON is defined as the first frame in which the metacarpal position changes. *See* **Note 16**. After tracing the movement frame by frame (Fig. 3b), the XY coordinates are extrapolated

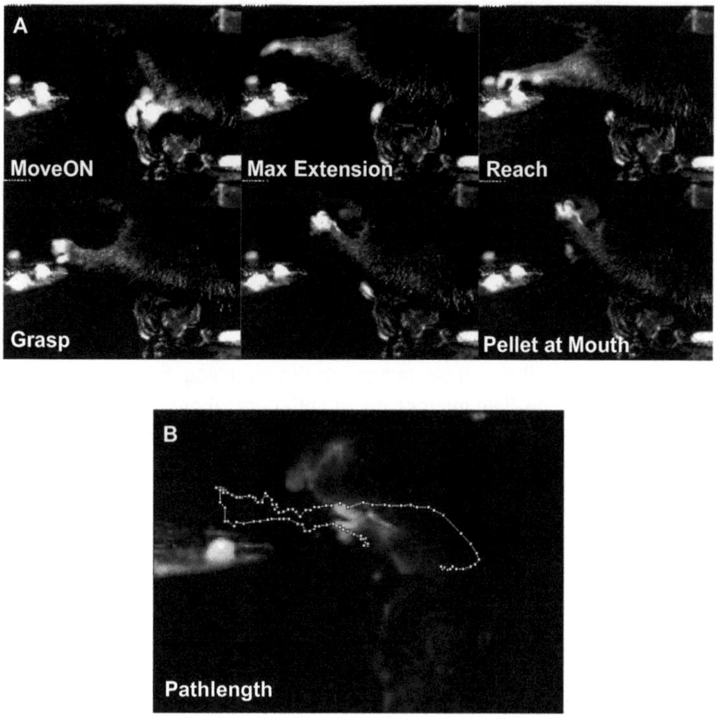

Fig. 3 Illustration of reach-to-grasp movement phases and movement trajectories: (**a**) Sequence of images showing the basic steps of the reach-to-grasp movement. *MoveON*: corresponds to the first frame in which a change in the position of the metacarpus from the support rod is observed. *Max Extension*: corresponds to the maximum extension of the animal's paw during movement execution. *Reach*: reaching the target (chocolate pellets) on the rotating table. *Grasp*: the animal grasps the target and through the rotation of the paw brings the pellet towards the mouth (*Pellet at Mouth*) at which point the movement ends. (**b**) Example of trajectory drawn manually

to plot the trajectories of the RtG movement and analyze the kinematic data as mean speed (mm/s), max speed (mm/s), duration (s), pathlength (mm), number of peaks, mean acceleration (mm/s^2), and max acceleration (mm/s^2). These kinematic parameters are then correlated with the neuronal activity recorded from both cortical hemispheres.

3.9 Data Analysis of Fluorescent Calcium Images

Images whose gray levels represent the fluorescent light captured by the microscope are acquired as image stacks and saved in *".tif"* format.

Image preprocessing: Before proceeding with the method of choice to analyze the data, there are a series of preprocessing steps that need to be performed. In particular, we usually perform three preprocessing steps: (1) alignment of the image stacks across sessions and animals by using the anatomical landmarks marked

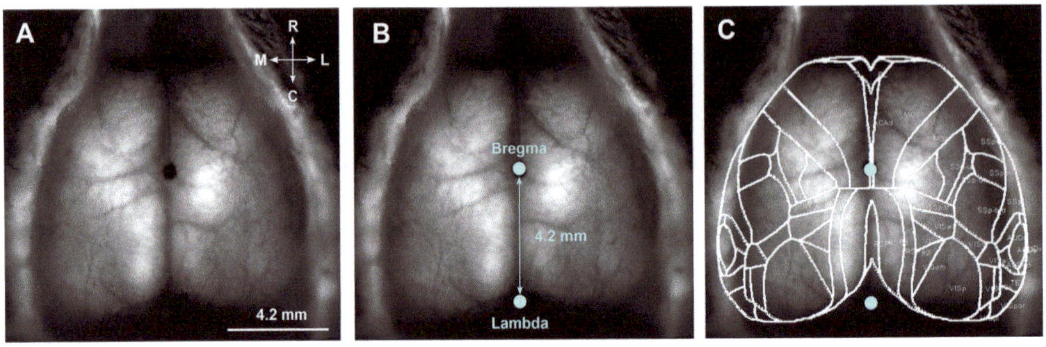

Fig. 4 Alignment of the cortical image stacks by using the anatomical landmarks. (**a**) Grayscale image acquired with the wide-field microscope of both cortical hemispheres. Scale bar: 4.2 mm. (**b**) Identification of the two anatomical landmarks (bregma and lambda) used for alignment and overlap of the Allen Brain Atlas parcellation mask. (**c**) Image of the cortex with the parcellation mask applied for identification of different cortical areas

during the surgery, (2) normalization of the fluorescence signal, and (3) removal of movement and hemodynamic artifacts.

1. Image alignment (Fig. 4a) is necessary to compare the neural activity from specific cortical regions from recordings acquired on separate days and from individual animals. This step is also necessary to parcel the cortex into regions according to a reference atlas [31]. To register our images, we employ a custom-made Python tool based on the PyQtGraph library and the Allen atlas Python API (Allen SDK). In particular, the tool registers images employing an affine transformation that is estimated by mapping the coordinates of bregma and lambda to that of the Allen Brain Atlas. Since, generally, the anatomical distance between bregma and lambda is 4.2 mm in the alignment phase these points are used as a reference to apply the parcellation mask on the image plane of the cortex (Fig. 4b). *See* **Notes 17** and **18**. Once the images are registered, the parcellation mask can be applied to segment the entire cortex into different regions (Fig. 4c).
2. Normalization of the raw fluorescence signal is necessary to compensate for the intrinsic variability of the measurement system. These can be caused by variation or oscillation in the illumination intensity of the LED, spatial inhomogeneity in the illumination, or associated variations in the expression of the fluorescent indicator in the cortical layer of the animal. To compensate for these factors and to obtain comparable values among animals on different days, the fluorescence signal is expressed as $\Delta F/F_0$. This value is calculated using the following formula: $\frac{\Delta F}{F_0} = \frac{(F - F_0)}{F_0}$, where F is the raw value of the fluorescence signal for the given pixel, while F_0 represents the

reference fluorescence that is the mean fluorescence per pixel over time. Generally this ratio is reported as a percentage.

3. In general, the fluorescence recorded with GCaMP indicators can be contaminated by nonneural phenomena such as hemodynamic response or movement artifacts that can lead to a wrongful interpretation of the fluorescence signal [32–34]. To remove the hemodynamic response, one can illuminate the sample at 530 nm, corresponding to the isosbestic point in the absorption spectrum of the hemoglobin [32, 33]. At this wavelength, the two forms, oxy and deoxyhemoglobin, exhibit the same absorption value allowing the reflectance signal to be measured regardless of the oxygen concentration contained in the hemoglobin. Therefore, with an additional light source, both the fluorescence and reflectance signal can be acquired by strobing illumination between the 485 nm and the 525 nm LEDs (blue to excite the GCaMP6f and green for the reflectance signal) it is possible to obtain a stroboscopic illumination of the sample and record the fluorescence signal (relating to the calcium activity) and the reflectance signal (relating to the hemodynamic contribution) by training a single line of transgenic animals. To eliminate the hemodynamic contribution it is sufficient to perform the ratio between the two recorded signals [33, 34]. Alternatively, one can employ postprocessing techniques that rely on decomposition algorithms and can identify the contribution of latent sources to the recorded fluorescence. In this context, it is possible to discriminate the neuronal activity from the hemodynamic contribution using the nonnegative matrix factorization (NMF) source separation technique [17, 35, 36]. To discriminate the hemodynamic contribution, there are other methods of signal decomposition such as the PCA/ICA technique, which allows one to manually exclude the components dependent on the blood flow activity or the motion artifacts caused by the vaso-constriction/dilation of the vascular structures [15].

Finally, to reduce the respiratory and heart rate contribution, it is advisable to filter the calcium signal with a 9 Hz low-pass filter.

After the preprocessing phases, although the analysis depends on the scientific question that the experimenter wants to address, the core of such analyses should be devoted to the investigation of the relationship between the fluorescence signal and the kinematics measures from the behavioral task. In our work, first, we wanted to know how many cortical areas are involved in the generation of the RtG movement. To answer this question, for each trial, we select several frames before and after the time of the movement onset and average these frames across trials to obtain an activation map [17]. The number of frames before and after the onset should be big enough to show that the activity in response to the movement starts from baseline and returns to it after the completion of the motor task. By generating these maps, one can directly observe all areas that are involved in the generation of the task, separate them into those areas active before and after movement onset, or identify the areas activated by correct or incorrect execution of the movement.

To understand how the neural activity in the different regions of the cortex is related to the motor output, we parcel the cortex according to the Allen reference atlas (Fig. 4) and extract a calcium fluorescence trace for each region by averaging the signal over all pixels of the region (Fig. 5a) *see* **Note 19**. Features of this signal are then directly correlated with the kinematics parameters. For example, in our work, we found a significant negative association between the PeakMax (the highest value of the fluorescence signal in a time window around the movement onset) and the pathlength of the movement (the total length of the path performed in one RtG) in the contralateral Caudal Forelimb Area (CFA) (Fig. 5b). In this way, it is possible to directly link features of the calcium signal dynamics with the kinematics parameters suggesting that the neural activity of a particular cortical region could participate in the encoding of the motor output.

Moreover, it is possible to investigate how the motor output can influence the neural dynamics between cortical regions (for example, functional connectivities (FC)). To this extent, one can compute pairwise cross-correlation matrices (Fig. 5c) by using measures of association among time-series, such as the Pearson correlation coefficient, between fluorescence signals of pairs of parceled cortical regions at the single-trial level. By repeating these measurements in different time-windows ranging from before to after the movement onset, it is possible to investigate how the FC rearranges or prepares with respect to the generation of the movement output. Finally, to extract commonalities in the way the regions associate with one another it is possible to use successive analysis directly on the cross-correlation matrices by rearranging the rows of the matrix using clustering algorithms such as the Hierarchical Clustering Analysis.

4 Notes

1. The application is optional; it depends on the required image resolution.
2. In case a liquid reward is provided, water restriction in rodents is well tolerated and is used as a motivational mechanism in motor learning experiments of different tasks such as the RtG [14, 15, 26, 27, 37]. The advantage of providing this type of reward lies in being able to adjust the quantity both in terms of

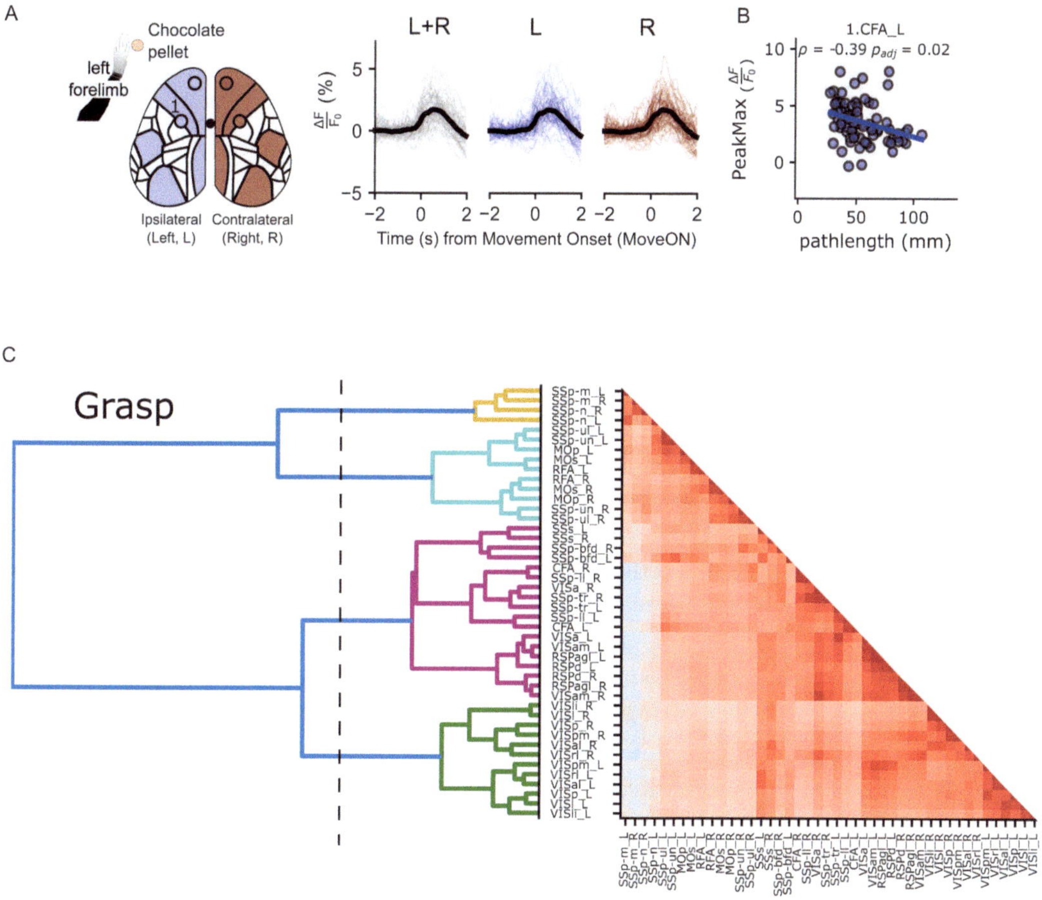

Fig. 5 Example of analyses of wide-field calcium imaging during the execution of a skilled voluntary movement: (**a**) Calcium fluorescence traces extracted from the cortical hemispheres. Image of the whole cortex with Allen's parcellation mask. Calcium fluorescence trace extracted from both (gray) left (blue) and right (red) hemispheres. (**b**) Feature of the calcium signal correlated with the kinematic parameters (PeakMax vs. Pathlength) in the contralateral Caudal Forelimb Area (CFA). (**c**) Functional Connectivity: cross-correlation matrices to study how motor output influences neural dynamics between different cortical regions. (Image modified from Quarta et al. [17])

the size of the drop and in terms of palatability, acting on the caloric intake by changing the type of liquid (water, sugared water, milk, or sweetened condensed milk). Considering a daily water consumption of about 5.8 ± 0.2 mL [26], and a water reward during the task of about 5–10 μL per drop, it is possible to provide a high number of rewards obtaining a considerable amount of repetition of movement in each recording session. Delivering a reward of this magnitude requires syringe delivery systems that have a thin spout with very fine control of the amount of liquid to deliver the exact amount of reward.

3. Compared to liquid rewards, tasks using pellets may result in fewer repetitions of the motor task. Despite this, the pellet-based solid reward is a widely used method for investigating RtG and it allows for investigating the object component of RtG behavior [17, 27, 38–42].
4. It is important to monitor the weight of the animals daily, both before and after training. Monitoring the weight, in addition to controlling the weight loss and preventing it from falling below 20% of the original weight, has the purpose of evaluating the animal's motivation.
5. Suppression of the nociceptive reflex following paw pinch is evaluated to determine the optimal depth of general anesthesia.
6. When fixing the animal's head, make sure that the pressure exerted by the two ear bars is not excessive to cause exophthalmos. If this occurs, change the position of the two ear bars.
7. Before the start of the surgery, the apparatus was calibrated to correctly align the animal's head to the rostral-caudal and mid-lateral planes.
8. Gently scratch the skull with a blade. This procedure facilitates the partial penetration of the dental cement into the skull bone allowing it to obtain a stable lining of the skull and avoid possible infections.
9. Identifying the two anatomical landmarks, bregma and lambda, is important for the alignment phase when analyzing calcium signal data.
10. During these steps, it is important to pay attention to two steps: avoid forming air bubbles below the coverslip glasses and apply the head-post in a position parallel to the surface of the skull. The presence of air bubbles would compromise image quality during calcium imaging, and a distortion of the aluminum support bar would cause an incorrect position of the animal's head with respect to the behavioral apparatus; this could complicate learning the RtG task during training sessions.
11. The cages for housing (Tecniplast S.p.A., Varese, Italy), used in our laboratory, are made of transparent plastic and equipped with a lid consisting of a frame covered with a polyester filter sheet with high filtering efficiency towards atmospheric dust (on particles between 8 and 10 μm). Although the animals are housed individually, the shape of the cage allows the animals to maintain eye contact and to receive the olfactory stimuli of the other subjects inside the enclosure.
12. Considering that single housing could increase the animal's stress levels, it is important to create an enriched environment inside the cage, making sure to use soft materials (such as

nesting material) and avoiding the introduction of small-sized plastic tubes or plastic houses where they could get stuck with the head-post causing it to detach.

13. The 470 nm LED to illuminate the dorsal surface of the skull and excite the GCaMP indicator should be set at an intermediate intensity to visualize the visual field exhaustively. In particular, at very low intensity the LED emits a flickering light, while at maximum it may lead to saturation of the fluorescence signal which over time can cause photobleaching of the indicator. To be sure that you have not damaged the indicator due to photobleaching, it is advisable to periodically acquire a recording of the calcium signal and check that there are no excessive variations in the raw fluorescence values between the start and the end of the recording or between recordings.
14. If the animal loses interest in performing the licking or RtG attempts, it may be helpful to encourage the animal to complete the activity by placing a pellet of chocolate close to the animal's nose or mouth by using a plastic tweezer.
15. Since in the early stages of learning, the animal will not immediately perform licking or RtG attempts, instead of starting with the automated version of the training, to facilitate learning, it may be useful to start with the manual version. In our setup, in the first days of training, the behavioral task was controlled by the operator using a button. Pressing the button results in the emission of the sound stimulus simultaneous with the lighting of the red LED followed by the rotation of the table (200 milliseconds later) to advance a single chocolate pellet at a time. In this way, there will be no further rotation until the operator presses the button again. The manual version makes it possible to control the learning of the task at every stage, correcting (if necessary) the execution of the movement. Since the immobility of the animal, with both paws on the perch, was a fundamental condition before starting the task, the manual version gave the operator the opportunity to wait and ensure that the animal met these behavioral requirements. This controlled version allows the motor task to be learned correctly without the introduction of incorrect behavioral habits (e.g. paw resting on the table, paws not resting on the perch, paw movement on the perch, grooming). When the movement has been learned, it is possible to switch to the automated version and continue with the animal's training. Using this dual version, fewer invalid or incorrect movements will be discarded during the automated version.
16. Tracing movements by hand requires a great amount of time depending on the total number of trials to be traced. In general, depending on the movement/behavior to be analyzed,

there exist commercial and noncommercial software that can automatically track the behavioral outcome. Above all, DeepLabCut is widely employed for tracking anatomical landmarks [43–46]. However, besides the hardware requirements, for the software to work very well, it is important to provide a good number of images as examples and it is also important to compare the results of DeepLabCut to that of an operator to ensure reproducible and valid tracking. This means that a good deal of trials needs to be scored by a skilled human operator.

17. Since bregma and lambda are used in the alignment phase, it is important to be as precise as possible in the identification and labeling of these two points during the surgery.
18. During the alignment, in addition to the two anatomical landmarks, it is advisable to use additional points as reference (e.g. the vasculature can be used to check that the alignment is correct).
19. An alternative anatomical atlas or decomposition method, like Spatial Independent Component investigation (sICA), could be used to segment the cortex for this kind of investigation [15]. Both methods offer the advantage of data dimensionality reduction compared to a pixel wise analysis. Using decomposition algorithms such as sICA has the advantage of providing functional parcels with the drawback that it is highly dependent on the activity observed in the experiment. Instead, employing an anatomical atlas has the advantage of being activity-independent. However, it is important to check anatomical regions that span over large portions of the cortex as the single pixels of the area could show heterogeneous behavior.

References

1. Evarts EV (1968) Relation of pyramidal tract activity to force exerted during voluntary movement. J Neurophysiol 31:14–27. https://doi.org/10.1152/jn.1968.31.1.14
2. Georgopoulos A, Kalaska J, Caminiti R, Massey J (1982) On the relations between the direction of two-dimensional arm movements and cell discharge in primate motor cortex. J Neurosci 2:1527–1537. https://doi.org/10.1523/JNEUROSCI.02-11-01527.1982
3. Iwaniuk AN, Whishaw IQ (2000) On the origin of skilled forelimb movements. Trends Neurosci 23:372–376. https://doi.org/10.1016/S0166-2236(00)01618-0
4. Whishaw IQ (2003) Did a change in sensory control of skilled movements stimulate the evolution of the primate frontal cortex? Behav Brain Res 146:31–41. https://doi.org/10.1016/j.bbr.2003.09.027
5. Graziano MSA, Aflalo TN (2007) Mapping behavioral repertoire onto the cortex. Neuron 56:239–251. https://doi.org/10.1016/j.neuron.2007.09.013
6. Wise SP, Evarts EV (1981) The role of the cerebral cortex in movement. Trends Neurosci 4:297–300. https://doi.org/10.1016/0166-2236(81)90093-X
7. Akerboom J, Chen T-W, Wardill TJ et al (2012) Optimization of a GCaMP calcium indicator for neural activity imaging. J Neurosci 32:13819–13840. https://doi.org/10.1523/JNEUROSCI.2601-12.2012
8. Chen T-W, Wardill TJ, Sun Y et al (2013) Ultrasensitive fluorescent proteins for imaging neuronal activity. Nature 499:295–300. https://doi.org/10.1038/nature12354
9. Dana H, Chen T-W, Hu A et al (2014) Thy1-GCaMP6 transgenic mice for neuronal

population imaging in vivo. PLoS One 9: e108697. https://doi.org/10.1371/journal.pone.0108697

10. Nakai J, Ohkura M, Imoto K (2001) A high signal-to-noise Ca2+ probe composed of a single green fluorescent protein. Nat Biotechnol 19:137–141. https://doi.org/10.1038/84397
11. Ohkura M, Matsuzaki M, Kasai H et al (2005) Genetically encoded bright Ca2+ probe applicable for dynamic Ca2+ imaging of dendritic spines. Anal Chem 77:5861–5869. https://doi.org/10.1021/ac0506837
12. Tallini YN, Ohkura M, Choi B-R et al (2006) Imaging cellular signals in the heart in vivo: cardiac expression of the high-signal Ca2+ indicator GCaMP2. Proc Natl Acad Sci USA 103: 4753–4758. https://doi.org/10.1073/pnas.0509378103
13. Tian L, Hires SA, Mao T et al (2009) Imaging neural activity in worms, flies and mice with improved GCaMP calcium indicators. Nat Methods 6:875–881. https://doi.org/10.1038/nmeth.1398
14. Allegra Mascaro AL, Conti E, Lai S et al (2019) Combined rehabilitation promotes the recovery of structural and functional features of healthy neuronal networks after stroke. Cell Rep 28:3474–3485.e6. https://doi.org/10.1016/j.celrep.2019.08.062
15. Makino H, Ren C, Liu H et al (2017) Transformation of cortex-wide emergent properties during motor learning. Neuron 94:880–890.e8. https://doi.org/10.1016/j.neuron.2017.04.015
16. Montagni E, Resta F, Conti E et al (2018) Wide-field imaging of cortical neuronal activity with red-shifted functional indicators during motor task execution. J Phys D Appl Phys 52: 074001
17. Quarta E, Scaglione A, Lucchesi J et al (2022) Distributed and localized dynamics emerge in the mouse neocortex during reach-to-grasp behavior. J Neurosci 42:777–788. https://doi.org/10.1523/JNEUROSCI.0762-20.2021
18. Ren C, Komiyama T (2021) Wide-field calcium imaging of cortex-wide activity in awake, head-fixed mice. STAR Protoc 2:100973. https://doi.org/10.1016/j.xpro.2021.100973
19. Allen WE, Kauvar IV, Chen MZ et al (2017) Global representations of goal-directed behavior in distinct cell types of mouse neocortex. Neuron 94:891–907.e6. https://doi.org/10.1016/j.neuron.2017.04.017
20. Salkoff DB, Zagha E, McCarthy E, McCormick DA (2020) Movement and performance explain widespread cortical activity in a visual detection task. Cereb Cortex 30:421–437. https://doi.org/10.1093/cercor/bhz206
21. Donello JE, Loeb JE, Hope TJ (1998) Woodchuck hepatitis virus contains a tripartite post-transcriptional regulatory element. J Virol 72: 5085–5092
22. Skinner BF (1938) The behavior of organisms: an experimental analysis. Appleton-Century, Oxford
23. Galea JM, Mallia E, Rothwell J, Diedrichsen J (2015) The dissociable effects of punishment and reward on motor learning. Nat Neurosci 18:597–602. https://doi.org/10.1038/nn.3956
24. Wächter T, Lungu OV, Liu T et al (2009) Differential effect of reward and punishment on procedural learning. J Neurosci 29:436–443. https://doi.org/10.1523/JNEUROSCI.4132-08.2009
25. Schultz W (2015) Neuronal reward and decision signals: from theories to data. Physiol Rev 95:853–951. https://doi.org/10.1152/physrev.00023.2014
26. Bachmanov AA, Reed DR, Beauchamp GK, Tordoff MG (2002) Food intake, water intake, and drinking spout side preference of 28 mouse strains. Behav Genet 32:435–443. https://doi.org/10.1023/A:1020884312053
27. Guo ZV, Hires SA, Li N et al (2014) Procedures for behavioral experiments in head-fixed mice. PLoS One 9. https://doi.org/10.1371/journal.pone.0088678
28. Suckow MA, Danneman P, Brayton C (2001) The laboratory mouse. CRC Press, Boca Raton
29. Resta F, Montagni E, de Vito G et al (2022) Large-scale all-optical dissection of motor cortex connectivity shows a segregated organization of mouse forelimb representations. Cell Rep 41:111627. https://doi.org/10.1016/j.celrep.2022.111627
30. Silasi G, Xiao D, Vanni MP et al (2016) Intact skull chronic windows for mesoscopic wide-field imaging in awake mice. J Neurosci Methods 267:141–149. https://doi.org/10.1016/j.jneumeth.2016.04.012
31. Dong HW (2008) The Allen reference atlas: a digital color brain atlas of the C57Bl/6J male mouse. Wiley, Hoboken
32. Ma Y, Shaik MA, Kozberg MG et al (2016) Resting-state hemodynamics are spatiotemporally coupled to synchronized and symmetric neural activity in excitatory neurons. Proc Natl Acad Sci USA 113:E8463–E8471. https://doi.org/10.1073/pnas.1525369113
33. Ma Y, Shaik MA, Kim SH et al (2016) Wide-field optical mapping of neural activity and

brain haemodynamics: considerations and novel approaches. Philos Trans R Soc B Biol Sci 371:20150360. https://doi.org/10.1098/rstb.2015.0360
34. Scott BB, Thiberge SY, Guo C et al (2018) Imaging cortical dynamics in GCaMP transgenic rats with a head-mounted widefield macroscope. Neuron 100:1045–1058. https://doi.org/10.1016/j.neuron.2018.09.050
35. Pnevmatikakis EA, Soudry D, Gao Y et al (2016) Simultaneous denoising, deconvolution, and demixing of calcium imaging data. Neuron 89:285–299. https://doi.org/10.1016/j.neuron.2015.11.037
36. Saxena S, Kinsella I, Musall S et al (2020) Localized semi-nonnegative matrix factorization (LocaNMF) of widefield calcium imaging data. PLoS Comput Biol 16:e1007791. https://doi.org/10.1371/journal.pcbi.1007791
37. Galiñanes GL, Bonardi C, Huber D (2018) Directional reaching for water as a cortex-dependent behavioral framework for mice. Cell Rep 22:2767–2783. https://doi.org/10.1016/j.celrep.2018.02.042
38. Whishaw IQ, Faraji J, Kuntz J et al (2017) Organization of the reach and grasp in head-fixed vs. freely-moving mice provides support for multiple motor channel theory of neocortical organization. Exp Brain Res 235:1919–1932. https://doi.org/10.1007/s00221-017-4925-4
39. Farr TD, Whishaw IQ (2002) Quantitative and qualitative impairments in skilled reaching in the mouse (Mus musculus) after a focal motor cortex stroke. Stroke 33:1869–1875. https://doi.org/10.1161/01.STR.0000020714.48349.4E
40. Lai S, Panarese A, Spalletti C et al (2015) Quantitative kinematic characterization of reaching impairments in mice after a stroke. Neurorehabil Neural Repair 29:382–392. https://doi.org/10.1177/1545968314545174
41. McKenna JE, Whishaw IQ (1999) Complete compensation in skilled reaching success with associated impairments in limb synergies, after dorsal column lesion in the rat. J Neurosci 19:1885–1894. https://doi.org/10.1523/JNEUROSCI.19-05-01885.1999
42. Nica I, Deprez M, Nuttin B, Aerts J-M (2017) Automated assessment of endpoint and kinematic features of skilled reaching in rats. Front Behav Neurosci 11:255. https://doi.org/10.3389/fnbeh.2017.00255
43. Bova A, Gaidica M, Hurst A et al (2020) Precisely timed dopamine signals establish distinct kinematic representations of skilled movements. eLife 9:e61591. https://doi.org/10.7554/eLife.61591
44. Mathis A, Mamidanna P, Cury KM et al (2018) DeepLabCut: markerless pose estimation of user-defined body parts with deep learning. Nat Neurosci 21:1281–1289. https://doi.org/10.1038/s41593-018-0209-y
45. Nath T, Mathis A, Chen AC et al (2019) Using DeepLabCut for 3D markerless pose estimation across species and behaviors. Nat Protoc 14:2152–2176. https://doi.org/10.1038/s41596-019-0176-0
46. Forys BJ, Xiao D, Gupta P, Murphy TH (2020) Real-time selective Markerless tracking of forepaws of head fixed mice using deep neural networks. eNeuro 7:ENEURO.0096-20.2020. https://doi.org/10.1523/ENEURO.0096-20.2020

Chapter 8

Technical Considerations for Widefield Optical Imaging During Visuomotor Behaviors

Daisuke Shimaoka

Abstract

In this chapter, we consider visuomotor behaviors in a broad sense as trained or natural motor actions based on external visual information. These behaviors are deemed as involving coordinated neural activity across brain areas, including the visual and motor system. Widefield optical imaging is a valuable tool for studying coordinated neural activity across these areas at the mesoscale. However, visuomotor behaviors can cause spurious imaging signal, making widefield imaging during visuomotor behavior technically challenging. This chapter is intended to provide an overview of the technical considerations for widefield optical imaging to study visuomotor behaviors. In Subheading 1, we discuss unique capabilities of widefield imaging that can be used to study coordinated brain activity within and across multiple areas. In Subheading 2, we survey experimental strategies for achieving widefield imaging during visuomotor behaviors, including recording and visual-stimulation systems. Subheading 3 discusses technical challenges that often arise during widefield imaging of visuomotor behaviors, including hemodynamic and mechanical artifacts that can distort the optical signal. We discuss how to reduce these imaging artifacts through experimental and data analysis approaches that are essential for improving the accuracy and reliability of imaging data.

Key words Visuomotor behavior, Multispectral imaging, Mesoscale, Hemodynamics, Head-fixation, Head-mounted microscope, Temporal multiplexing, Spectral multiplexing

1 Introduction: Why Widefield Optical Imaging for Visuomotor Behaviors

Visuomotor behaviors involve multistage brain processing, and are essential for daily activities such as foraging, escaping, and social interactions in the natural environment. They can require a combination of predicting upcoming visual information, detecting or discriminating external visual information, integrating information from vision and other sensory modalities, making decisions, and executing motor actions. In many cases, movements must occur within a few hundred milliseconds, and then it may be necessary to reevaluate these actions based on feedback. In primates, part of this process is mediated by the "dorsal visual stream," a series of hierarchically organized areas located in the dorsal part of the cortex,

James E. Niemeyer et al. (eds.), *Awake Behaving Mesoscopic Brain Imaging*, Neuromethods, vol. 214,
https://doi.org/10.1007/978-1-0716-4120-0_8,

but the existence and role of this stream in other species, such as rodents, has been debated [1, 2]. Hence, to study the neural dynamics underlying visuomotor behaviors, it is desirable to monitor millisecond-scale dynamics from multiple cortical areas in the dorsal visual stream and beyond. This combination of high temporal resolution and broad spatial sampling has been impossible to achieve with traditional electrophysiological methods, but is feasible with widefield optical imaging.

This section compares the advantages of widefield imaging for studying visuomotor behaviors with other recording techniques such as unit electrophysiology, multiphoton imaging, and functional magnetic resonance imaging (fMRI). Widefield imaging is often used to study "mesoscale" functional structures at a scale down to sub-millimeter and can also resolve temporal events down to the near-millisecond range depending on the selection of neural activity sensors and recording configuration [3]. These capabilities allow widefield imaging to study mesoscale functional structures such as columnar organization and inter-areal communication, as well as their temporal evolution over time, which are important for understanding visuomotor behaviors. This chapter focuses on widefield fluorescence imaging with activity sensors such as genetically encoded calcium indicators (GECIs), genetically encoded voltage indicators (GEVIs), genetically encoded glutamate indicators, and synthetic voltage-sensitive dyes (VSDs), as well as imaging of diffuse reflectance to probe hemodynamic activity.

1.1 Mesoscale Structures

Widefield optical imaging emerged as a potent tool for investigating the intricate functional composition of cortical regions. It plays a pivotal role in delineating the boundaries of visual and motor cortical domains, a prerequisite for comprehending their interplay during visuomotor activities. With its superior spatial resolution compared to other imaging modalities like fMRI, widefield imaging has contributed to resolving functional structures within an area at sub-millimeter scale of multiple species [4, 5]. Noteworthy instances of these architectures include cortical columns, intricate clusters of neurons forming a cylindrical structure across cortical layers and retinotopic organization, orchestrating the spatial arrangement of neurons according to their receptive field locations in the visual field. A recent breakthrough lies in its capacity to capture functional configurations during motor behavior, unearthing the dynamic patterns activated during spontaneous running [6] (Fig. 1).

In addition to resolving functional structures of cortical areas, widefield imaging has been employed to detect changes in the structure over time. For example, Kenet et al. have discovered populations of neurons in cat V1 encompassing a set of dynamically switching orientation columns [7]. It has been also used to study

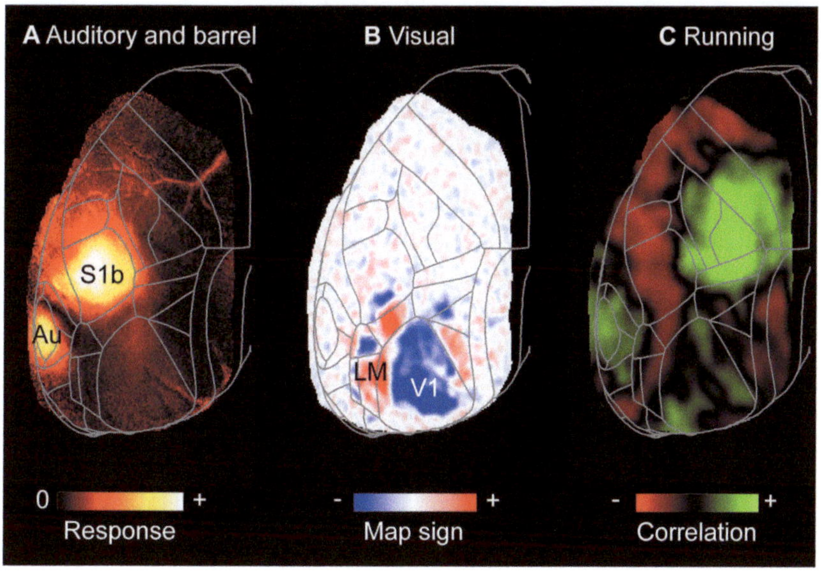

Fig. 1 Functional structures as unveiled through mesoscale optical imaging of a mouse cortical hemisphere. (**a**) Activation patterns in the S1 barrel cortex (S1b) and Auditory cortex (Au), triggered by contralateral whisker stimulation via air puff. (**b**) Higher and lower visual cortical areas, discerned via visual field sign analysis. (**c**) Distinctive depolarization patterns resulting from spontaneous treadmill running while the mouse's head is immobilized. (**a**–**c**, reproduced from [6], licensed under CC-BY 4.0)

the spread of activity across the cortical surface in response to spontaneous or evoked stimuli [8–13], and propagation of neural activation across multiple spatial scales (neuronal avalanches) [14]. By using genetic engineering techniques to selectively express activity sensors in specific cell types and layers (e.g., [15, 16]), it is now possible to study these functional signals in the defined population of neurons [17]. These observations, which can describe cell type- and layer-specific mesoscale dynamics, are closely aligned with neural mass models that describe neural population dynamics on a cortical sheet (e.g., [18]). On the other hand, neuroanatomical and neurohistological data are rapidly accumulating in some mammalian species (e.g., mouse: https://mouse.brain-map.org; marmoset: https://www.marmosetbrain.org; macaque: https://www.blueprintnhpatlas.org). Combining these data with brain imaging data, including that by widefield optical imaging, it is becoming possible to construct more realistic, quantitative models of neuronal activity. Such models, refined with data assimilation techniques, are expected to forecast upcoming mesoscale neural activity.

1.2 Inter-areal Communication

Visuomotor behavior involves coordinated activity across brain regions, such as motor and visual cortices. This coordinated activity may be quantified in terms of synchrony, coherence, and causality. Widefield imaging has been used to characterize these network properties (e.g., [19, 20]). In addition, these properties can be further analyzed with graph-theoretical approaches, to identify critical brain areas and sub-areas such as network hubs. Functional connectivity analysis has been applied to data of many recording modalities, such as electroencephalography (EEG) and electrocorticography (ECoG). With its advantageous spatial resolution, the connectivity analysis of widefield data is expected to better inform the location of the functionally connected brain areas. For instance, Kuroki et al., reported a confined cortical region between visual and somatosensory areas to exhibit hub-like characteristics in response to simultaneous visual, auditory, and somatosensory stimulation, which may play a critical role in multisensory integration [20].

The coordinated activity across brain regions can be also evaluated through quantifying spread of activity over time and across cortical space. Indeed, widefield imaging has revealed stereotypical spatiotemporal patterns, called motifs, across multiple cortical areas. These stereotypical motifs are shown to explain a more significant amount of widefield imaging signals than the functional connectivity does [21]. Several methods have been applied to extract these motifs from imaging data, including methods based on similarity to anatomical data [9], unsupervised clustering (k-means, hierarchical clustering) [13], and dimensionality reduction (e.g., Independent Component Analysis) [21]. Motifs were initially observed in sleep, anesthetized, and quiescent awake states, and their functional and behavioral roles have been debated. These activities may represent "mental imageries": spontaneous representation of the environment or expectation of the upcoming sensory information. More recently, motifs have been also observed during visuomotor behaviors, but current evidence suggests that they are not predictive of detection and discrimination behavior [21, 22].

1.3 Expandability

Widefield imaging is a recording technique that is relatively simple to set up and can be used while the space above the brain being studied is unobstructed. This makes it easy to use in conjunction with other recording or stimulation techniques. For example, it has been paired with single-unit electrophysiology to study the relationship between single-unit spikes and widespread activity across brain areas [23]. Barson et al. reported ways to achieve simultaneous widefield and two-photon imaging to reveal the relation between local network activities within an area and global activities across areas [24]. Widefield calcium imaging has also been paired with intrinsic-signal imaging via multispectral recording approaches to reveal neurovascular coupling [25]. Widefield imaging has also been combined with brain stimulation techniques to investigate causal interactions between distant brain areas [26, 27]. These combinations of widefield imaging with other techniques enhance its capabilities and can help to understand the neural network dynamics involved in visuomotor behaviors.

2 Materials

There are several experimental strategies for achieving widefield imaging during visuomotor behaviors. The most common method is the use of a tandem-lens imaging system in combination with head fixation (manual restriction of head movement), which stabilizes the head during in vivo imaging while still permitting the animal to express a range of behaviors. This section covers the standard imaging system adopted for widefield imaging, equipment for behavioral reporting under head fixation, and the recent development of head-mounted miniature microscopes, which allow for a wider range of behaviors in the animal.

2.1 Imaging in Head-Fixed Condition

2.1.1 Imaging System

The standard optical design used for widefield imaging employs a tandem-lens design [28] (Fig. 2a), which consists of two lenses (typically macroscope lenses)—one facing the imaging sensor housed in a camera (the reversed or objective lens), and the other facing the brain sample (called primary or imaging lens). The sample at the focal plane of the primary lens is projected to infinity, and the reversed lens focuses the image to its focal plane on the imaging sensor. Between the two tandem lenses lies an infinity-focused image path. The focal length of the two lenses decides the effective magnification. This tandem-lens design efficiently collects photons from the brain samples and allows low-intensity light illumination (e.g., light-emitting diode (LED)), reducing the bleaching of the neural activity sensor. The infinity-focused image path between the lenses can accommodate spectral filters and dichroic mirrors, enabling the collection of photons in a specific spectral range by the imaging sensor. This path can be further split into multiple horizontal paths by adding a dichroic mirror, enabling multi-camera imaging for Fluorescence Resonance Energy Transfer (FRET)-based activity sensors (e.g., [29]) and uniform illumination of the brain from above (epi-illumination). Köhler illumination is often incorporated to improve the homogeneity of illumination on the sample further [28].

There have been two main approaches to choosing the depth of field of the imaging system. Traditionally, shallow (< 100 um) depth of field was preferred and achieved with moderate-to-high (typically >0.3) numerical aperture (NA) lenses, which are suitable for samples with brain signals at a limited range of depths, such as brain slices. However, for samples with brain signals occurring across depths, imaging systems with deep (>1 mm) depth of field are more appropriate, such as for large ($> 2 \times 2$ mm), curved cortical surfaces or cortical tissue with target fluorophores expressed across multiple layers. In addition to using low (<0.1) NA lenses, the effective depth of field can also be increased through the use of light polarization [30, 31]. This involves shining a

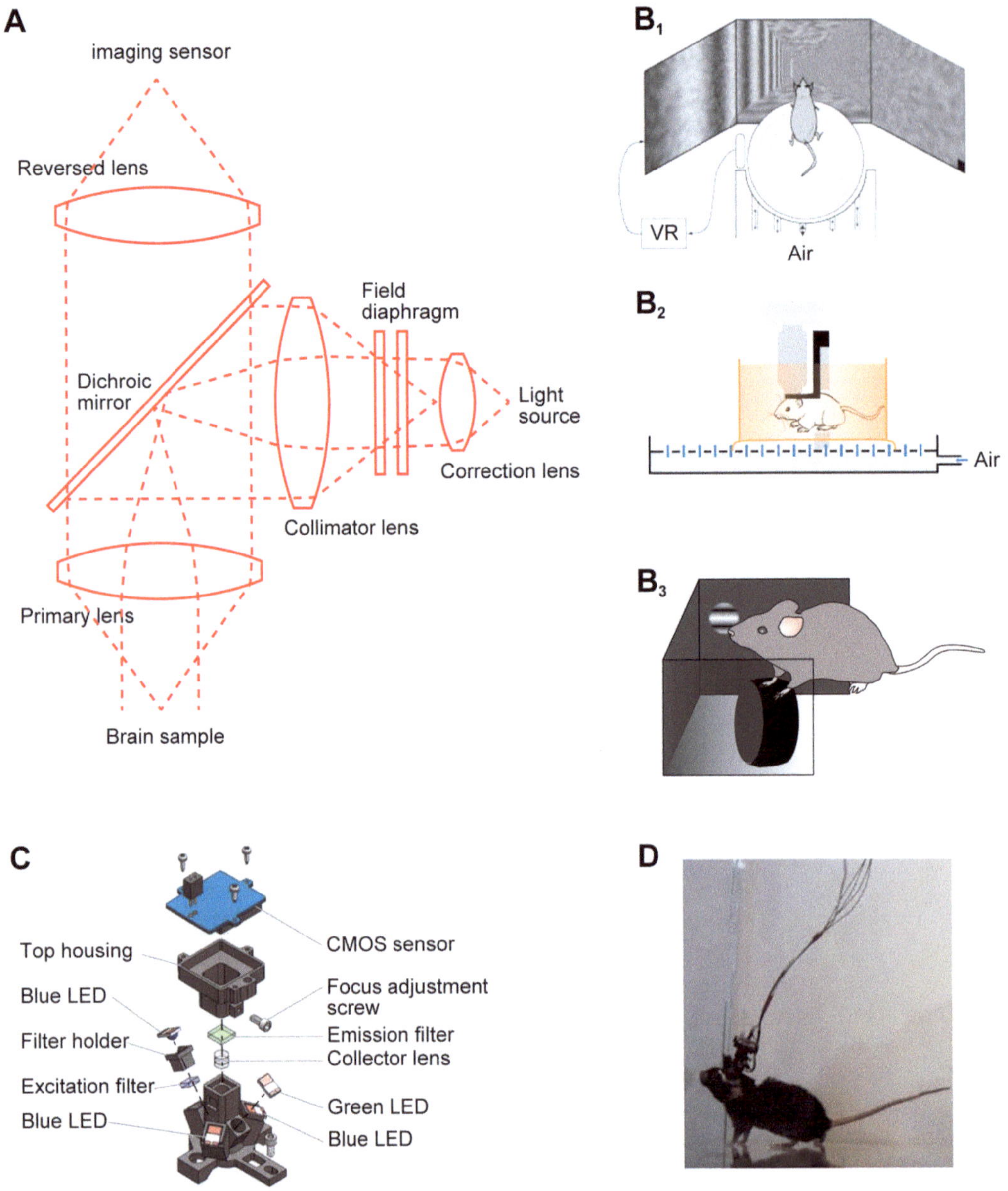

Fig. 2 Equipment for widefield optical imaging during visuomotor behavior. (**a**) Schematic diagram of tandem-lens macroscope with Köhler epi-illumination, often paired with head-fixation. (**b**) Examples of apparatus for behavioral reporting during head-fixation. B_1. Spherical treadmill, paired with virtual-reality. B_2. Flat-floored air-lifted platform. B_3. Steering wheel. (**c**) Imaging system with head-mounted microscope for mice. (**d**) A mouse bearing the head-mounted microscope engaging in natural behavior. (B_1 reproduce from [71], licenced under CC-BY 4.0, B_2 reproduced from [35] with permission from MyJOVE Corporation, c, d reproduced from [49] with permission from Springer Nature)

linearly polarized beam perpendicular to the tissue and detecting the "cross-linear" photons that have lost their initial polarization and tend to reach deeper structures by using a linear polarizing filter in the infinity-focused image path. However, this method can also introduce contamination of optical signals from depths beyond the target layer, so the depth of field should be chosen case by case based on the distribution of the target fluorophore.

2.1.2 Apparatus for Behavioral Report

Here we survey the apparatus used for visuomotor behaviors under head-fixation, the standard condition for widefield imaging.

2.1.3 Treadmill (Fig. 2B$_1$)

Locomotion is a type of movement that occurs in many animals and can be studied using treadmills. There are several types of treadmills, such as belts [32], rotating disks [33], cylinders [34], and floating spheres with constant air flow ([32]). These treadmills can be combined with visual feedback or virtual reality systems, and they can have one or two degrees of freedom. The 1D version only detects forward and backward movements, while the 2D version also detects movements to the left and right. Once the distance from the head to the treadmill floor is appropriately adjusted, the head-fixed animal typically starts locomoting without needing to acclimate. The 2D version paired with a virtual reality system can provide a more immersive environment, but it may require more acclimation time. There are a few potential drawbacks of this approach: (1) Locomotion activates multiple motor and sensory systems, leading to widespread activation in various brain regions. To study the neural activity for each specific modality, it is essential to monitor each body part separately and regress out each body movement. (2) There is a mismatch between visual and vestibular cues. (3) In case of the 1D apparatus, there may be a large mismatch between the animal's motor plan and how it can actually navigate.

2.1.4 Flat-Floored Air-Lifted Platform (Fig. 2B$_2$)

The flat-floor system is similar to the floating ball system, but instead of a ball, the floor itself is made to float using air [35]. This allows the animal to explore the floor by moving the air-lifted floor itself. Unlike the floating ball system, the flat-floor system is in the real world and can be furnished with additional sensory cues such as odors and sounds.

2.1.5 Button/Lever

Utilizing buttons or levers as experimental tools is a prevalent technique when investigating animal behavior including visuomotor behavior. This approach becomes particularly pertinent for species possessing advanced hand dexterity, such as primates and rodents.

For primates, Friedman et al., employed a reach-to-grasp task to probe cortical regions that govern sequential motor phases in macaques [36]. Furthermore, lever pressing has been instrumental in designing visuomotor tasks for marmosets [37]. For rodents, the lever pressing paradigm has facilitated the elucidation of sequential motor phases, providing insights into the microcircuitry for the motor behavior [38]. Additionally, this methodology has been pivotal in exploring how sensory cues influence motor actions [39].

2.1.6 Steering Wheel (Fig. 2B$_3$)

In a steering wheel setup, a head-fixed animal can turn the wheel either to the left or right to report its choice behaviors based on the visual information on the screen. This system has been used for visual discrimination tasks in mice [40]. However, this method is prone to the animals' development of choice bias. The exact cause of this is not fully understood, but it is known that head position relative to the wheel is one of the influencing factors.

2.1.7 Eye-Tracking System

Eye tracking, particularly in animals like primates with advanced eye movement systems, offers insights into an animal's internal state. Researchers achieve this by training animals to fixate on specific locations within their visual field [41–43].

To track gaze position, two primary methods have been used: magnetic coil implantation in the eye and infrared camera systems that capture pupil movements. However, a limitation with conventional eye trackers is their reduced capability to detect microsaccades—swift eye movements smaller than 1 degree occurring several times per second. Addressing this, the digital Dual Purkinje Imaging (dDPI) eye tracker emerges as a breakthrough, offering unparalleled precision of 0.005 deg RMS. This advancement ensures the reliable detection of microsaccades and fixational drifts.

Recent applications of high-precision eye tracking, particularly in free-viewing scenarios, have leveraged the encoding modeling analysis to discern the receptive fields of individual cells [44]. Integrating this advanced eye-tracking technology with widefield imaging and the encoding modeling analysis [45] would facilitate resolving the population receptive fields (pRF) from each camera pixel.

2.2 Imaging with Head-Mounted Microscope

Brain recording experiments, including widefield imaging, have widely employed the head-fixation system combined with the behavioral reporting apparatus we have discussed above. However, there are still many research areas that cannot be accessed with this method, including the stimulation of an animal's vestibular system, exploration in three-dimensional space, and naturalistic social interactions with other animals. Moreover, head-fixation is known to induce stress, leading to an alteration of natural behavior [46].

To address these limitations, some researchers have developed head-mounted microscope systems that allow animals to move their heads with fewer constraints. One of the first of its kind is a fiber-based microscope [47]. This scope was capable of imaging

2×2 mm field of view at a single wavelength. Scott et al. [48] developed a miniaturized imaging system that brought the power of tabletop microscopes to the head-mounted domain. This system was composed of a 3D-printed plastic macroscope body that holds optics, multiple LEDs for fluorescence and reflectance imaging, a CMOS imaging sensor, and readout electronics [48]. It enabled the imaging of the dorsal part of one cortical hemisphere (7.8×4 mm field of view) in rats while they performed an evidence accumulation task, in which they report their perception by inserting their nose into one of three ports based on the flow of visual information over time. The system, including the headplate and cable, weighs 33 g, which is less than 10% of the typical weight of an adult rat, and is suitable for use with animals as heavy as rats, but not with lighter animals such as mice. More recently, Rynes et al. (2021) developed an even lighter imaging system (<4 g) that can image an 8×10 mm field of view (Fig. 2c, d) [49]. It encapsulates three LEDs for multispectral imaging, a single collimator lens, a CMOS sensor, and readout electronics. With this system, the two cortical hemispheres of mice were imaged during interactions with another mouse and during transitions from wakefulness to sleep. These miniaturized head-mounted microscopes are expected to be used for the study of visuomotor behaviors in the near future.

3 Methods

This section addresses the technical challenges that often arise when imaging the brain during visuomotor behaviors. These challenges include isolating the excitation and emission light from ambient light, and eliminating hemodynamic and mechanical artifacts that can distort the optical signals produced by neuronal activity. Potential solutions for these challenges are also discussed.

3.1 Light Contamination (Fig. 3, Note 5.1)

Visuomotor behaviors and widefield imaging exploit light for different purposes: the former is for stimulation of the photoreceptor in the retina, while the latter is for measuring brain activity through collection of fluorescence or reflectance lights. To accurately measure brain activity using light, it is essential to ensure that the imaging sensor does not detect the photons from the visual stimuli. In other words, photons from the two light sources should not be contaminated. There are three main strategies for preventing light contamination in this context.

3.1.1 Physical Segregation (Fig. 3A$_1$)

The photons collected at the imaging sensor for visual stimulation can be separated from other photons by using a physical screening device, such as a cone placed on the skull surrounding the imaging window. This cone blocks radiating and scattering light from the visual stimulation device, allowing only the light needed for

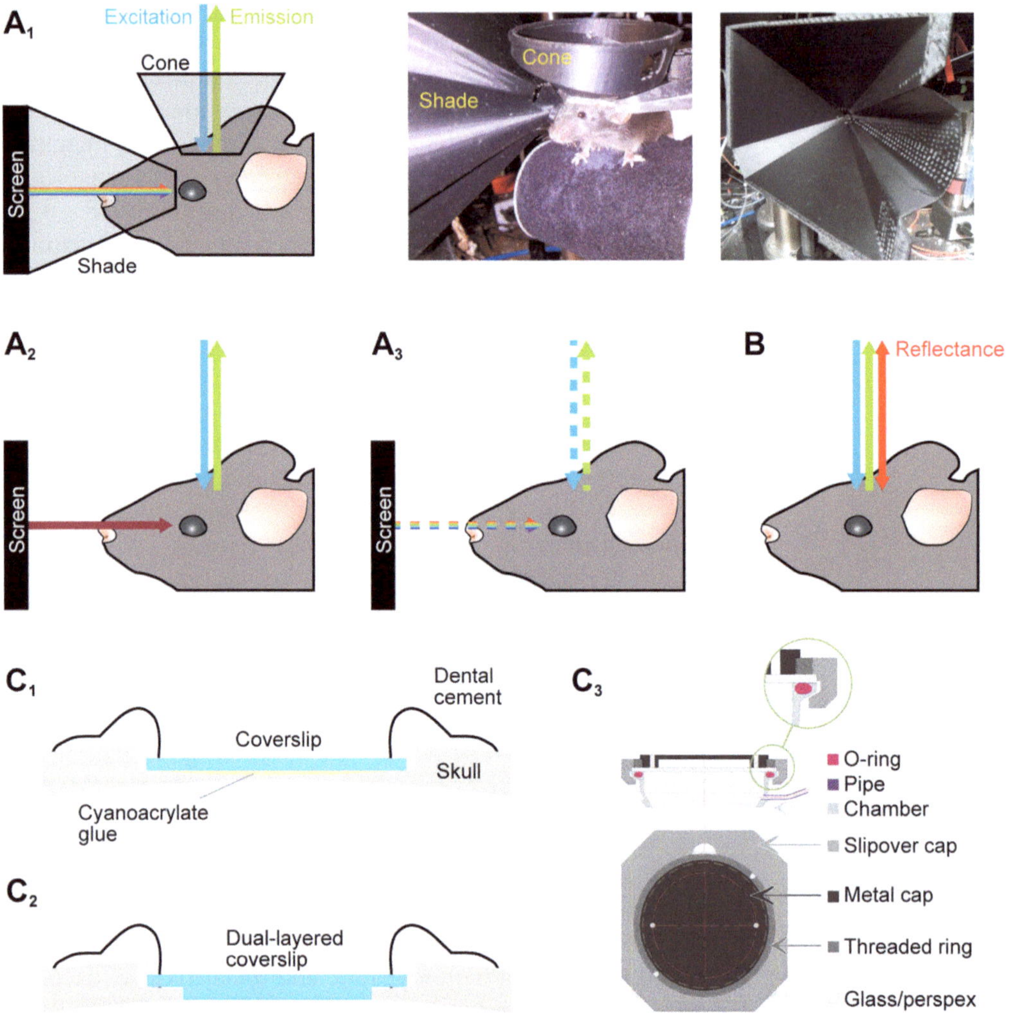

Fig. 3 Strategies for reducing imaging artifacts. (**a**) Strategies to reduce light contamination. A_1. (*left*) Physical segregation. (*middle*) An application of a light-isolation cone around the cranial window, and a light-isolation shade around the screen. A hole on the side of the cone allows direct access to the brain for simultaneous electrophysiological recording. A slit on the shade allows the fitting of an optical filter, further reducing spectral contamination. (*right*) The light-isolation shade, viewed from the side of the screen. A_2. In spectral segregation, the wavelength for visual stimulation is altered to be not overlapping with those for imaging. A_3. In temporal segregation, times for visual stimulation is altered to be not overlapping with those for imaging. (**b**) Multispectral imaging for simultaneous neuronal and hemodynamic imaging. For hemodynamic imaging, additional light source(s) is added (red). (**c**) Ways to suppress brain suppression. C_1. Permanent sealing of coverslip glued onto a thinned skull, secured with dental cement. The layer of dental cement is covered with black pigments to reduce light scattering. C_2. Permanent sealing of double-layered coverslip onto a cranial window with dental cement. C_3. Temporal sealing of coverslip with sliding-top cranial window. (C_3 reproduced from [72] with permission from Elsevier B. V.)

imaging to pass through. However, this approach is not immediately applicable when the imaging is combined with techniques that require access to the brain, such as extracellular unit recording. In these cases, a small port opened on the lateral wall of the cone may provide access to the brain. Alternatively, it is possible to prevent the light contamination with a shade covering the entire stimulation device. Such a device can preserve access to the brain, but it can cause distress in the animal being studied and may result in altered behavior and eye discharge.

3.1.2 Spectral Segregation (Fig. 3A$_2$)

In widefield fluorescence imaging, only photons within the emission wavelengths of the fluorophore are captured by the imaging sensor. In case of GCaMP imaging, for instance, photons at 530 nm are collected at the sensor. Photons outside of this wavelength range can be utilized for visual stimulation. One way to achieve segregation in the spectral domain is to limit the spectral range of the visual stimulation devices. These devices, such as liquid crystal displays and cathode-ray tubes, usually have red, green, and blue light emitters. If the target wavelength for imaging falls within the emission spectrum of one of these emitters, it may be possible to eliminate spectral overlap by removing that emitter. However, the emission spectra of these emitters often overlap, making this approach less effective. Using an additional optical filter in front of the visual stimulation device may help to reduce the spectral overlap further. One disadvantage of these spectral segregation strategies is that they alter the character of the visual stimuli, which can impact the visual system of the animal being studied. In some species, photoreceptor intensity is known to distribute heterogeneously across retinal surface [50]. Some brain areas are also known to be particularly responsive to specific wavelengths, e.g., color-sensitive columns in primates [51, 52], which may contribute to visually oriented motor actions [53]). Thus, it is vital to consider the effect of the altered spectrum in the visual stimuli on photoreceptors and brain areas that are particularly sensitive to specific wavelengths when implementing them. On the other hand, this approach allows for an unobstructed space between the brain and the primary lens, unlike the physical segregation approach (Fig. 3A$_1$).

3.1.3 Temporal Multiplexing (Fig. 3A$_3$)

Photons for the visual stimulation and for the widefield imaging can be also separated with a temporal multiplexing approach (similar to the technique employed for multiphoton imaging [54]). In this approach, the imaging light and the visual stimulation light are emitted intermittently, alternating with each imaging frame. The emission times of the two light sources are synchronized with the frame-grabbing times at the imaging sensor, and the imaging frames corresponding to times when only the excitation light is emitted are analyzed to probe brain activity. For this method to work, the light sources for visual stimulation and optical imaging must be capable of rapidly turning on and off in response to external control signal, and the imaging camera must support outputs for frame-grabbing times. Alternation of the two light sources can be achieved with a single-board microcontroller (e.g., Arduino) that can operate at megahertz range. One possible way to implement the temporal multiplexing is as follows: The camera is operated to run at a prespecified frame rate and produce camera acquisition times, which are sent to a microcontroller. The microcontroller is also connected to the light sources for visual stimulation and for imaging. Every time the microcontroller receives the signal corresponding to the end of a frame acquisition, it flips the polarity of the output to the two light sources. One advantage of this temporal multiplexing approach is that it clears the space between the primary lens and the brain being studied, but it also halves the camera frame rate and visual stimulation frequency. If the effective visual stimulation frequency is too low, the flickering of the screen itself can evoke visual responses in the retina and subsequent processing in the brain.

3.2 Hemodynamic Artifact

3.2.1 Imaging Hemodynamics

The optical signal acquired through widefield optical imaging is often contaminated by signals from sources other than the target fluorophores, such as flavoproteins that are naturally occurring fluorophores in the brain. These intrinsic fluorescence signals can be negligible when the target fluorophores are strongly expressed. Another source of contamination is the hemodynamic signal, which results from the absorption of emission light by the hemoglobin in the brain tissue along the optical path between the excitation light source and the target fluorophore. This hemoglobin-induced absorption affects the fluorescence in a multiplicative manner, indicating that the hemodynamic contamination of the fluorescence signal persists regardless of the level of fluorophore expression. Furthermore, hemodynamic signals are known to be enhanced during awake, behaving conditions, highlighting the need for accurately removing the hemodynamic artifact in the widefield signal [55]. The hemodynamic signal consists of two main components: concentrations of oxygenated hemoglobin (HbO) and deoxygenated hemoglobin (HbR). These two components are generally correlated, but they have different dynamic characteristics, such as onset latency [56]. Hence, measuring these multiple hemodynamic signals is crucial for accurately estimating the net neural activity from the target fluorophore. One way to measure these hemodynamic signals is to image the back-scattered light from the brain tissue exposed to specific wavelengths. Hemoglobin absorbs light with different efficiency at different wavelengths, which is summarized in its absorption spectra. This efficiency also depends on whether the hemoglobin is oxygenated or not, and the absorption spectra for HbO and HbR do not overlap at most wavelengths, with a few exceptions called isosbestic points (586 nm and 808 nm). Light at one of these wavelengths reflects the total volume of hemoglobin (HbT) regardless of its oxygenation state. Light at other wavelengths reflects a mixture of HbO and HbR, with the mixture ratio depending on the wavelength. HbO has been also estimated using the wavelength (630 nm) at which the absorption ratio between HbO and HbR is small (~0.1). More recently, for GCaMP imaging, ultraviolet light (~410 nm) has been utilized to estimate hemodynamic signals. At this wavelength, GCaMP is not calcium-dependent, so the fluorescence signal is thought to reflect nonneuronal signals, including hemodynamic signals [17]. This method can correct hemodynamic contaminants in the emission path (but not the excitation path), and cause phototoxic damage to brain tissue over long periods.

3.2.2 Simultaneous Neuronal and Hemodynamics Imaging (Fig. 3b)

To remove the hemodynamic components from the imaging signal, it is necessary to image fluorescence from the target fluorophore and the hemodynamic signals from the same brain region simultaneously. There are two main approaches for achieving simultaneous imaging: temporal and spectral multiplexing. Temporal multiplexing involves alternating the emission times of the different light sources, while spectral multiplexing uses light sources at different wavelengths. Both approaches allow for the simultaneous imaging of the target fluorophore and the hemodynamic signals.

In the temporal multiplexing approach, multiple light sources with different wavelengths are switched on and off during each imaging frame while the imaging sensor continuously acquires images at different wavelengths [17]. This switching can be controlled through a single-board microcontroller triggered by the camera acquisition signals (in the same way implemented in Subheading 3.1). The temporal multiplexing approach only requires one imaging sensor, making it more easily adoptable, but it also reduces the effective sampling frequency.

In the spectral multiplexing approach, light with multiple wavelengths is continuously emitted, and the back-scattered light of different wavelengths is captured by multiple imaging sensors or a single multispectral camera. Alternatively, multiple cameras can be placed at the end of diverted optical paths to collect photons from different wavelengths [22, 57]. This approach does not compromise the sampling frequency but requires more complex equipment than temporal multiplexing to align the multiple images.

3.3 Mechanical Artifact (Note 5.2)

All functional imaging techniques, including widefield imaging, rely on the assumption that the position of the pixel on an imaging sensor relative to the brain is stably fixed during the recording period. Any movement of the brain relative to the pixel position can distort the imaging results, especially during visuomotor behavior. Even if the animal being studied is held or trained to be still, subtle movements caused by the autonomic nervous system, such as breathing and heartbeats, can move the brain periodically, affecting the imaging results. There are two ways to address these issues: the first minimizes the brain movement itself during the recording, while the second resorts to image analysis techniques to remove the artifacts from the recorded signals. Here we discuss the former strategy, and will address the latter in Subheading 3.4.

3.3.1 Body Movement

Body movement is a common cause of the displacement of the brain with respect to the imaging sensor, particularly when it affects the distance between the brain and the imaging sensor. For example, head nodding can cause displacement when imaging from above the head. To minimize this displacement, it is recommendable to adopt a head-fixation system that is resistant to these movements, such as one with three clamping points located on the sides and back of the head rather than just the sides (e.g., [58]). On the other hand, translational or rotational movements that do not change the distance between the brain and the imaging sensor can be corrected using image registration algorithms, which will be addressed in Subheading 3.4. Some studies with multiphoton imaging have adopted head-fixation systems that allow for head rotation while still maintaining a stable distance between the brain and the imaging sensor [59], which is readily adoptable for widefield imaging.

3.3.2 Brain Pulsation (Fig. 3c)

Widefield imaging has been widely conducted through thinned skull [60] (Polished and reinforced thinned skull, PoRTS, Fig. $3C_1$) or even through the intact skull in mice. However, if a cranial window is made for some purposes, such as expressing genetically encoded activity indicators or for applying synthetic dyes, the portion of the brain over the craniotomy can pulsate extensively due to the heartbeat, especially in primates. This pulsation causes periodic displacement of the brain, leading to spurious imaging signal. It can be suppressed by sealing the cranial window with a coverslip permanently glued to the surrounding skull (Fig. $3C_2$) [61]. Alternatively, the cranial window can be sealed by a coverslip pressed against an O-ring sunk into the surface of the chamber, and the cranial window is filled with a substance such as agar, silicone oil, cerebrospinal fluid (CSF), or phosphate-buffered saline (PBS) [62] (Fig. $3C_3$). The latter mechanism allows removal of the coverslip, thus ideal if it is necessary to access the brain directly, for staining with artificial fluorescence dyes or for maintenance of the surface condition. In species with less-pronounced pulsation, these rigid seals are overkill. Instead, softer materials such as polydimethylsiloxane (PDMS) can sufficiently suppress the brain pulsation. Widefield imaging has been successfully performed using agar in rats (e.g., [63]) and PDMS in mice [64].

We have discussed sources of artifactual optical signals during visuomotor behaviors, and ways to reduce them through experimental approaches. The following section discusses data analysis techniques that are used to reduce imaging artifacts, which is essential for improving the accuracy and reliability of imaging data.

3.4 Offline Artifact Removal

Widefield imaging data collected during visuomotor behavior is susceptible to two primary sources of artifacts that do not accurately reflect brain activity: brain motion and hemodynamics. Brain motion can be caused by the animal's ballistic movements or by rhythmic movements such as heartbeats and respiration. Hemodynamic activity, including changes in blood volume and oxygenation, can act like a "shutter," attenuating fluorescence signal from the target fluorophore. This section reviews strategies for minimizing these artifacts in acquired imaging data.

3.4.1 Brain Movement and Pulsation

The imaging data collected during visuomotor behaviors can contain spurious signals that stem from displacement of the brain relative to the imaging sensor. These artifacts can be reduced by mechanically stabilizing the optical path or by deploying image registration algorithms to compensate for the displacement. Image registration typically involves three steps: first, creating a template image from a period when the animal is stationary; second, computing the distance between each frame and the template image (either in real space or Fourier space); and third, registering each frame to the template. These procedures are usually done offline as the first step of the data analysis pipeline, but it is becoming increasingly possible to do them in real time during image acquisition [65]. Image registration can reduce brain movement artifacts for most visuomotor behaviors, but it is not a universal solution as it cannot correct motion along the z-axis. Therefore, it is vital to minimize motion along the z-axis during image acquisition, as discussed in Subheading 3.3.

If the displacement is caused by cyclic motions such as breathing and heartbeat, the spurious signal caused by such movement can be reduced with a data resampling technique. During each cycle of these movements, the brain's position relative to the imaging sensor often moves in a predictable manner, meaning that the brain's position is fixed at the same temporal phase across cycles. As a result, imaging data resampled at the same phase should be free of artifacts caused by cyclic motions. This method is most effective for imaging slow brain activity, such as hemodynamic signals.

3.4.2 Hemodynamics

In Subheading 3.2, we discussed the importance of multispectral imaging in obtaining the net neuronal activity in widefield signal. This section reviews strategies for removing hemodynamic signals from the multispectral data.

Spectral Unmixing

Spectral imaging simultaneously produces two types of frames: (**A**) a signal from the target fluorophore contaminated by hemodynamic components, and (**B**) a signal from sources outside the target fluorophore. To reduce the hemodynamic artifact in signal **A**, the hemodynamic signal B is scaled by the regression coefficient with signal **A**. The regression is typically performed on signals from individual pixels or on the average of signals across pixels. This simple regression procedure has been widely used in widefield imaging studies [66].

As we discussed in Subheading 3.2, the hemodynamic signal consists of two main components: the concentrations of oxygenated and deoxygenated hemoglobin (HbO and HbR). Although these two components are generally correlated, they have different dynamic characteristics. To accurately extract the net neuronal signal from the observed widefield signal, it is essential to remove the both hemodynamic components from the contaminated signal. To do this, signal **B** must be sampled at least two different wavelengths. The resulting multiple signal **B**s are generally highly correlated because they almost inevitably contain HbO and HbR. Therefore, it is not effective to remove signal **A** by regressing it against multiple signal **B**s. Instead, it is necessary to first estimate HbO and HbR from the multiple signal **B**s, and then subtract these components from the contaminated fluorescence signal A. The different light propagation properties of HbO and HbR in scattering tissue (as described by the Beer-Lambert Law) have been used to successfully estimate HbO and HbR from multiple signal **B**s [25, 56, 57].

Data Compression

It has been reported that data compression algorithms to widefield imaging data can identify spatial components with prominent vascular patterns (e.g., [58]). These components may be considered hemodynamic artifacts and can be subtracted away to reveal the net neural activity. This approach can be applied to any imaging data and does not require specialized equipment, unlike the spectral unmixing approach. However, it does require manual inspection of spatial components, which can be time-consuming and subject to inspector bias. Additionally, there is a risk of removing actual neuronal signals due to the correlation between neuronal spiking and hemodynamic signals known as neurovascular coupling [67, 68]. This coupling can lead to a violation of the assumptions for data compression algorithms. These drawbacks make this approach less compelling compared to the spectral unmixing approach.

3.5 Example Application: Widefield Calcium/Voltage Imaging During Visual Detection

As an example of widefield imaging during visuomotor behavior, here we discuss one study that investigated the impact of the ongoing bilateral fluctuation on visual response and perception [22]. It was well documented that ongoing activity can display a striking degree of bilateral symmetry between the cortical hemispheres (e.g., [13, 69]). It was not clear, however, to what extent this bilateral ongoing activity can influence sensory-evoked activity.

For the experimental setup, transgenic mice expressing GECI or GEVI were implanted with a head-fixation bar and a glass coverslip over a thinned skull (Fig. $3C_1$), ensuring stable imaging over an extended period exceeding 3 months covering the task training period. Within this setup, the mice were positioned in a head-fixed manner and engaged in a visuomotor task, utilizing a miniature wheel (Fig. $2B_3$).

The widefield imaging was conducted over the dorsal portions of both cortical hemispheres (Fig. 4a) using a tandem-design widefield scope (Fig. 2a). To minimize ambient light contamination in the imaging signal, light-isolation techniques were implemented through the temporal unmixing approach (Fig. $3A_3$).

Both widefield calcium and voltage signal unveiled significant ongoing bilateral fluctuations, particularly evident in areas proximal to the midline (Fig. 4b). Further analysis of these fluctuations revealed their dominant influence on unilateral visual responses (Fig. 4c). The bilateral fluctuations dominated unilateral and localized visual responses, revealed after averaging signals across >100 trials (Fig. 4d). These findings posit that to fully comprehend neuronal activity within one hemisphere, it might be imperative to incorporate recordings from the analogous areas in the other hemisphere.

Interestingly, despite the prevalence of fluctuations in critical visual areas such as PM (Fig. 4e), their presence did not hinder visual detection behavior (Fig. 4f). This observation led to the hypothesis that a downstream mechanism potentially nullifies bilateral fluctuations by contrasting activity between the left and right visual cortices, calling for further research on the role of the bilateral ongoing activity.

4 Conclusion

To summarize, widefield optical imaging is a method that allows researchers to observe mesoscale spatiotemporal dynamics within and between brain areas, including functional structures and their evolution over time. It can be also employed to study coordinated inter-areal dynamics, which are important for understanding visuomotor behaviors. Additionally, widefield imaging is easily combined with other neurostimulation and recording techniques. However, there are several sources of artifactual signals that can impact the

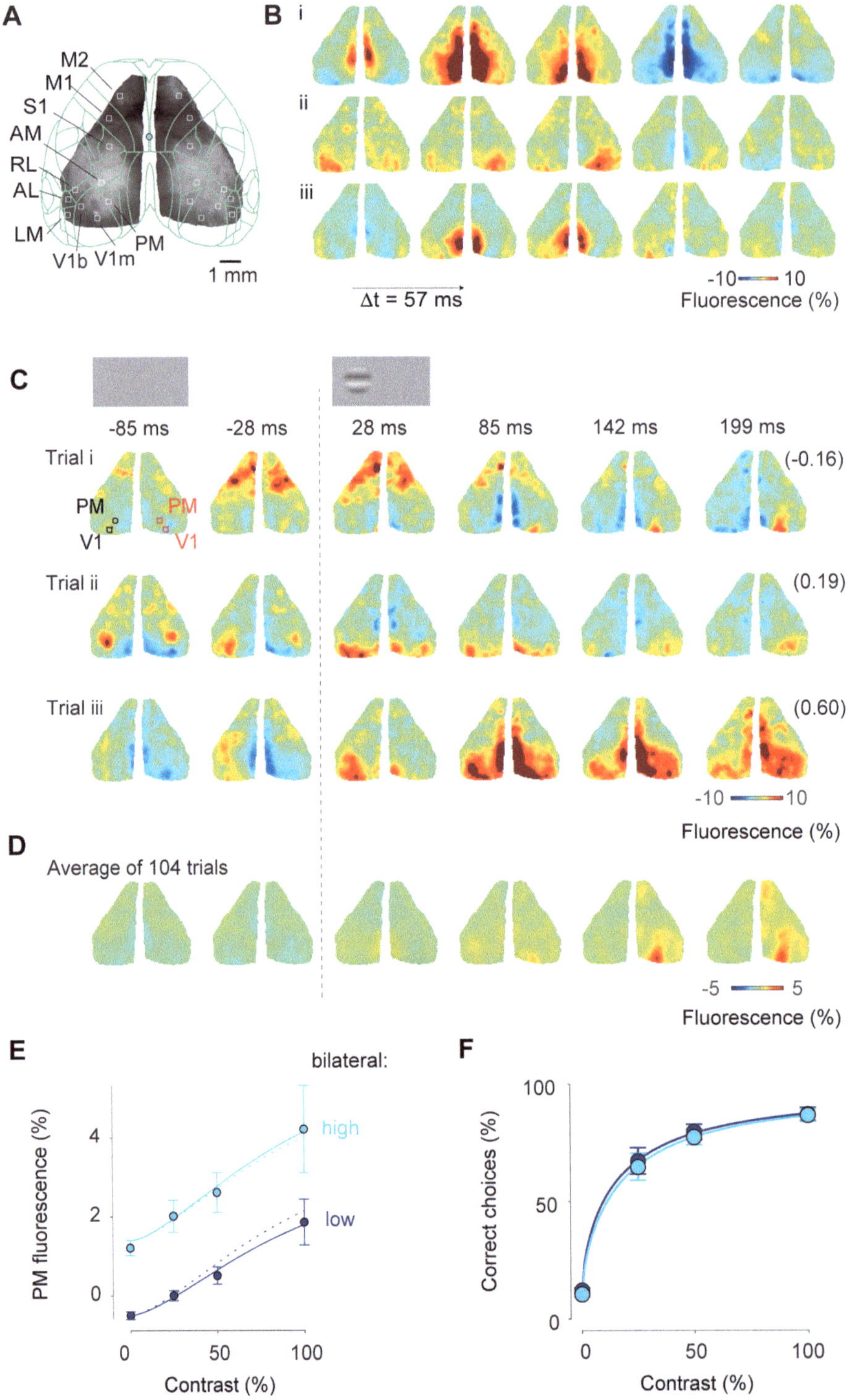

Fig. 4 Widefield calcium imaging during visual detection task. (**a**) Location of somatosensory and motor areas based on the anatomical database by Allen Brain Atlas. The background image indicates a raw fluorescence signal obtained from an example mouse expressing GCaMP6. Superimposed green contour lines indicate borders registered in the Common Coordinate Framework of the Allen Brain Atlas. (**b**) Three example sequences of calcium signals during the period before visual stimulation when a mouse was holding the wheel without detectable eye movements. (**c**) Three example sequences of calcium signal during the period

accuracy of widefield imaging during visuomotor behavior. One is the contamination of photons from the visual stimulation device into the imaging signal. This contamination can be reduced through physical segregation, temporal multiplexing, and spectral multiplexing. The second major source of spurious signal stems from the movement of the brain, which can be minimized either by mechanically stabilizing and suppressing the brain movement or through the image registration procedure. The third major sources are hemodynamic artifact, which impacts imaging signal in a multiplicative manner. This artifact can be minimized by simultaneous neuronal and hemodynamic imaging, combined with spectral unmixing approaches. With these issues adequately addressed, widefield imaging has the potential to provide valuable insights into the brain processes involved in visuomotor behaviors.

5 Notes

5.1. *Strategies to Prevent Light Contamination* (Fig. 3a)

To effectively mitigate light contamination, it's essential to evaluate the advantages and limitations of each prevention method. Physical segregation (Fig. 3A$_1$) offers robust light shielding but can overly constrain the animal's movement. On the other hand, temporal multiplexing (Fig. 3A$_3$) offers a noninvasive approach but necessitates specialized equipment like cameras and visual stimulation devices. This method also poses potential challenges, such as inducing brain responses due to light flickering. Hence, the chosen strategy should align with the extent of light contamination relative to the authentic optical signals originating from brain activity.

To quantify light contamination, one can assess the brain's optical signal while toggling the visual stimulation device on and off. Additionally, illuminating the imaging window vicinity with a flashlight can mimic stray light effects. An optimal imaging setup should exhibit consistent signal readings irrespective of the device's activation or the

Fig. 4 (continued) after presentation of 50% contrast to the left monocular visual field. In each example, the right-most number in parentheses indicates correlation coefficient of the sequence to the average across repeats. (**d**) Average neural activity across repeats. (**e**) Contrast responses from trial groups of high (cyan) and low (blue) contralateral ongoing activity averaged across animals. Curves indicate contrast-response functions where the scale parameter is free (solid curve) or fixed (dotted curve) between the high and low ongoing activities. Error bars represent S.E. across seven animals. (**f**) Probability of turning the wheel to the correct direction as a function of stimulus contrast, averaged across animals. The probability is computed separately in trial groups of low (blue) and high (cyan) ongoing activity measured in area PM. Error bars represent S.E. across seven animals. Curves indicate psychometric functions obtained by fitting a probabilistic observer model [40] separately to the two sets of responses flashlight's illumination. Ideally, light contamination artifacts should be minimized to levels below the brain signal's inherent fluctuations, which can vary based on the imaging modality and sensor characteristics.

Opting for physical segregation typically involves crafting light-isolation components (cones and shades), which may be greatly facilitated by the use of 3D printer. This approach offers agility in prototyping various designs cost-effectively. Although 3D-printed materials may lack durability, they can offer excellent light-shielding capabilities when fabricated with 100% infill. If concerns arise regarding light scattering from the printed surface, enhancing the material's surface with black flocked paper (e.g., BFP1, Thorlabs), could be a viable option.

5.2. *Imaging Window Preparation* (Fig. 3c)

For optimal spatial resolution in widefield imaging, the use of the PoRTS window (Fig. $3C_1$) is highly recommended. When executed correctly, this method offers excellent optical clarity while minimizing the potential activation of microglia and astrocytes, which could lead to an inflammatory response and subsequent alterations in neuronal physiology. The creation of the PoRTS window typically involves the use of a high-speed hand-held drill (e.g., Volvere i7, NSK). Alternatively, a scalpel blade can be employed to gently scrape off the surface of the skull. To avoid overheating the brain tissue, it's essential to thin the skull gradually, working in short intervals. Clear away any skull debris either by rinsing with water or using an air pump. For optimal clarity, thin the skull down to the compact bone layer, ensuring removal over the spongy bone layer. To complete the PoRTS window, reinforce the thinned area with a cover glass using either cyanoacrylate adhesive [60] or an adhesive resin cement overlaid with a thin layer of clear nail polish [70].

The cranial window preparation (Fig. $3C_2$) is indispensable for achieving the desired optical clarity in widefield imaging; however, it demands meticulous attention to detail. To remove the skull segment corresponding to a cranial window, a dental drill is commonly utilized to etch a defined window perimeter. In cases where the skull presents a flat surface, a biopsy punch may serve as a viable alternative for etching the window. Throughout this procedure, the appearance of small cracks and moisture indicative of cerebrospinal fluid signifies successful penetration of the skull. A vital aspect of this preparation is minimizing bleeding beneath the skull to preserve the brain's optical signal integrity. To achieve this, immerse the exposed area in artificial cerebrospinal fluid (aCSF), ensuring coverage up to the entire dura mater beneath the skull flap. Subsequent removal of the dura from the skull flap should be performed slowly, with a fine tweezers (e.g., jeweler's forceps) or a disposal needle. Should bleeding occur post-flap removal, refrain from using cotton applicators or sponges to prevent potential long-term inflammatory responses within the brain. Instead, delicately apply aCSF to the brain's surface to cleanse and mitigate bleeding.

Acknowledgments

The author would like to thank Nicholas Price and Mehdi Adibi for their comments and suggestions on an early version of the manuscript.

References

1. Gămănuţ R, Shimaoka D (2021) Anatomical and functional connectomes underlying hierarchical visual processing in mouse visual system. Brain Struct Funct 1–19. https://doi.org/10.1007/s00429-021-02415-4
2. Saleem AB (2020) Two stream hypothesis of visual processing for navigation in mouse. Curr Opin Neurobiol 64:70–78. https://doi.org/10.1016/j.conb.2020.03.009
3. Grinvald A, Hildesheim R (2004) VSDI: a new era in functional imaging of cortical dynamics. Nat Rev Neurosci 5:874–885. https://doi.org/10.1038/nrn1536
4. Shimaoka D, Wong YT, Rosa MG, Seow N, Price C (2024) Naturalistic movies and encoding analysis define areal borders in marmoset third-tier visual cortex. Progress in Neurobiology, 102657. https://doi.org/10.1016/j.pneurobio.2024.102657
5. Ho CLA, Zimmermann R, Weidinger JDF et al (2021) Orientation preference maps in microcebus murinus reveal size-invariant design principles in primate visual cortex. Curr Biol 31:733–741.e7. https://doi.org/10.1016/j.cub.2020.11.027
6. Shimaoka D, Harris KD, Carandini M (2018) Effects of arousal on mouse sensory cortex depend on modality. Cell Rep 22:3160–3167. https://doi.org/10.1016/j.celrep.2018.02.092
7. Kenet T, Bibitchkov D, Tsodyks M et al (2003) Spontaneously emerging cortical representations of visual attributes. Nature 425:954–956
8. Jancke D, Chavane F, Naaman S, Grinvald A (2004) Imaging cortical correlates of illusion in early visual cortex. Nature 428:423–426. https://doi.org/10.1038/nature02396
9. Mohajerani MH, Chan AW, Mohsenvand M et al (2013) Spontaneous cortical activity alternates between motifs defined by regional axonal projections. Nat Neurosci 16:1426–1435. https://doi.org/10.1038/nn.3499
10. Muller L, Reynaud A, Chavane FEDER, Destexhe A (2014) The stimulus-evoked population response in visual cortex of awake monkey is a propagating wave. Nat Commun 5:1–14. https://doi.org/10.1038/ncomms4675
11. Ferezou I, Haiss F, Gentet LJ et al (2007) Spatiotemporal dynamics of cortical sensorimotor integration in behaving mice. Neuron 56:907–923. https://doi.org/10.1016/j.neuron.2007.10.007
12. Stroh A, Adelsberger H, Groh A et al (2013) Making waves: initiation and propagation of corticothalamic Ca2+ waves in vivo. Neuron 77:1136–1150. https://doi.org/10.1016/j.neuron.2013.01.031
13. Shimaoka D, Song C, Knöpfel T (2017) State-dependent modulation of slow wave motifs towards awakening. Front Cell Neurosci 11:108. https://doi.org/10.3389/fncel.2017.00108
14. Scott G, Fagerholm ED, Mutoh H et al (2014) Voltage imaging of waking mouse cortex reveals emergence of critical neuronal dynamics. J Neurosci 34:16611–16620. https://doi.org/10.1523/JNEUROSCI.3474-14.2014
15. Madisen L, Garner AR, Shimaoka D et al (2015) Transgenic mice for intersectional targeting of neural sensors and effectors with high specificity and performance. Neuron 85:942–958. https://doi.org/10.1016/j.neuron.2015.02.022

16. Daigle TL, Madisen L, Hage TA et al (2018) A suite of transgenic driver and reporter mouse lines with enhanced brain-cell-type targeting and functionality. Cell 174:465–480.e22. https://doi.org/10.1016/j.cell.2018.06.035
17. Allen WE, Kauvar IV, Chen MZ et al (2017) Global representations of goal-directed behavior in distinct cell types of mouse neocortex. Neuron 94:891–907.e6. https://doi.org/10.1016/j.neuron.2017.04.017
18. Markounikau V, Igel C, Grinvald A, Jancke D (2010) A dynamic neural field model of mesoscopic cortical activity captured with voltage-sensitive dye imaging. PLoS Comput Biol 6: e1000919. https://doi.org/10.1371/journal.pcbi.1000919.s001
19. Vanni MP, Murphy TH (2014) Mesoscale transcranial spontaneous activity mapping in GCaMP3 transgenic mice reveals extensive reciprocal connections between areas of Somatomotor cortex. J Neurosci 34:15931–15946. https://doi.org/10.1523/JNEUROSCI.1818-14.2014
20. Kuroki S, Yoshida T, Tsutsui H et al (2018) Excitatory neuronal hubs configure multisensory integration of slow waves in association cortex. Cell Rep 22:2873–2885. https://doi.org/10.1016/j.celrep.2018.02.056
21. MacDowell CJ, Buschman TJ (2020) Low-dimensional spatiotemporal dynamics underlie cortex-wide neural activity. Curr Biol 1–25. https://doi.org/10.1016/j.cub.2020.04.090
22. Shimaoka D, Steinmetz NA, Harris KD, Carandini M (2019) The impact of bilateral ongoing activity on evoked responses in mouse cortex. elife. https://doi.org/10.7554/eLife.43533
23. Clancy KB, Orsolic I, Mrsic-Flogel TD (2019) Locomotion-dependent remapping of distributed cortical networks. Nat Neurosci 503:1–13. https://doi.org/10.1038/s41593-019-0357-8
24. Barson D, Hamodi AS, Shen X et al (2019) Simultaneous mesoscopic and two-photon imaging of neuronal activity in cortical circuits. Nat Methods 1–13. https://doi.org/10.1038/s41592-019-0625-2
25. Ma Y, Shaik MA, Kozberg MG et al (2016) Resting-state hemodynamics are spatiotemporally coupled to synchronized and symmetric neural activity in excitatory neurons. Proc Natl Acad Sci USA 113:E8463–E8471. https://doi.org/10.1073/pnas.1525369113
26. Lim DH, Mohajerani MH, LeDue J et al (2012) In vivo large-scale cortical mapping using channelrhodopsin-2 stimulation in transgenic mice reveals asymmetric and reciprocal relationships between cortical areas. Front Neural Circuits 1–19. https://doi.org/10.3389/fncir.2012.00011/abstract
27. Resta F, Montagni E, de Vito G et al (2022) Large-scale all-optical dissection of motor cortex connectivity shows a segregated organization of mouse forelimb representations. Cell Rep 41:111627. https://doi.org/10.1016/j.celrep.2022.111627
28. Ratzlaff EH, Grinvald A (1991) A tandem-lens epifluorescence macroscope: hundred-fold brightness advantage for wide-field imaging. J Neurosci Methods 36:127–137
29. Akemann W, Mutoh H, Perron A et al (2010) Imaging brain electric signals with genetically targeted voltage-sensitive fluorescent proteins. Nat Methods 7:643–649. https://doi.org/10.1038/nmeth.1479
30. White BR, Bauer AQ, Snyder AZ et al (2011) Imaging of functional connectivity in the mouse brain. PLoS One 6:e16322. https://doi.org/10.1038/nn.2620
31. Song X, Guo Y, Li H et al (2022) Mesoscopic landscape of cortical functions revealed by through-skull wide-field optical imaging in marmoset monkeys. Nat Commun 13:2238–2219. https://doi.org/10.1038/s41467-022-29864-7
32. Durand S, Heller GR, Ramirez TK et al (2022) Acute head-fixed recordings in awake mice with multiple Neuropixels probes. Nat Protoc. https://doi.org/10.1038/s41596-022-00768-6
33. Kislin M, Mugantseva E, Molotkov D et al (2014) Flat-floored air-lifted platform: a new method for combining behavior with microscopy or electrophysiology on awake freely moving rodents. JoVE 1–11. https://doi.org/10.3791/51869
34. Saleem AB, Diamanti EM, Fournier J et al (2018) Coherent encoding of subjective spatial position in visual cortex and hippocampus. Nature 562:124–127. https://doi.org/10.1038/s41586-018-0516-1
35. Niell CM, Stryker MP (2010) Modulation of visual responses by behavioral state in mouse visual cortex. Neuron 65:472–479. https://doi.org/10.1016/j.neuron.2010.01.033
36. Friedman RM, Chehade NG, Roe AW, Gharbawie OA (2020) Optical imaging reveals functional domains in primate sensorimotor cortex. NeuroImage 221:117188. https://doi.org/10.1016/j.neuroimage.2020.117188
37. Ebina T, Masamizu Y, Tanaka YR et al (2018) Two-photon imaging of neuronal activity in motor cortex of marmosets during upper-limb

movement tasks. Nat Commun 1–16. https://doi.org/10.1038/s41467-018-04286-6

38. Isomura Y, Harukuni R, Takekawa T et al (2009) Microcircuitry coordination of cortical motor information in self-initiation of voluntary movements. Nat Neurosci 1–9. https://doi.org/10.1038/nn.2431
39. Kondo M, Matsuzaki M (2021) Neuronal representations of reward-predicting cues and outcome history with movement in the frontal cortex. Cell Rep 34:108704. https://doi.org/10.1016/j.celrep.2021.108704
40. Burgess CP, Lak A, Steinmetz NA et al (2017) High-yield methods for accurate two-alternative visual psychophysics in head-fixed mice. Cell Rep 20:2513–2524. https://doi.org/10.1016/j.celrep.2017.08.047
41. Slovin H (2002) Long-term voltage-sensitive dye imaging reveals cortical dynamics in behaving monkeys. J Neurophysiol 88:3421–3438. https://doi.org/10.1152/jn.00194.2002
42. Seidemann E, Arieli A, Grinvald A, Slovin H (2002) Dynamics of depolarization and hyperpolarization in the frontal cortex and saccade goal. Science 295:862–865
43. Chen SC-Y, Benvenuti G, Chen Y et al (2022) Similar neural and perceptual masking effects of low-power optogenetic stimulation in primate V1. elife. https://doi.org/10.7554/eLife.68393
44. Yates JL, Coop SH, Sarch GH et al (2023) Detailed characterization of neural selectivity in free viewing primates. Nat Commun 14: 3656–3611. https://doi.org/10.1038/s41467-023-38564-9
45. Shimaoka D, Wong YT, Rosa MG, Price NSC (2023) Naturalistic movies and encoding analysis redefine areal Borders in primate visual cortex. bioRxiv 1–21. https://doi.org/10.1101/2023.12.25.573299
46. Juczewski K, Koussa JA, Kesner AJ et al (2022) Stress and behavioral correlates in the head-fixed method: stress measurements, habituation dynamics, locomotion, and motor-skill learning in mice. Sci Rep 1–19. https://doi.org/10.1038/s41598-020-69132-6
47. Ferezou I, Bolea S, Petersen CCH (2006) Visualizing the cortical representation of whisker touch: voltage-sensitive dye imaging in freely moving mice. Neuron 50:617–629. https://doi.org/10.1016/j.neuron.2006.03.043
48. Scott BB, Thiberge SY, Guo C et al (2018) Imaging cortical dynamics in GCaMP transgenic rats with a head-mounted Widefield macroscope. Neuron 100:1045–1058.e5. https://doi.org/10.1016/j.neuron.2018.09.050
49. Rynes ML, Surinach DA, Linn S et al (2021) Miniaturized head-mounted microscope for whole-cortex mesoscale imaging in freely behaving mice. Nat Methods 1–14. https://doi.org/10.1038/s41592-021-01104-8
50. Nadal-Nicolás FM, Kunze VP, Ball JM et al (2020) True S-cones are concentrated in the ventral mouse retina and wired for color detection in the upper visual field. elife. https://doi.org/10.7554/eLife.56840
51. Nasr S, Polimeni JR, Tootell RBH (2016) Interdigitated color- and disparity-selective columns within human visual cortical areas V2 and V3. J Neurosci 36:1841–1857. https://doi.org/10.1523/JNEUROSCI.3518-15.2016
52. Liu Y, Li M, Zhang X et al (2020) Hierarchical representation for chromatic processing across macaque V1, V2, and V4. Neuron 108:538–550.e5. https://doi.org/10.1016/j.neuron.2020.07.037
53. Denman DJ, Luviano JA, Ollerenshaw DR et al (2018) Mouse color and wavelength-specific luminance contrast sensitivity are non-uniform across visual space. elife. https://doi.org/10.7554/eLife.31209
54. Leinweber M, Zmarz P, Buchmann P et al (2014) Two-photon calcium imaging in mice navigating a virtual reality environment. JoVE 1–6. https://doi.org/10.3791/50885
55. Pisauro MA, Dhruv NT, Carandini M, Benucci A (2013) Fast hemodynamic responses in the visual cortex of the awake mouse. J Neurosci 33:18343–18351. https://doi.org/10.1523/JNEUROSCI.2130-13.2013
56. Kozberg MG, Chen BR, DeLeo SE et al (2013) Resolving the transition from negative to positive blood oxygen level-dependent responses in the developing brain. Proc Natl Acad Sci USA 110:4380–4385. https://doi.org/10.1073/pnas.1212785110/-/DCSupplemental
57. Valley MT, Moore MG, Zhuang J et al (2020) Separation of hemodynamic signals from GCaMP fluorescence measured with wide-field imaging. J Neurophysiol 123:356–366. https://doi.org/10.1152/jn.00304.2019
58. West SL, Aronson JD, Popa LS et al (2022) Wide-field calcium imaging of dynamic cortical networks during locomotion. Cereb Cortex 32:2668–2687. https://doi.org/10.1093/cercor/bhab373
59. Voigts J, Harnett MT (2020) Somatic and dendritic encoding of spatial variables in Retrosplenial cortex differs during 2D navigation.

Neuron 105:237–245.e4. https://doi.org/10.1016/j.neuron.2019.10.016

60. Drew PJ, Shih AY, Driscoll JD et al (2010) Chronic optical access through a polished and reinforced thinned skull. Nat Methods 7:981–984. https://doi.org/10.1038/nmeth.1530
61. Goldey GJ, Roumis DK, Glickfeld LL et al (2014) Removable cranial windows for long-term imaging in awake mice. Nat Protoc 9:2515–2538. https://doi.org/10.1038/nprot.2014.165
62. Arieli A, Grinvald A, Slovin H (2002) Dural substitute for long-term imaging of cortical activity in behaving monkeys and its clinical implications. J Neurosci Methods 114:119–133. https://doi.org/10.1016/s0165-0270(01)00507-6
63. Petersen CC, Grinvald A, Sakmann B (2003) Spatiotemporal dynamics of sensory responses in layer 2/3 of rat barrel cortex measured in vivo by voltage-sensitive dye imaging combined with whole-cell voltage recordings and neuron reconstructions. J Neurosci 23:1298–1309
64. Heo C, Park H, Kim Y-T et al (2016) A soft, transparent, freely accessible cranial window for chronic imaging and electrophysiology. Sci Rep 6:27818. https://doi.org/10.1038/srep27818
65. Mitani A, Komiyama T (2018) Real-time processing of two-photon calcium imaging data including lateral motion artifact correction. Front Neuroinform 12:98. https://doi.org/10.3389/fninf.2018.00098
66. Wekselblatt JB, Flister ED, Piscopo DM, Niell CM (2016) Large-scale imaging of cortical dynamics during sensory perception and behavior. J Neurophysiol 115:2852–2866. https://doi.org/10.1152/jn.01056.2015
67. Hillman EMC (2014) Coupling mechanism and significance of the BOLD signal: a status report. Annu Rev Neurosci 37:161–181. https://doi.org/10.1146/annurev-neuro-071013-014111
68. Logothetis NK, Pauls J, Augath M et al (2001) Neurophysiological investigation of the basis of the fMRI signal. Nature 412:150–157
69. Mohajerani MH, McVea DA, Fingas M, Murphy TH (2010) Mirrored bilateral slow-wave cortical activity within local circuits revealed by fast Bihemispheric voltage-sensitive dye imaging in anesthetized and awake mice. J Neurosci 30:3745–3751. https://doi.org/10.1523/JNEUROSCI.6437-09.2010
70. Sofroniew NJ, Vlasov YA, Hires SA, Freeman J (2015) Neural coding in barrel cortex during whisker-guided locomotion. elife. https://doi.org/10.7554/eLife.12559.001
71. Krumin M, Lee JJ, Harris KD, Carandini M (2018) Decision and navigation in mouse parietal cortex. elife. https://doi.org/10.7554/eLife.42583
72. Arieli A, Grinvald A (2002) Optical imaging combined with targeted electrical recordings, microstimulation, or tracer injections. J Neurosci Methods 116:15–28. https://doi.org/10.1016/s0165-0270(02)00022-5

Chapter 9

Mesoscale Ca^{2+} Imaging During Locomotion

Sarah L. West, Russell E. Carter, Martha L. Streng, and Timothy J. Ebner

Abstract

Locomotion is one of the most basic of animal behaviors, providing animals with the ability to actively explore and interact with the external world. While the basic cyclic pattern of locomotion is created in the spinal cord by central pattern generators, virtually all regions of the nervous system are not only engaged but play important roles in locomotion. During locomotion, there is widespread modulation of neurons in the cerebral cortex, including premotor, primary motor, primary sensory, parietal, and retrosplenial cortices. Cerebral cortical activity during locomotion serves to adjust the parameters of locomotion and to process somatosensory, visual, and auditory stimuli more effectively. Therefore, full comprehension of locomotion requires monitoring the widespread modulation of neuronal activity throughout the cerebral cortex to understand the neural dynamics and changing interactions between regions. A similar approach is needed for other brain regions involved in controlling locomotion such as the cerebellar cortex. With the development of new generations of genetically encoded Ca^{2+} indicators and improved tools for optical imaging, mesoscale Ca^{2+} imaging provides a powerful approach for monitoring neuronal activity over large regions at relatively high spatial and temporal resolution. Our recent mesoscale Ca^{2+} imaging in the cerebral cortex demonstrates the dynamic interactions between cortical regions and changing states from rest to walk. This chapter describes our methods for mesoscopic Ca^{2+} imaging in the cerebral cortex during locomotion and the extension of the technique to the cerebellar cortex.

Key words Wide-field Ca^{2+} imaging, Cerebral cortex, Cerebellum, Locomotion, Neural dynamics, Polymer window, GCaMP6, Functional connectivity, Independent component analysis

1 Introduction

Brains and behavior are complex. Sensation, movement, and cognition involves processing and integrating information within and across brain regions. Even the simplest behaviors widely engage the nervous system. To achieve an understanding of how the brain plans, executes, and acquires behaviors necessitates descriptions of the neuronal activity at multiple temporal and spatial scales [1–7]. As hypothesized by many investigators, behavior is the

Sarah L. West and Russell E. Carter contributed equally with all other contributors.

James E. Niemeyer et al. (eds.), *Awake Behaving Mesoscopic Brain Imaging*, Neuromethods, vol. 214,
https://doi.org/10.1007/978-1-0716-4120-0_9,

emergent property of the interactions among ensembles of neurons, neural circuits, and brain areas [8–11]. An important neural scale is the mesoscopic, the level between single neurons and larger brain regions. With strong parallels to physical systems, in which the collective interactions of atoms and molecules give rise to system properties such as magnetism or temperature, the mesoscopic level in the nervous system represents many of the properties of behaviors and brain states. Therefore, this chapter reviews our approach to mesoscale neuronal recordings in both the cerebral and cerebellar cortices.

Traditionally, monitoring population neural activity has relied on electrophysiological recordings, either single neuron activity and/or local field potentials. While there has been considerable progress in developing multi-contact electrodes and electrode arrays, electrode recordings typically provide spatial resolution in a small region(s) and are not easily scaled to record across large regions. In humans, meso- and macroscopic investigations into brain networks have relied on functional magnetic resonance imaging (fMRI) and magnetoencephalography (MEG). Additionally, these two important technologies have limitations in either their spatial (MEG) or temporal (fMRI) resolution. Recent advances in optical sensing tools allow cellular resolution monitoring at multiple scales. Genetically encoded Ca^{2+} indicators (GECI, such as GCaMP6) have enabled in vivo high-resolution monitoring of the activity of hundreds to thousands of neurons [12–17] and the simultaneous imaging of neuronal activity over large regions at relatively high spatial and temporal resolution. Mesoscale wide-field optical imaging studies of the dorsal cerebral cortex in head-fixed mice has been used to study the complex spatiotemporal patterns of activity during a spectrum of behaviors [18–23].

Locomotion is one of the most fundamental of behaviors across the animal kingdom. The basic cyclic pattern of locomotion is generated by spinal cord central pattern generators (for reviews, *see* [24, 25]). However, virtually all regions of the nervous system are engaged during locomotion. The brainstem plays a critical role in starting and stopping locomotion, the basal ganglia is essential in the selection of locomotion related motor patterns, and the cerebellum in motor learning. Further, the cerebral cortex plays a critical role in monitoring the constantly changing external world and adjusting the output of central pattern generators accordingly (for review, *see* [24, 25]). Electrophysiological recordings show that parietal and motor cortical areas regulate locomotion in response to obstacles and sensory stimuli [26]. Primary sensory cortical computations are altered in the somatosensory, visual, and auditory cortices, presumably to interpret relevant stimuli more effectively during locomotion [27, 28]. The retrosplenial cortex is also engaged [29]. Therefore, a full understanding of locomotion requires investigations into all these structures, as championed by

Grillner and colleagues [25]. Our recent mesoscale Ca^{2+} imaging findings highlight this widespread cortical modulation and demonstrates the dynamical exchange of information between cortical regions and the distinct cortical functional states occurring from rest to walk and on return to rest [20]. This chapter details our methods for mesoscale Ca^{2+} in the cerebral and cerebellar cortices during locomotion.

2 Materials

2.1 Cranial and Cerebellar Polymer Window Fabrication

For either cerebral cortical or cerebellar imaging, the surface curvature of the corresponding mouse skull was mapped using a modified CNC mill as described previously [30]. This surface profile was then used to create a 3D profile of the skull surface to make a structural frame of the cranial implant, with the fabrication and assembly steps shown in Fig. 1a. Files for cerebral cortical imaging are freely available for download (*see* [30]). The frames were 3D-printed on a Form 2 printer (Formlabs) using their Black v4

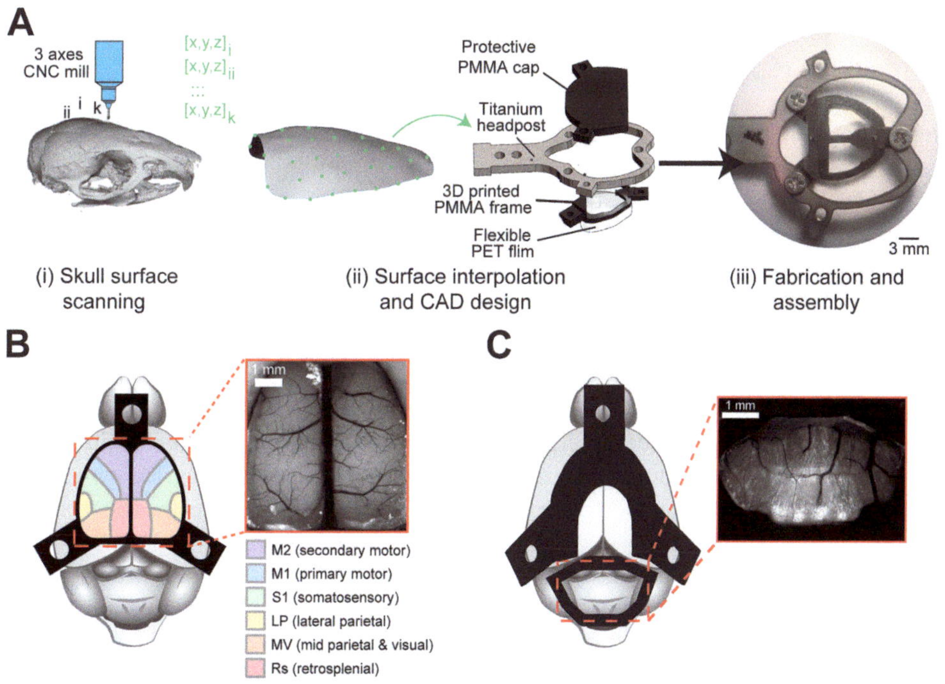

Fig. 1 (**a**) Fabrication of cerebral cortical window implants. (i) Skulls of mice are scanned using a modified CNC mill and combined to create an average skull surface profile for the mouse line. (ii) Skull profiles are used to 3D print a conformant PMMA frame. (iii) PMMA frame is assembled with flexible PET film and a titanium headpost to form the complete implant. (Adapted from Ghanbari et al. [30] with permission.) (**b**) Fluorescence image of the cerebral cortex viewed through the cortical window, with approximate functional regions. (Adapted from West et al. [20] with permission.) (**c**) Fluorescence image of the cerebellar cortex viewed through the cerebellar window

resin (RS-F2-GPBK-04, Formlabs). The three holes in the implant were tapped using a 0–80 hand tap (#15J611, Grainger). For cerebellar imaging, implants consist of two components: the window itself, which is implanted over the cerebellum, and a separate implant for head fixation that is affixed to the skull over the cerebral cortex, and similarly tapped with three holes as for the cerebral cortical windows. A small section of 50 μm transparent polyethylene terephthalate (PET) was bonded to the frame using a clear, two-part epoxy (Scotch-WeldTM DP100 Plus Clear, 3M Inc.). Small relief cuts were made into the PET before bonding, with care taken to not have the cuts enter inside of the frame boundary, to help the PET film curve to the contour of the implant. To aid in bonding, the frame and PET were placed into a custom two-part mold to hold the two pieces together during the bonding process inside of a table-top clamp. Implants were kept in the clamp overnight, and then the excess PET was carefully cut away from the outside implant edges. The PET was then cleaned with ethanol and low-lint cleaning tissue (KimWipes, Kimtech Inc.).

2.2 Mouse Model

Mice expressing genetically encoded Ca^{2+} indicators, GCaMP6f and GCaMP6s, are extensively used for imaging neuronal Ca^{2+} dynamics in awake animal preparations. For cerebral cortical imaging, we utilize Thy1-GCaMP6f mice purchased from Jackson Laboratory (#024339) and bred with C57BL/6 mice in our animal facility. Thy1-GCaMP6f predominantly express GCaMP in excitatory neurons in layers II/III of the cerebral cortex [31]. For cerebellar imaging, we utilize PcP2-GCaMP6s mice, bred by crossing PcP2-Cre mice (JDHU, Jackson Laboratory #010536) with Ai162 mice (Jackson Laboratory #031562), resulting in the expression of GCaMP6s in cerebellar Purkinje cells. Mice between 2 and 6 months old are ideal for this type of implant, as their skulls have finished growing by this time [32] and the dura tends to adhere to the skull more frequently in older mice, making removal of the skull without causing extensive dural damage extremely difficult. Mice are housed in a 12/12-h reverse light–dark cycle, with all experiments performed in the dark phase to not disrupt the normal sleep cycle.

2.3 Fluorescence Microscope Setup

Awake Ca^{2+} fluorescence imaging was performed under an epifluorescence microscope (e.g., Nikon AZ100, or a custom Thorlabs microscope). For cerebral cortical imaging we focus down on layers II/III (~150 μm below the brain surface) and for the cerebellar cortex we focus down to a sufficient depth to resolve both Purkinje cell somata and dendrites, approximately ~180 μm to the Purkinje cell layer. The microscope stage has a platform to support our custom freely moving treadmill (*see* Subheading 2.4 below). Single-photon fluorescence imaging was performed using a high-speed CMOS camera (Andor Zyla 4.2, Oxford Instruments) controlled by Micromanager imaging software [33]. Using a 1× objective and the digital zoom on the microscope, the field-of-view was adjusted so that the cranial window filled as much of the image as possible. The field-of-view within the window for the cerebral cortex was approximately 6.2 × 6.2 mm (Fig. 1b) and approximately 5.0 × 3.5 mm for the cerebellar cortex (Fig. 1c). A dual-wavelength approach was used to image both Ca^{2+}-dependent (470 nm) and Ca^{2+}-independent (405 nm) GCaMP6f signals in an alternating fashion [20, 34–36]. This was performed using a fast LED switcher (OptoLED, Cairn). Images were acquired at 40 fps (resulting in 20 fps for each channel), using an 18 ms exposure time and binned at 256 × 256 pixels. This resulted in a pixel resolution of ~24 × 24 μm. For the cerebellar cortex we routinely image at 1024 × 1024 pixels, with the corresponding increase in pixel resolution. A light-proof collar was attached to the end of the objective and extended down to meet with the titanium headplate of the implant. This was used to prevent any strobing light from being detected by the mouse (also *see* **Note 1**).

2.4 Horizontal Disk

Mice were head-fixed via the titanium headplate of the implant to a freely moving horizontal disk that allows the mice to rest or walk spontaneously (Fig. 2a, b). The disk was custom laser-cut from 1/4-inch acrylic and attached to a central axel on a low-friction ball bearing (McMaster-Carr). The outer edge of the disk was printed with ridges so a rotary encoder could be attached (*see* Subheading 3.3). A custom stainless steel head bar extended from the central axel over the disk. Two holes at the midpoint of the head bar were compatible with two holes in the titanium headplate allowing for head fixation of the mouse with two 2–56 screws. Adhesive plastic traction mats were applied to the surface of the disk to improve purchase during mouse locomotion. These mats were black to create a dark background for paw tracking from the recorded behavior videos (*see* Subheading 3.8).

2.5 Disk Displacement Collection

Displacement of the disk treadmill was collected with an optic rotary encoder controlled by an Arduino Uno microcontroller (Arduino) running custom code (Fig. 2a, b). A custom 3D-printed cog was attached to the rotary encoder and aligned to the outer edge of the horizontal disk. Disk displacement was recorded using the serial port of the microcontroller and saved with PuTTY software.

2.6 Infrared Behavior Video Recordings

Mouse behavior activity was recorded with two high speed infrared-sensitive CMOS cameras (DMK 33UP1300, The Imaging Source; or Flea3, Point Grey, Fig. 2b). Video of full-body behavior was captured by a camera placed in front of the mice approximately 45° from center. This captured the movement of three of the paws during rest and locomotion. A second camera positioned 90° to

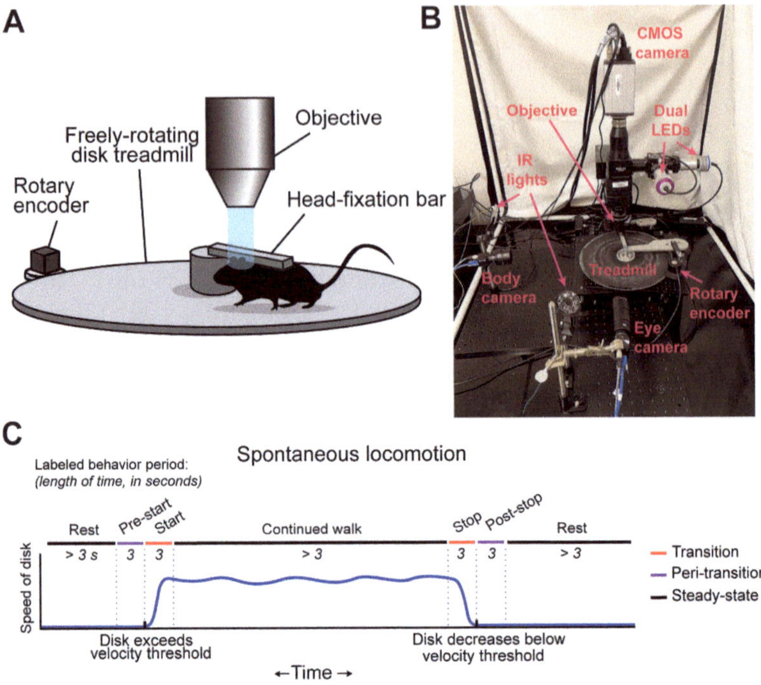

Fig. 2 (**a**) Diagram of spontaneous locomotion behavior setup. (Adapted from Ghanbari et al. [30] with permission.) (**b**) Photograph of mesoscale Ca^{2+} imaging experimental setup. (**c**) Diagram of spontaneous locomotion behavior tasks segmentation. (Adapted from West et al. [20] with permission)

the mouse captured eye movement and pupil diameter. Spinnaker SDK software (FLIR Systems) was used to both control the behavior cameras and save the video. A USB3 or faster connection was required to deliver images to the recording computer at the required speeds without dropping images. The behavior setup was lit with LED infrared light arrays (Acxico) to avoid contaminating the fluorescence imaging and to maximize the comfort of the mice.

2.7 Motorized Treadmill Locomotion

Mesoscale Ca^{2+} imaging during motorized treadmill locomotion required materials in addition to those used in the spontaneous behavior setup. A Nema 17 stepper motor (17HS4401, Motion King; 1.8° step angle, 1.7 A rated current, 40 N cm torque) replaced the rotary encoder. A custom 3D printed cog connected the motor to the horizontal disk. A Texas Instruments DRV8825 motor driver was used to drive the stepper motor, with the reference voltage set to no higher than 0.5 V. A 100 μF capacitor was wired between the motor driver and the motor power source input to prevent any rapid fluctuations in current as the power is turned on or off from causing jolting of the motor. To maintain smooth operation, the stepper motor was run at 1/32 step size. An

Arduino Due microcontroller (Arduino) controlled the motor driver using the AccelStepper code library (Adafruit Industries) in addition to custom code. Audio tones were played by two piezo buzzers (TDK, PS1240P02CT3) wired in parallel to indicate to mice the upcoming actions of the stepper motor. Note that the high processing rate of the Arduino Due (84 MHz) was required to play high-frequency tones while simultaneously controlling the stepper motor at a smooth rate (*see* **Note 2**).

2.8 Time Synchronization of Data Collection Modalities

Custom Spike2 code (Cambridge Electronic Design Limited, version 5.21) generated temporally coordinating trigger signals to cameras and microcontrollers via a CED 1401 mk II interface (Cambridge Electronic Design Limited). Fluorescence and behavior cameras were triggered at each frame, while microcontrollers began collection or control paradigms after a single trigger occurring simultaneously with the first imaging frames.

3 Methods

3.1 Cranial Window Implantation

Mice were weighed and administered 2.0 mg/kg Buprenorphine ER (s.c.) and 2.0 mg/kg Meloxicam (s.c.) for analgesia and inflammation, respectively. These injections were performed at least 2 h before the surgery start time, as it takes approximately 4 h to reach peak concentration in the circulation [37]. Mice were anesthetized using 5% isoflurane in an induction box to a deep anesthetic state, and the scalp was shaved and cleaned using standard aseptic techniques (*see* **Note 3**). The mice were then transferred to, and head-fixed in, a stereotax using ear bars and a nose cone. The level of isoflurane was reduced to 2% just prior to this transfer. Ophthalmic eye ointment was used to cover the eyes, as mice do not close their eyes under anesthesia. A feedback-regulated heating pad was used to maintain the body temperature at 37 °C throughout the entire surgical procedure. Depth of anesthesia was assessed every 15 min via toe pinch. If no response was observed, the isoflurane level was decreased by 0.1%. If a response was detected, the isoflurane level was increased by 0.5% and surgical procedures were halted for 15 min until there was no toe pinch response. Warmed lactated Ringers solution was administered subcutaneously every 2 h to prevent dehydration during the procedure. The scalp was removed to expose the skull over the dorsal surface of the cerebral and cerebellar cortices, and the skull's surface was scraped to remove the fascia and to increase the surface area for dental cement contact.

Cyanoacrylate glue (Vetbond, 3M) is applied over the boundary of the skin incision and the skull to seal the incision. After allowing this to cure, we flood the area with sterile saline and gently scrub the area with a sterile cotton-tipped applicator to prevent the cyanoacrylate from frosting over. A high-speed drill was used to manually mill a path in the dorsal surface of the skull to match the internal profile of the implant frame. This allows for the edges of the implant to rest on the skull surface while allowing visual access to the exposed brain surface through the PET window. Periodically flood the area with sterile saline to aid in heat dissipation and cleaning off of any bond dust generated during the milling process. An important step is to acquire a picture of the skull before removing the skull flap of the craniotomy to allow for correct image alignment during later data analysis steps. To aid in the bone flap removal, soak the skull in saline for a few minutes before carefully inserting a fine forceps (e.g., #5 straight, Fine Science Tools) to gently lift up the four corners of the bone flap. The saline is allowed to run under the bone during this process, to help to keep the dura from getting caught on the underside of the skull flap. Care must be taken when lifting up the skull flap to prevent it from depressing into the opposite side during the lifting process. Steps to avoid this can include using two forceps on opposite sides of the skull flap, gluing a wooden stick to the center of the skull flap (before removal) to aide in lifting the bone flap straight up, or if the bone in the drill path has completely separated from the rest of the skull, the skull flap may be pulled off horizontally in one direction (using care not to cut the brain surface with the skull flap during the removal process). Once the bone flap is removed, place a piece of sterile gel foam or gauze pad soaked in sterile saline over the brain. Absorption of the saline with a cotton-tipped applicator or Kim-Wipe on the gel foam/gauze will apply gentle pressure to the brain and will aide in stopping any bleeding that may occur.

Following removal of the bone flap, a line of gel superglue (Locktight) is applied over the border of the cerebral cortex and olfactory bulbs and the lambdoid suture. The cranial window is then placed over the two lines of gel superglue to adhere the cranial window to the skull. A small drop of sterile saline applied to the underside of the implant will help prevent the PET from sticking to the brain surface and reduce the risk of tearing a surface blood vessel if any movement of the window happens during the adhesion step. Once the implant is secured in the gel superglue, the implant is further sealed in place using dental cement (C&B Metabond, Parkell Inc.). The dental cement should cover all of the exposed skull and slightly go over the skin. A custom titanium headplate is attached to the cranial window by three 0–80 screws into the previously tapped holes. A second application of dental cement should fill all of the space between the cranial window and the titanium headplate. Ensure that no dental cement runs onto or falls on the PET as that will obstruct the imaging field. A 3D-printed cap is attached to the titanium headplate to protect the PET surface.

Implantation of cerebellar windows largely follow the general procedures for cerebral cortical window implantation, with a large craniotomy performed over the dorsal cerebellum, exposing lobules IV–V, VI, and VII of both vermis, simplex, and crus I/II regions. The separate head fixation implant is affixed to the skull over the cerebral cortex, in a similar position as for the cerebral cortical windows, secured using dental cement, and attached to the custom titanium headplate as outlined above.

3.2 Mouse Habituation and Training

Ideally, mice should be housed in a reversed light–dark (12 h–12 h) room with experiments performed, as the dark (night) period is the normal walking and high activity phase of the mouse circadian rhythm. After at least 7 days of recovery from the polymer window implantation surgery, mice should be habituated to head fixation on the horizontal disk in increasing time increments (e.g. 5 min, 15 min, 40 min, 1 h, etc. per day) until mice stand in a relaxed posture or locomote spontaneously during the majority of the head fixation period.

3.3 Spontaneous Locomotion Behavior

Locomotion kinematics are calculated from the angular displacement of the rotary encoder attached to the horizontal disk. Displacement is recorded at 1 kHz by an Arduino Uno microcontroller (Arduino). The velocity of the disk is calculated from the differential of the displacement and converted to the desired units using the radius from the center of the horizontal disk to the point the mouse is head-fixed. Noise in the velocity should be removed by a sliding average window (100 ms window, 1-ms step size). With this smoothing, periods that remain within velocities of −0.25 and 0.05 cm/s are labeled as rest, and locomotion can be defined as periods of movement when the disk reaches a velocity of 0.25 cm/s or more. Locomotion onset is defined as the time the wheel velocity first exceeds 0 cm/s, while offset is the time velocity returns to 0 cm/s. Periods of velocity that remain between 0.05 and 0.25 cm/s were labeled "fidgeting," while those dipping below −0.25 cm/s were labeled as backwards movement.

3.4 Behavior Video Analysis DeepLabCut

Behavior video was analyzed using a residual neural network (DeepLabCut™, [38]). Markers were placed on each visible paw in the recorded full-body video to track individual paw movement and step cycle. High spatial (700 × 240 pixels for an image of the mouse's entire body) and temporal (40 frames per second) resolution videos were required for successful paw tracking. In the eye video, eight markers were placed around the outer edge of the pupil. A circle was fit to the eight pupil markers to estimate pupil diameter (an indicator of general arousal) using the Pratt method [39]. Tracked behavior was then segmented into behavior periods based on the simultaneous rotary encoder data, as described above.

Table 1
Mouse training paradigm for the motorized treadmill experiments

Step	Description
1	Rest for 10–30 s. Warning tone 2 s before motor changes. Motor starts at low acceleration, runs at 1 of 3 randomly selected slow speeds. Returns to rest after 10–30 s.
2	Introduce faster acceleration rate (s).
3	Increase speeds.
4	Introduce deceleration periods in middle of walk periods.
5	Introduce acceleration periods, increase speeds.
6	Increase speeds.
7	Introduce maintaining cues.
8	Randomize order of rest and walk periods. Warning tone 3 s before motor changes.
9	Warning tone 4 s before motor changes.
10	Warning tone 5 s before motor changes.

Training took place over at least 10 days, with each training step intended to be implemented for 1 h on subsequent training days. If a mouse showed distress or resistance to a training step, training was regressed to the previous training step until the mouse showed comfort and competence before continuing to the next step

3.5 Externally Driven Locomotion via a Motorized Treadmill

The behavior task was composed of 5-min trials divided into stages of randomly varying length in time (3–25 s; Fig. 2b). Each trial began and ended with the motor at rest (speed = 0 cm/s). Intermediary stages ran the treadmill at a randomly selected speed from the range of comfortable walking speeds of the head-fixed mice (below 4 cm/s) or returned to rest. The comfortable walking speed of the mice on the motorized treadmill is slower than that of mice walking on the disk spontaneously. Changes in speed occurred at randomly selected comfortable acceleration rates (below 1.25 cm/s). Auditory warning cues were played on piezoelectric buzzers (TDK, PS1240P02CT3) 5 s before the treadmill changed speed to reduce surprise or startle in the mice. The pitch of the tones fell within reported mouse auditory ranges [40] and were played with the Arduino *tone* library. Two piezoelectric buzzers (TDK, PS1240P02CT3) were wired in parallel. As above, the higher processing rate of the Arduino Due (84 MHz) was required over that of a more standard microcontroller processor to play these high-frequency tones while simultaneously controlling the stepper motor at a smooth rate. Mice were trained in the motorized treadmill task using the paradigm outlined in Table 1.

3.6 Preprocessing of Fluorescence Videos

The whole preprocessing was performed using custom code in MATLAB. The blue and violet fluorescence channels of each stack were automatically deinterleaved by detecting the mean brightness of a region over the brain for both even and odd images, then labeling the brighter set of images as the blue channel and the darker set the violet channel.

Motion artifacts were removed with spatial registration. A representative blue-channel image was selected from each imaging session (e.g., first image of first stack), and the clearest and most centered image from this set was chosen for each mouse as the reference image for spatial registration across sessions and the drawing of brain masks (see below). All blue channel images were registered to the respective recording session's representative image using the *dftregistration* MATLAB function [41]. This function uses the two-dimensional Fourier space to quickly align images with relatively small spatial differences. The resulting transformation matrices for each blue image were applied to the corresponding violet images, as the differences between the blue-channel and violet-channel images could potentially introduce errors if a direct registration of a violet-channel image to a blue-channel image was performed. Next, images were registered across recording sessions within mice to remove variability in brain placement in relation to the fluorescence microscope in any given session. Since these placement differences could be too large to be corrected with the *dftregistration* function, these registrations were done using the MATLAB Image Processing Toolbox function *imregtform*. Each registration was calculated between the given recording session's representative image to the mouse's overall representative image, and those transformations were applied to all images using the MATLAB Image Processing Toolbox function *imwarp* with an affine transformation. The *imregtform* function performed exceedingly well in registering images across recording sessions, but occasional registrations required manual adjustments. In these cases, four or more points corresponding to visible anatomical landmarks, such as blood vessel branch points, were manually selected on the representative images and registered. Registration transformations were calculated with the MATLAB Image Processing Toolbox function *fitgeotrans* and applied to all images in a recording session. Brain masks were drawn for each mouse so only the pixels corresponding to brain were kept for further analysis, and pixels depicting other features (such as the edges of the surgical implant, dura thickening at the periphery of the cortical window) were discarded. This removes non-brain artifacts as well as reduces computation time in later processing steps. Brain masks were drawn manually on each mouse's representative image using custom MATLAB code, then applied to all spatially registered images.

A low-pass temporal filter was applied to remove all activity above 7 Hz (fifth-order Butterworth filter, high cutoff frequency 7 Hz) with the MATLAB Signal Processing Toolbox functions *butter* and *filtfilt*. No high-pass filtering was applied, as locomotion produces physiologically relevant, large offsets in cortical activity of the cortex that may last for the full duration of locomotion.

However, any nonphysiologic drifts in fluorescence signal are expected to have been detected in the violet imaging channel, and those fluctuations were removed in the hemodynamic correction step described below. The effects of hemodynamics and other nonneuronal sources were removed from the fluorescence signal using the violet-channel images. Within each stack, the activity of each pixel in the blue channel was regressed against the activity of the corresponding pixel in the violet channel, using a simple linear regression with intercepts via the MATLAB built-in function *regress*. The residuals of these regressions were kept as the hemodynamic-corrected signal. This hemodynamic corrected signal was used in all subsequent steps of analysis. Fluorescence traces were converted to % $\Delta F/F$ using the formula:

$$\%\Delta F/F = 100 \times \left(\frac{F+\mu}{u} - 1\right)$$

Where F = fluorescence timeseries and μ = mean of the blue channel data minus the mean of the violet channel data across all stacks. An example of the modulation in Ca^{2+} fluorescence across the cerebral cortex reveals the extensive modulation prior to the start, during and the end of spontaneous locomotion (Fig. 3a), with almost all regions of interest engaged throughout (Fig. 3b). Also *see* **Note 4**.

3.7 Fluorescence Spatial Segmentation Using Independent Component Analysis

Spatial segmentation of the dorsal cerebral cortex was performed with singular value decomposition (SVD) compression and independent component analysis (ICA). This method provides a blind source segmentation (BSS) of the cerebral cortex based only on statistical properties of the Ca^{2+} activity and does not use any prior assumptions regarding cerebral function or architecture. For each mouse, images from all trials were concatenated and compressed to reduce noise and data size, using the Fast Singular Value Decomposition (fSVD) algorithm [42]. Only the first 200 singular values were used to recreate the spatial dimension of the data [23]. Spatial independent component analysis (sICA) was applied to the compressed data to identify a catalog of functionally relevant cortical regions. We computed the first 50 spatial independent components (ICs) using the Joint Approximation Diagonalization of Eigenmatrices (JADE) algorithm that decomposes mixed signals into ICs by minimizing the mutual information with a series of Givens rotations [43, 44] (Fig. 4a). Masks of ICs were made by setting intensity values below 3.5 to 0 (Fig. 4b). Masks covering less than 150 contiguous pixels or that, upon visual inspection, corresponded to artifacts not associated with cortical activity were discarded, including vascular artifacts that survived the regression step above. An IC that included multiple discontinuous areas, such as homotopic cortical regions, was separated into individual ICs, and

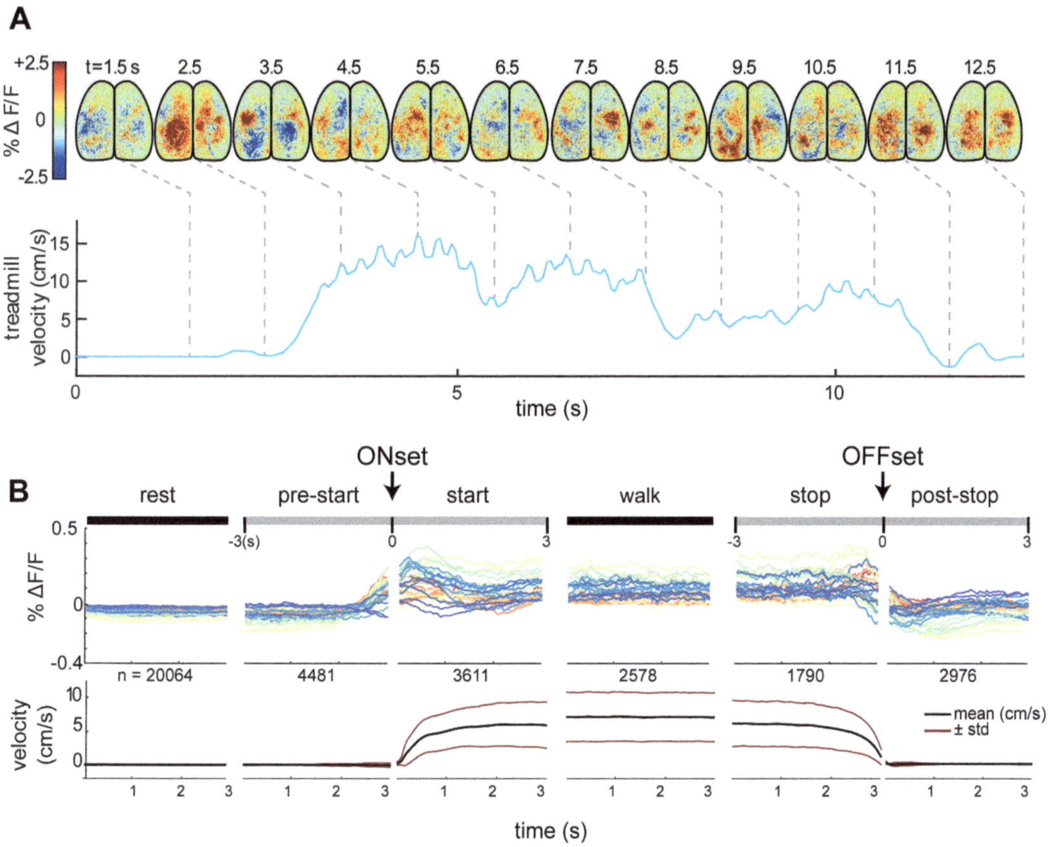

Fig. 3 (**a**) Change in mesoscale Ca^{2+} fluorescence during an example bout of locomotion after preprocessing. (**b**) Mean Ca^{2+} fluorescence time series from regions of interest during spontaneous locomotion. (Adapted from West et al. [20] with permission)

these individual ICs were used in subsequent analyses. To group data across animals, ICs in each mouse catalog were manually assigned to nodes of interest that were present in the majority of mice and corresponded, approximately, to known cortical regions based on the Common Coordinate Framework [45]. If more than one IC in a mouse appeared to belong to the same node, both ICs were given that label.

The dual-wavelength hemodynamic correction procedure described above may not remove all artifacts in the fluorescence signal due to the fluctuation of blood vessel diameter, particularly during the physical exertion of locomotion. To avoid such signal contamination in the remaining ICs, pixels in each IC that overlapped with blood vessels were manually removed. Mean fluorescence traces were calculated for each node by cross-multiplying the weighted masks of the corresponding IC with the preprocessed fluorescence stacks and dividing by the summed amplitude of the IC mask.

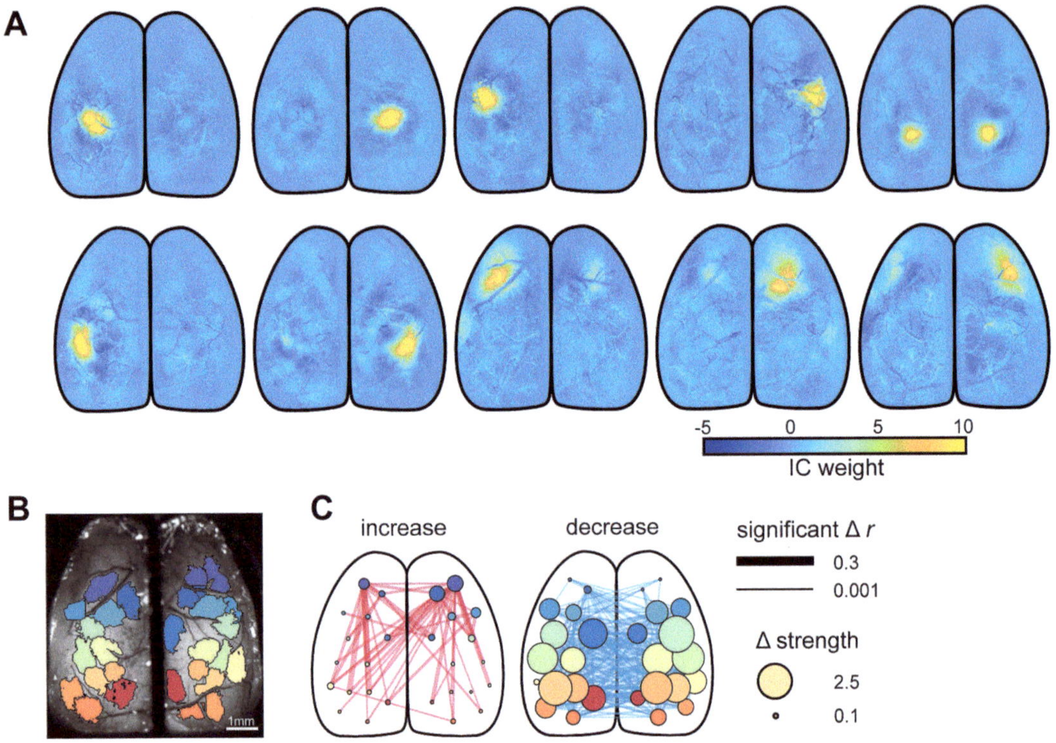

Fig. 4 (**a**) Example subset of raw independent components (ICs) from data from a single mouse. (**b**) Example spatial segmentation of the cerebral cortex obtained through thresholding (weight > 3.5) of the complete set of calculated ICs shown in (**a**). (**c**) Example average change in functional connectivity between spatially segmented regions as measured by Pearson correlation (*r*) across locomotion onset. The size of each graph node corresponds to the strength of connectivity of that region (sum of all *r* values). (All panels adapted from West et al. [20] with permission)

Mesoscale Ca^{2+} imaging lends itself to studies of the functional connectivity of regions across the cortex, which reveals the large-scale organization of the brain during behaviors including locomotion (*see* [20, 29]; Fig. 4c). Functional connectivity may be calculated as the correlation of activity between a predefined set of regions of interest found via spatial segmentation [20]. Alternatively, the correlation between individual pixels within and across regions may each be calculated and summarized [29]. Additional measures of connectivity may be used to fit the experimental question including directed measures such as Granger causality or transfer entropy [2, 20] (also *see* **Note 5**).

3.8 Conclusions

In this chapter, we provide experimental and analytical methods on the use of mesoscale Ca^{2+} imaging during locomotion in head-fixed mice. Recording neuronal activity at the mesoscale level fills an important spatial level of understanding, bridging the gap between micro- and macroscale neural dynamics. Mesoscopic Ca^{2+} imaging has rapidly become a major approach to study neural network activity in a spectrum of behaviors, as described in other chapters. Importantly, mesoscopic optical imaging is in its infancy with new developments and capabilities appearing at a rapid pace. These advances will ensure that mesoscale optical imaging continues to be an essential tool to investigate nervous system function and dysfunction.

4 Notes

1. *Imaging*

   During dual wavelength imaging, the 405 nm (Ca^{2+}-independent, blue channel) fluorescent output needs to be lower than the 470 nm (Ca^{2+}-dependent, violet channel) signal. We have found that if the power of the 405 nm light source is too strong, Ca^{2+}-dependent signals can be observed in what should be the Ca^{2+}-independent signal. Therefore, when doing the hemodynamic correction, it is possible to remove true Ca^{2+} activity from the signal.

   As the caps of the implants do not fully seal around the titanium headplate, dust and debris can accumulate on top of the PET film. Following head-fixation and removal of the cap, the PET surface should be carefully cleaned using several applications of sterile saline and a KimWipe or other lint-free absorbent applicator. Care must be taken to avoid pushing on the PET, as this could detach the PET from the frame and introduce an opening, increasing the chance of an infection. In some cases, the dura or bone may regrow between the brain and underside of the PET, obscuring brain regions beneath the distal edges of the window.

2. *Behavior Setup*

   Several difficulties presented themselves while using head-fixed mice on the horizontal disk treadmill. The first came with ensuring the animals' comfort while on the disk. Notably, the size of the animals varied greatly across individuals, making it difficult to determine an optimal height of the head-bar from the surface of the disk. A head bar with easily adjustable height and head angle is recommended. Another problem presented was the variation in time spent locomoting across mice. Some mice show little interest in walking on the horizontal disk, and it must be decided if those mice should be encouraged to walk (i.e. by manually moving the disk or through reward-based training) or removed from the experimental study. Additionally, animal waste is often deposited on the disk throughout an imaging session and must be cleaned frequently. Urine can reduce traction on the disk and discourage walking, while feces can interfere with automated behavior video analysis or be flung into difficult to reach corners of the experimental setup. Finally, the left-right curve of the disk can create moderately asymmetric limb movement. Other treadmill configurations, such as a vertical wheel [48, 49] or a three-dimensional sphere capable of spinning in any direction [50, 51], may alleviate some of these concerns. However, these alternatives may be too tall in the vertical plane to fit under all fluorescence microscopes.

3. *Surgical*

   The craniotomy and brain window implantation are complex and challenging surgical procedures. A new surgeon may take up to several months before becoming proficient in the procedure, whereas a more experienced surgeon may only need to perform a few surgeries to become proficient. Here, we outline several important steps to be aware of during the surgical procedure. When removing the skull flap, it is possible that the central sinus, or areas along the craniotomy edge may start to bleed. If the bleeding does not stop after application of gel foam/gauze, and it is near the edge of the craniotomy, a small amount of sterile bone wax may be applied directly over the bleed. If the sagittal sinus tears open in the middle of the brain, we have found that there is little that can be done to stop the bleed before extensive blood loss occurs. Animals that have had this happen tend to have a harder time during recovery and a lower quality brain image during recordings, and it may be necessary to terminate the surgery if this occurs. If access to a skilled surgeon is not available, an alternative could be the use of a recently developed automated milling robot [46, 47]. This open-source solution can allow for complex drilling/milling procedures that do not require the extensive training needed for manual drilling/milling.

4. *Data Analysis*

   These mesoscale imaging experiments generate very large datasets, as much as 20 GB/h of recording, requiring extensive reserves of hard drive data storage. Accordingly, processing of these datasets is computationally intensive. Data files should be limited in size by limiting the length of single recordings, subdividing single recordings into multiple files, and/or compressing files without losing important subtleties of the data. A machine with sufficient computer random-access memory (RAM) to store entire data files in working memory is sometimes required to perform certain computations. Therefore, access to high-performance computing resources may be necessary. The use of graphics processing units (GPUs; in contrast to central processing units (CPUs)), can accelerate some processes that are otherwise impractically slow or resource intensive. Various programming languages including Python and MATLAB implement specialized routines to run computations on a GPU.

While we used dual wavelength imaging to remove the hemodynamic contamination, other methods have been used, for example, reflection imaging with green light [52]. Additional approaches to remove hemodynamic artifacts include using independent component analysis (ICA) that extracts blood vessels, which can then be removed before subsequent analysis [20, 36, 52]. Also, manual removal of the blood vessels is another approach [35]. A recent review of the question of hemodynamics correction concluded that consensus on the optimal strategy does not yet exist [53].

Also, the analytical methods used to normalize the hemodynamic contribution include precedents for both linear [54–57] and divisive methods [52, 58, 59]. Several papers have dealt with the equivalence of the division and subtraction methods of normalizing fluorescence; a common conclusion is that the two methods are essentially indistinguishable [4, 57, 60]. While hemodynamic correction is important, virtually every mesoscopic Ca^{2+} imaging study that compared their results using hemodynamic corrected versus noncorrected data found that the hemodynamic contribution to GCaMP6 Ca^{2+} epifluorescence imaging is limited and does not have a significant impact on the results [4, 19, 23, 35, 61, 62]. Importantly for this chapter on mesoscopic imaging during locomotion, we show in a recent paper that the functional connectivity results were essentially the same on corrected and non-hemodynamic data [20].

Finally, the preprocessing of mesoscopic Ca^{2+} imaging data differ greatly, for example, the degree of spatial and temporal filtering (for reviews, *see* [63, 64]). While we prefer to use filters sparingly, their application will be dictated by the specific questions being addressed. Similarly, as discussed above, a variety of analysis methods have been developed for mesoscale Ca^{2+} imaging, each with its own advantages and drawbacks [63].

5. *Other Analysis Approaches to Mesoscale Imaging Data*

The rich, multidimensional nature of both the Ca^{2+} data and the behaviors offers great opportunities as well as analytical challenges. While our analysis of mesoscale imaging data during locomotion initially relied on ICA and functional connectivity as discussed above, other approaches have been utilized for the analysis of Ca^{2+} neuronal dynamics and their relation to behavior [63, 64]. An important step in mesoscale imaging analysis is defining the regions and decomposing the signals into their component parts. Segmentation approaches range from manual parcellation by placing regions of interest (ROIs) placed over the imaging field and analysis of the change in fluorescence relative to the baseline [65, 66]. Another approach is to utilize existing functional and or anatomical maps, with many using the Allen Brain Atlas Common Cortical Framework (CCF) [61]. Spike-triggered mapping [67], seed-based correlation methods [68, 69], and clustering algorithms [70] have also been used to develop activity based spatial maps, all based on identifying common features of the fluorescence signals.

Blind source separation (BSS) techniques, such as the ICA discussed above, are unsupervised and data-driven methods that strive to decompose the Ca^{2+} activity into functionally relevant sources. Principal component analysis (PCA) and singular value decomposition (SVD) are BSS techniques used in the analysis of mesoscale Ca^{2+} imaging data [23, 71]. However, one weakness is that the extracted PCA/SVD components are orthogonal, with uncertain physiological implications. In contrast, some BSS methods, for example, ICA, decompose the data into maximally statistically independent components (ICs) [72–74]. ICA can be applied in the spatial domain to reveal independent spatial components (spatial ICs, Fig. 4a, b). Also, ICA can be applied to the temporal domain to reveal independent temporal patterns and their associated locations [36, 72, 75]. Nonnegative matrix factorization (NMF), a BSS approach used for Ca^{2+} imaging [76, 77], requires the components to be nonnegative [78]. However, the nonnegative restriction can be relaxed, as in localized semi-NMF, which confines the spatial component to CCF [79]. As for all parcellation methods, the spatial components can be used to investigate region-based neuronal Ca^{2+} activity in a hypothesis-driven manner. The advantages and limitations of these various analyses has been discussed [21, 63].

Acknowledgments

This work was supported in part by was supported in part by National Institutes of Health grants, R01 NS111028 to T.J.E., P30 DA048742 to T.J.E., and K99 NS121274 to M.L.S.

References

1. Fox MD, Snyder AZ, Vincent JL, Corbetta M, Van EDC, Raichle ME (2005) The human brain is intrinsically organized into dynamic, anticorrelated functional networks. Proc Natl Acad Sci USA 102(27):9673–9678
2. Avena-Koenigsberger A, Misic B, Sporns O (2017) Communication dynamics in complex brain networks. Nat Rev Neurosci 19(1): 17–33
3. Gilad A, Gallero-Salas Y, Groos D, Helmchen F (2018) Behavioral strategy determines frontal or posterior location of short-term memory in neocortex. Neuron 99(4):814–828.e817. https://doi.org/10.1016/j.neuron.2018.07.029

4. Allen WE, Kauvar IV, Chen MZ, Richman EB, Yang SJ, Chan K, Gradinaru V, Deverman BE, Luo L, Deisseroth K (2017) Global representations of goal-directed behavior in distinct cell types of mouse neocortex. Neuron 94(4): 891–907
5. Ferezou I, Haiss F, Gentet LJ, Aronoff R, Weber B, Petersen CC (2007) Spatiotemporal dynamics of cortical sensorimotor integration in behaving mice. Neuron 56(5):907–923
6. Dipoppa M, Ranson A, Krumin M, Pachitariu M, Carandini M, Harris KD (2018) Vision and locomotion shape the interactions between neuron types in mouse visual cortex. Neuron 98(3):602–615
7. Saleem AB, Ayaz A, Jeffery KJ, Harris KD, Carandini M (2013) Integration of visual motion and locomotion in mouse visual cortex. Nat Neurosci 16(12):1864–1869
8. Buzsaki G (2010) Neural syntax: cell assemblies, synapsembles, and readers. Neuron 68(3):362–385
9. Harris KD (2005) Neural signatures of cell assembly organization. Nat Rev Neurosci 6(5):399–407
10. Carrillo-Reid L, Yang W, Kang Miller JE, Peterka DS, Yuste R (2017) Imaging and optically manipulating neuronal ensembles. Annu Rev Biophys 46:271–293
11. Carrillo-Reid L, Yang W, Bando Y, Peterka DS, Yuste R (2016) Imprinting and recalling cortical ensembles. Science 353(6300):691–694
12. Abe T, Maeda Y, Iijima T (1988) Transient increase of the intracellular Ca2+ concentration during chemotactic signal transduction in Dictyostelium discoideum cells. Differentiation 39(2):90–96
13. Miyawaki A, Llopis J, Heim R, McCaffery JM, Adams JA, Ikura M, Tsien RY (1997) Fluorescent indicators for Ca2+ based on green fluorescent proteins and calmodulin. Nature 388(6645):882–887
14. Nagai T, Yamada S, Tominaga T, Ichikawa M, Miyawaki A (2004) Expanded dynamic range of fluorescent indicators for Ca(2+) by circularly permuted yellow fluorescent proteins. Proc Natl Acad Sci USA 101(29): 10554–10559
15. Zhao Y, Araki S, Wu J, Teramoto T, Chang YF, Nakano M, Abdelfattah AS, Fujiwara M, Ishihara T, Nagai T, Campbell RE (2011) An expanded palette of genetically encoded Ca(2) (+) indicators. Science 333(6051):1888–1891
16. Tian L, Hires SA, Mao T, Huber D, Chiappe ME, Chalasani SH, Petreanu L, Akerboom J, McKinney SA, Schreiter ER, Bargmann CI, Jayaraman V, Svoboda K, Looger LL (2009) Imaging neural activity in worms, flies and mice with improved GCaMP calcium indicators. Nat Methods 6(12):875–881
17. Chen TW, Wardill TJ, Sun Y, Pulver SR, Renninger SL, Baohan A, Schreiter ER, Kerr RA, Orger MB, Jayaraman V, Looger LL, Svoboda K, Kim DS (2013) Ultrasensitive fluorescent proteins for imaging neuronal activity. Nature 499(7458):295–300
18. Zatka-Haas P, Steinmetz NA, Carandini M, Harris KD (2021) Sensory coding and the causal impact of mouse cortex in a visual decision. eLife 10:e63163. https://doi.org/10.7554/eLife.63163
19. Makino H, Ren C, Liu H, Kim AN, Kondapaneni N, Liu X, Kuzum D, Komiyama T (2017) Transformation of cortex-wide emergent properties during motor learning. Neuron 94(4):880–890
20. West SL, Aronson JD, Popa LS, Feller KD, Carter RE, Chiesl WM, Gerhart ML, Shekhar AC, Ghanbari L, Kodandaramaiah SB, Ebner TJ (2022) Wide-field calcium imaging of dynamic cortical networks during locomotion. Cereb Cortex 32(12):2668–2687. https://doi.org/10.1093/cercor/bhab373
21. Nietz AK, Streng ML, Popa LS, Carter RE, Flaherty E, Aronson JD, Ebner TJ (2022) To be and not to be: wide-field Ca2+ imaging reveals neocortical functional segmentation combines stability and flexibility. bioRxiv:2022.2009.2016.508301. https://doi.org/10.1101/2022.09.16.508301
22. Orsolic I, Rio M, Mrsic-Flogel TD, Znamenskiy P (2021) Mesoscale cortical dynamics reflect the interaction of sensory evidence and temporal expectation during perceptual decision-making. Neuron 109(11): 1861–1875.e1810. https://doi.org/10.1016/j.neuron.2021.03.031
23. Musall S, Kaufman MT, Juavinett AL, Gluf S, Churchland AK (2019) Single-trial neural dynamics are dominated by richly varied movements. Nat Neurosci 22(10):1677–1686. https://doi.org/10.1038/s41593-019-0502-4
24. Rossignol S, Dubuc R, Gossard JP (2006) Dynamic sensorimotor interactions in locomotion. Physiol Rev 86(1):89–154
25. Grillner S, El MA (2020) Current principles of motor control, with special reference to vertebrate locomotion. Physiol Rev 100(1): 271–320
26. Drew T, Marigold DS (2015) Taking the next step: cortical contributions to the control of locomotion. Curr Opin Neurobiol 33:25–33. https://doi.org/10.1016/j.conb.2015.01.011

27. Parker PRL, Brown MA, Smear MC, Niell CM (2020) Movement-related signals in sensory areas: roles in natural behavior. Trends Neurosci 43(8):581–595
28. Schneider DM (2020) Reflections of action in sensory cortex. Curr Opin Neurobiol 64:53–59
29. Clancy KB, Orsolic I, Mrsic-Flogel TD (2019) Locomotion-dependent remapping of distributed cortical networks. Nat Neurosci 22:778–786
30. Ghanbari L, Carter RE, Rynes M, Dominguez J, Chen G, Naik A, Hu J, Sagar MAK, Halton L, Mossazaghi N, Gray MM, West SL, Eliceiri KW, Ebner TJ, Kodandaramaiah SB (2019) Cortex-wide neural interfacing via transparent polymer skulls. Nat Commun 10:1500
31. Dana H, Chen TW, Hu A, Shields BC, Guo C, Looger LL, Kim DS, Svoboda K (2014) Thy1-GCaMP6 transgenic mice for neuronal population imaging in vivo. PLoS One 9(9):e108697
32. Vora SR, Camci ED, Cox TC (2015) Postnatal ontogeny of the cranial base and craniofacial skeleton in male C57BL/6J mice: a reference standard for quantitative analysis. Front Physiol 6:417. https://doi.org/10.3389/fphys.2015.00417
33. Edelstein AD, Tsuchida MA, Amodaj N, Pinkard H, Vale RD, Stuurman N (2014) Advanced methods of microscope control using μManager software. J Biol Methods 1(2):e10. https://doi.org/10.14440/jbm.2014.36
34. Jacobs EAK, Steinmetz NA, Carandini M, Harris KD (2020) Cortical state fluctuations during sensory decison making. Curr Biol 30:4944–4955
35. MacDowell CJ, Buschman TJ (2020) Low-dimensional spatiotemporal dynamics underlie cortex-wide neural activity. Curr Biol 30(14):2665–2680
36. Cramer SW, Haley SP, Popa LS, Carter RE, Scott E, Flaherty EB, Dominguez J, Aronson JD, Sabal L, Surinach D, Chen CC, Kodandaramaiah SB, Ebner TJ (2023) Wide-field calcium imaging reveals widespread changes in cortical functional connectivity following mild traumatic brain injury in the mouse. Neurobiol Dis 176:105943. https://doi.org/10.1016/j.nbd.2022.105943
37. Foley PL, Liang H, Crichlow AR (2011) Evaluation of a sustained-release formulation of buprenorphine for analgesia in rats. J Am Assoc Lab Anim Sci 50(2):198–204
38. Mathis A, Mamidanna P, Cury KM, Abe T, Murthy VN, Mathis MW, Bethge M (2018) DeepLabCut: markerless pose estimation of user-defined body parts with deep learning. Nat Neurosci 21(9):1281–1289
39. Pratt V (1987) Direct least-squares fitting of algebraic surfaces. ACM SIGGRAPH Comput Graph 21(4):145–152. https://doi.org/10.1145/37402.37420
40. Grothe B, Pecka M (2014) The natural history of sound localization in mammals--a story of neuronal inhibition. Front Neural Circuits 8:116. https://doi.org/10.3389/fncir.2014.00116
41. Guizar-Sicairos M, Thurman ST, Fienup JR (2008) Efficient subpixel image registration algorithms. Opt Lett 33(2):156–158
42. Halko N, Martinsson P-G, Shkolnisky Y, Tygert M (2010) An algorithm for the principal component analysis of large data sets. arXiv:1007.5510. https://doi.org/10.48550/arXiv.1007.5510
43. Cardoso JF (1999) High-order contrasts for independent component analysis. Neural Comput 11(1):157–192
44. Sahonero-Alvarez G, Calderon H (2017) A comparison of SOBI, FastICA, JADE and infomax algorithms. In: Proceedings of the 8th International Multi-Conference on Complexity, Informatics and Cybernetics (IMCIC 2017)
45. Wang Q, Ding SL, Li Y, Royall J, Feng D, Lesnar P, Graddis N, Naeemi M, Facer B, Ho A, Dolbeare T, Blanchard B, Dee N, Wakeman W, Hirokawa KE, Szafer A, Sunkin SM, Oh SW, Bernard A, Phillips JW, Hawrylycz M, Koch C, Zeng H, Harris JA, Ng L (2020) The Allen mouse brain common coordinate framework: a 3D reference atlas. Cell 181(4):936–953.e920. https://doi.org/10.1016/j.cell.2020.04.007
46. Rynes ML, Ghanbari L, Schulman DS, Linn S, Laroque M, Dominguez J, Navabi ZS, Sherman P, Kodandaramaiah SB (2020) Assembly and operation of an open-source, computer numerical controlled (CNC) robot for performing cranial microsurgical procedures. Nat Protoc 15(6):1992–2023
47. Ghanbari L, Rynes ML, Hu J, Schulman DS, Johnson GW, Laroque M, Shull GM, Kodandaramaiah SB (2019) Craniobot: a computer numerical controlled robot for cranial microsurgeries. Sci Rep 9(1):1023
48. Rasmussen R, Nicholas E, Petersen NC, Dietz AG, Xu Q, Sun Q, Nedergaard M (2019) Cortex-wide changes in extracellular potassium ions parallel brain state transitions in awake

behaving mice. Cell Rep 28(5):1182–1194. e1184. https://doi.org/10.1016/j.celrep.2019.06.082
49. Adam EM, Johns T, Sur M (2022) Dynamic control of visually guided locomotion through corticosubthalamic projections. Cell Rep 40(4):111139. https://doi.org/10.1016/j.celrep.2022.111139
50. Leinweber M, Ward DR, Sobczak JM, Attinger A, Keller GB (2017) A sensorimotor circuit in mouse cortex for visual flow predictions. Neuron 95(6):1420–1432.e1425. https://doi.org/10.1016/j.neuron.2017.08.036
51. Thurley K, Ayaz A (2017) Virtual reality systems for rodents. Curr Zool 63(1):109–119. https://doi.org/10.1093/cz/zow070
52. Ma Y, Shaik MA, Kim SH, Kozberg MG, Thibodeaux DN, Zhao HT, Yu H, Hillman EM (2016) Wide-field optical mapping of neural activity and brain haemodynamics: considerations and novel approaches. Philos Trans R Soc Lond B Biol Sci 371(1705):20150360
53. Couto J, Musall S, Sun XR, Khanal A, Gluf S, Saxena S, Kinsella I, Abe T, Cunningham JP, Paninski L, Churchland AK (2021) Chronic, cortex-wide imaging of specific cell populations during behavior. Nat Protoc 16(7): 3241–3263. https://doi.org/10.1038/s41596-021-00527-z
54. Frostig RD, Lieke EE, Ts'o DY, Grinvald A (1990) Cortical functional architecture and local coupling between neuronal activity and the microcirculation revealed by in vivo high-resolution optical imaging of intrinsic signals. Proc Natl Acad Sci USA 87(16):6082–6086
55. Kim CK, Yang SJ, Pichamoorthy N, Young NP, Kauvar I, Jennings JH, Lerner TN, Berndt A, Lee SY, Ramakrishnan C, Davidson TJ, Inoue M, Bito H, Deisseroth K (2016) Simultaneous fast measurement of circuit dynamics at multiple sites across the mammalian brain. Nat Methods 13(4):325–328
56. Spors H, Grinvald A (2002) Spatio-temporal dynamics of odor representations in the mammalian olfactory bulb. Neuron 34(2):301–315. https://doi.org/10.1016/s0896-6273(02)00644-x
57. Wekselblatt JB, Flister ED, Piscopo DM, Niell CM (2016) Large-scale imaging of cortical dynamics during sensory perception and behavior. J Neurophysiol 115(6):2852–2866. https://doi.org/10.1152/jn.01056.2015
58. Akemann W, Mutoh H, Perron A, Park YK, Iwamoto Y, Knopfel T (2012) Imaging neural circuit dynamics with a voltage-sensitive fluorescent protein. J Neurophysiol 108(8): 2323–2337. https://doi.org/10.1152/jn.00452.2012
59. Kramer RS, Pearlstein RD (1979) Cerebral cortical microfluorometry at isosbestic wavelengths for correction of vascular artifact. Science 205(4407):693–696. https://doi.org/10.1126/science.223243
60. Cai Y, Wu J, Dai Q (2022) Review on data analysis methods for mesoscale neural imaging in vivo. Neurophotonics 9(4):041407. https://doi.org/10.1117/1.NPh.9.4.041407
61. Vanni MP, Murphy TH (2014) Mesoscale transcranial spontaneous activity mapping in GCaMP3 transgenic mice reveals extensive reciprocal connections between areas of somatomotor cortex. J Neurosci 34(48): 15931–15946
62. Murphy TH, Boyd JD, Bolanos F, Vanni MP, Silasi G, Haupt D, LeDue JM (2016) High-throughput automated home-cage mesoscopic functional imaging of mouse cortex. Nat Commun 7:11611
63. Nietz AK, Popa LS, Streng ML, Carter RE, Kodandaramaiah SB, Ebner TJ (2022) Wide-field calcium imaging of neuronal network dynamics in vivo. Biology (Basel) 11(11): 1601. https://doi.org/10.3390/biology11111601
64. Ren C, Komiyama T (2021) Characterizing cortex-wide dynamics with wide-field calcium imaging. J Neurosci 41(19):4160–4168. https://doi.org/10.1523/JNEUROSCI.3003-20.2021
65. Busche MA, Kekus M, Adelsberger H, Noda T, Forstl H, Nelken I, Konnerth A (2015) Rescue of long-range circuit dysfunction in Alzheimer's disease models. Nat Neurosci 18(11): 1623–1630
66. Murakami T, Yoshida T, Matsui T, Ohki K (2015) Wide-field Ca2+ imaging reveals visually evoked activity in the retrosplenial area. Front Mol Neurosci 8:20. https://doi.org/10.3389/fnmol.2015.00020
67. Xiao D, Vanni MP, Mitelut CC, Chan AW, LeDue JM, Xie Y, Chen AC, Swindale NV, Murphy TH (2017) Mapping cortical mesoscopic networks of single spiking cortical or sub-cortical neurons. eLife 6:e19976
68. Vanni MP, Chan AW, Balbi M, Silasi G, Murphy TH (2017) Mesoscale mapping of mouse cortex reveals frequency-dependent cycling between distinct macroscale functional modules. J Neurosci 37(31):7513–7533
69. Mohajerani MH, Chan AW, Mohsenvand M, Ledue J, Liu R, McVea DA, Boyd JD, Wang YT, Reimers M, Murphy TH (2013) Spontaneous cortical activity alternates between

motifs defined by regional axonal projections. Nat Neurosci 16(10):1426–1435

70. Pinto L, Rajan K, DePasquale B, Thiberge SY, Tank DW, Brody CD (2019) Task-dependent changes in the large-scale dynamics and necessity of cortical regions. Neuron 104(4): 810–824.e819. https://doi.org/10.1016/j.neuron.2019.08.025
71. Kondo M, Matsuzaki M (2021) Neuronal representations of reward-predicting cues and outcome history with movement in the frontal cortex. Cell Rep 34(5):108704. https://doi.org/10.1016/j.celrep.2021.108704
72. Brown GD, Yamada S, Sejnowski TJ (2001) Independent component analysis at the neural cocktail party. Trends Neurosci 24(1):54–63
73. Stone JV (2002) Independent component analysis: an introduction. Trends Cogn Sci 6(2):59–64. https://doi.org/10.1016/s1364-6613(00)01813-1
74. Calhoun VD, Adali T (2006) Unmixing fMRI with independent component analysis. IEEE Eng Med Biol Mag 25(2):79–90
75. Calhoun VD, Adali T, Pearlson GD, Pekar JJ (2001) Spatial and temporal independent component analysis of functional MRI data containing a pair of task-related waveforms. Hum Brain Mapp 13(1):43–53
76. Yang W, Miller JE, Carrillo-Reid L, Pnevmatikakis E, Paninski L, Yuste R, Peterka DS (2016) Simultaneous multi-plane imaging of neural circuits. Neuron 89(2):269–284
77. Quarta E, Scaglione A, Lucchesi J, Sacconi L, Allegra Mascaro AL, Pavone FS (2022) Distributed and localized dynamics emerge in the mouse neocortex during reach-to-grasp behavior. J Neurosci 42(5):777–788. https://doi.org/10.1523/jneurosci.0762-20.2021
78. Mirzal A (2017) NMF versus ICA for blind source separation. Adv Data Anal Classif 11(1):25–48. https://doi.org/10.1007/s11634-014-0192-4
79. Saxena S, Kinsella I, Musall S, Kim SH, Meszaros J, Thibodeaux DN, Kim C, Cunningham J, Hillman EMC, Churchland A, Paninski L (2020) Localized semi-nonnegative matrix factorization (LocaNMF) of widefield calcium imaging data. PLoS Comput Biol 16(4):e1007791. https://doi.org/10.1371/journal.pcbi.1007791

Part IV

Mesoscale Imaging of Brain Disease

Chapter 10

Mesoscale Ca^{++} Imaging of Seizures and Interictal Activity in Models of Chronic Epilepsy

William F. Tobin and Matthew C. Weston

Abstract

Epilepsy is defined by the presence of recurrent, spontaneous seizures that involve brain-wide networks. Brain activity outside of seizures (interictal activity) is also often abnormal in epilepsy, but the relationship between these two phenomena is not well understood. In this chapter, we describe a method for imaging brain activity across the surface of the dorsal cortex in mouse models of epilepsy using Ca^{++} indicators, as well as how to analyze this activity to study both spontaneous seizures and interictal activity. The data generated from this approach can be used to (1) identify cortical areas that are susceptible to seizures and abnormal activity patterns, (2) investigate spontaneous seizure initiation, spread, and termination, (3) relate features of interictal abnormalities to seizures, and (4) evaluate the efficacy of therapies in normalizing cortical activity. The low cost, usefulness, and flexibility of the approach should encourage those investigating animal models of epilepsy and other neurological diseases to adopt it as part of their phenotyping repertoire.

Key words Epilepsy, Mouse models, Genetics, Mesoscale, Ca^{++} imaging, Spontaneous seizures

1 Introduction

The epilepsies are a diverse group of disorders in which normal brain function is disrupted leading to recurrent spontaneous seizures as well as a host of associated psychiatric and neurological symptoms. Although brain activity during seizures has been intensively studied for years, far less is known about how epileptic disease processes reshape brain activity outside of seizures and the brain states from which seizures emerge [1, 2]. This information is critical for determining how changes at the molecular and cellular levels translate into the abnormal large-scale brain activities that underlie symptoms and how brain activity might be modified to restore normal function [3, 4].

The greatest impediment to collecting this information from human patients is the lack of appropriate controls. Outside of overtly epileptic events, differences in background genetics and

James E. Niemeyer et al. (eds.), *Awake Behaving Mesoscopic Brain Imaging*, Neuromethods, vol. 214,
https://doi.org/10.1007/978-1-0716-4120-0_10,

life history limit our ability to identify which features of a patient's brain activity reflect the disease state as opposed to other individual factors [1]. Additional challenges arise from the limitations of technologies currently used to record large-scale brain activity from humans, such as EEG and fMRI. While EEG offers high temporal resolution, the relationship between neural firing and the signals it records is highly nonlinear and difficult to interpret [5]. On the other hand, human functional brain imaging can collect data from large regions of tissue but suffers from poor resolution and reliance on proxy measures of neural activity [6]. Fortunately, the increasing number of animal models of chronic epilepsy has made it possible to compare epileptic and healthy individuals that share the same background genetics and life history. Because of its ability to simultaneously image multiple brain regions, mesoscopic Ca^{++} imaging (MeCI) is uniquely suited to answer fundamental questions about how epileptic disease processes reshape brain activity and the context in which seizures occur [7]. It offers better spatial resolution and a more linear relationship to neuronal activity relative to alternatives at this scale, such as EEG and fMRI, with temporal resolution that lies between these two modalities. A relatively small but growing number of studies have used MeCI to investigate seizure initiation and propagation in rodent brains. Key insights from these studies include that seizures propagate along synaptic pathways and to areas where synaptic inhibition is compromised, and that seizures and interictal spikes may share common initiation mechanisms [8–10].

To come to these conclusions, these studies leveraged the power of using focally applied chemoconvulsants to initiate epileptiform activity. This approach has clear advantages for investigating seizure initiation and spread because the abnormal activity begins at a constrained time and place. Chronic epilepsy models offer an attractive alternative to chemoconvulsant models because they place the epileptic seizures in the context of a brain that has undergone the process of epileptogenesis and experienced chronic seizures. Seizures in this context may differ in key ways from induced seizures [11, 12]. Additionally, non-seizure brain activity in these models can be imaged and compared to that of healthy controls. Between-genotype differences in this ongoing activity can give insight into the types of brain activity that give rise to seizures, and measure how severely activity in disease models deviates from normal in different regions.

In this chapter we describe a method for imaging large-scale brain activity in a genetic mouse model of childhood epilepsy at the mesoscale [13]. This approach can share many features of imaging neural activity in healthy mice or chemoconvulsant-induced epilepsy with MeCI, including the imaging hardware, basic image processing, and further analyses [14, 15]. The general rarity of spontaneous seizures poses a challenge to studying these events themselves and relating interictal activity to seizure activity in many models. However, even in these cases, a great deal can be learned from MeCI about how gene variants, epileptic disease processes, and treatment strategies impact basic features of brain activity [16]. Deriving this information from raw MeCI data requires careful considerations and analyses that we will discuss. This approach has the potential to advance our understanding of the features that distinguish the epileptic brain, how they relate to underlying causes at the cellular and molecular scale, and can potentially be used to determine where and how brain activity needs to be modulated to holistically restore normal function.

2 Materials

2.1 Experimental Animals

In principle, MeCI can be performed in any animal model of epilepsy. An early application of the tandem-lens epifluorescent macroscope, the most common instrument used for MeCI, was intrinsic imaging of the primate cortex [17]. Recent advances in techniques to introduce calcium indicator proteins into the neurons of a diverse array of species mean that MeCI can be performed in almost any model organism [18]. Yet among mammalian species, the mouse is unrivaled in the number of genetic lines available that carry both research tools and disease-causing gene variants. For this reason, this chapter will focus on mice but with some simple adaptations the techniques described here can be applied to other species. In fact, several recent studies have imaged seizures in zebrafish with excellent results [19].

An important issue that needs to be considered when planning MeCI studies of epileptic animals is the probability of imaging spontaneous seizures. Because these events are the hallmark of epilepsy in human patients, and therapeutic modulation of their occurrence remains the gold standard of treatment efficacy, many investigators will want to measure seizure activity. However, in most rodent epilepsy models, spontaneous seizures are rare and unpredictable [20], making it unlikely that they will occur during the relatively short course of MeCI imaging sessions. EEG recording in rodents typically span days, if not weeks, in order to capture the number of spontaneous seizures needed to power studies of their properties [21, 22]. Mesoscale imaging experiments, on the other hand, are usually conducted over a period of hours at most, due to the desire to avoid animal welfare issues associated with long-term head fixation and to keep the volume of data generated manageable. Thus, unless the chosen model has frequent or predictable seizures, an imaging approach that only samples for minutes or hours per day will likely capture few or no spontaneous seizures. Many animals that have numerous spontaneous seizures may also suffer from early lethality or poor health. Thus, the ideal choice of a model is one that has frequent, nonlethal seizures. Previously, we imaged mice with a human gain-of-function variant in the *KCNT1* gene to image spontaneous seizures [13, 23]. These mice can have several seizures per day, yet are generally long-lived and healthy. An interesting and largely unexplored alternative that may be possible in some models is to use a naturalistic trigger to cause seizures in a chronically epileptic animal. An example of this would be elevating body temperature to trigger febrile seizures in mouse models of Dravet syndrome [24]. Other physiological triggers such as stress or specific neuromodulatory states could also potentially be engineered into imaging experiments as seizure triggers depending on the phenotypic details of the model being studied.

Another factor that needs to be considered when designing MeCI studies is the age at which imaging needs to be performed. The vast majority of published MeCI studies have been performed in adolescent or adult mice, although slightly more demanding methods can be used to collect data from neonates. Important events in epileptogenesis likely occur early in development in many genetic models [20], making it more difficult, although by no means impossible, to study disease onset in these cases. On the other hand, models that rely on insults later in development to cause chronic epilepsy, such as trauma, induced genetic lesions, kindling, or pharmacology are well suited to studies of epileptogenesis using MeCI.

Finally, MeCI is largely restricted to the cortex where it can report on activity in all dorsal cortical areas simultaneously. Imaging deeper brain structures requires significantly more involved surgical techniques and a reduction in the area of imaged tissue. This makes the technique most relevant for models in which the cortex is involved in disease, although even in cases where seizures start in other brain regions MeCI can still potentially be used to study the propagation of epileptic activity, changes in basic, ubiquitous features of neurophysiology, and any disease-related changes that occur distant from the site of seizure initiation. In general, some knowledge of the spatial relationship and connectivity between the imaged tissue and the site of epileptic activity initiation is critical to interpreting MeCI data.

2.1.1 Genetically Encoded Calcium Indicators

Another important choice is how to introduce the desired Ca^{++} indicator. Ca^{++}-sensitive dyes have been used to monitor activity at the mesoscale, but for imaging chronically epileptic mice a genetically encoded indicator is preferable because it facilitates the longitudinal imaging needed to capture spontaneous seizures, as well as day-to-day variation in the severity of pathology, which can be quite high in epileptic animals. There are several transgenic mouse lines to introduce GCaMP and other Ca^{++} indicators, either into the germline or driven by Cre expressed under the control of a cell-type specific promoter. Some GCaMP-expressing lines have been reported to cause epileptiform activity on their own, possibly due to the presence of high levels of the Cre enzyme, or the presence of the Ca^{++}-buffering GCaMP molecule early in development [25]. This should be considered when choosing a GCaMP-expressing mouse line. Previously, we used a Snap-25 GCaMP line that expresses GCaMP6s pan-neuronally, as this line was reported not to cause epileptiform activity on its own [25]. Although the expressing level of GCaMP in this line is dim relative to other lines, in our experience high levels of GCaMP are not necessary for mesoscale imaging (*see* **Note 1**). If a brighter line is required, it's possible to suppress the expression of GCaMP until later in development to avoid these issues. However, this approach has, to our knowledge, only been demonstrated in wild-type mice, and care should be taken in any epileptic line to ensure the transgene is not altering or generating disease phenotypes.

Another option is to inject AAV virus that expresses one of many available Ca^{++} indicators. In the past, such viruses have most often been used to drive indicator expression in a restricted tissue volume following focal injection, but several recent advances have made it possible to achieve widespread expression in the CNS. One of these approaches relies on the use of a viral capsid that confers a tropism for CNS tissue (PHP.eB). These viruses are typically intravenously injected into adult mice, most often via the retro-orbital sinus [26]. Another promising option is to use the common serotype AAV9 which, when injected into the transverse sinus of P0 mouse pups, can deliver high levels of transgene expression throughout the brain [27]. All of these approaches have the potential to achieve widespread CNS expression of Ca^{++} indicators necessary for effective mesoscale imaging (*see* **Note 2**). The choice of using a transgenic line versus virus injection must be weighed by considering the complexity of a specific breeding scheme versus the extra work involved in performing the injections.

3 Methods

3.1 Head-Plate Surgery

Imaging under the mesoscope requires surgery to expose the skull for optical access and attach a headplate that is used to fix the animal under the objective lens. Anesthesia can be induced using 4% isoflurane gas and the levels maintained at 1–2% during surgery. Typically, mice are secured to a stereotaxic frame using ear bars, at least for the initial portion of surgery. Once the scalp has been removed, the dorsal skull exposed and cleaned, and the headplate securely attached to the skull, it is possible to shift head fixation to the headplate for greater stability. This can be useful if delicate operations such as craniotomy for electrodes, a glass imaging window, or drug delivery cannulae are incorporated into the surgery.

There are numerous published designs for headplates or headbars suitable for mesoscale imaging [28, 29]. For maximal stability, especially during imaging sessions when the animal is awake and moving, these are typically milled or cut out of a stiff metal such as aluminum or titanium. To achieve optical access, the skull can be removed and replaced with a glass window, thinned, or simply made transparent with cyanoacrylate application (Loctite 4305). Replacement with a glass window gives the best optical access and spatial resolution as well as the ability to perform 2-photon imaging in the same tissue as MeCI, but it also takes the most time and effort to learn. Furthermore, the longer surgery time and potential for trauma may exacerbate epilepsy or weaken already epileptic mouse strains. In our experience, imaging through a cyanoacrylate-coated skull gives signals that are more than adequate to study epileptic brain activity, which is already characterized by its high intensity relative to ongoing activity. It also is sufficient for detecting neural events associated with locomotion, sensory stimulation, and spontaneous activity. After surgery, the head plate can be filled with silicone elastomer (Kwik-Sil, WPI) for protection of the skull during recovery and in between imaging sessions.

3.2 Data Acquisition

The simplest mesoscale imaging setups use tandem-lens configurations with objective and image forming lenses, often consisting of consumer photographic camera lenses. To collect mesoscale fluorescence images, we constructed a tandem-lens epifluorescent macroscope (Fig. 1) [17]. It incorporated objective and image forming lenses with 50- and 105-mm focal lengths, respectively (Nikon Nikkor 50 mm f/1.4 and 105 mm f/1.8), and produced 2.1X magnification. A broadband LED light source (X-Cite 120 LED) was used for illumination with a GFP filter set (Chroma 49002) in 50-mm circular glass, mounted in Thorlabs SM2 hardware in the configuration illustrated in Fig. 1. A 530-nm LED (Thorlabs, M530F2, LEDD1B) coupled to an optical fiber pointed obliquely at the cortical surface was used to illuminate every other frame, and the resulting reflectance images were used to correct the GCaMP signal for hemodynamic artifacts [30]. To toggle illumination, both LEDs were driven using the camera frame exposure signals and routed with a microcontroller board (Arduino Uno). All images were captured with an Andor Zyla 5.5 sCMOS camera (Camera Link 10-tap) controlled by the open-source microscopy software μManager from within Matlab.

Mice were secured under the objective lens by positioning a pair of fixed threaded rods through the outermost holes of their head plate, which was clamped in place with thumb screws. The threaded rods were mounted on custom steel plates attached to 25-mm square optical rails (Thorlabs, 9 in long), themselves supported by post bracket platforms (Thorlabs, C1515) on 1.5″ optical posts (Thorlabs, P14). In this position, mice were free to run on

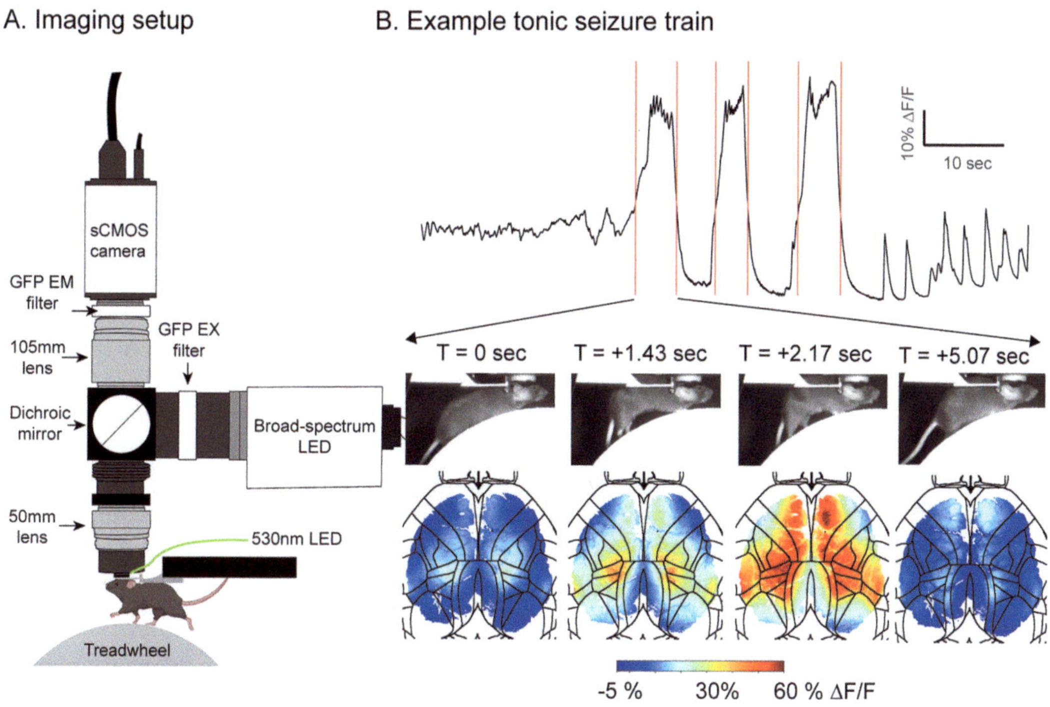

Fig. 1 Widefield imaging of dorsal cortex in Kcnt1$^{m/m}$ and WT mice. (**a**) A schematic of the tandem-lens, epifluorescent macroscope used for in vivo, awake widefield Ca^{++} imaging. Every other frame is illuminated with a 530 nm LED pointed obliquely at the skull surface; the resulting reflectance images are used to correct the GCaMP signal for hemodynamic artifacts. (**b**) An example seizure train. The trace shows average $\Delta F/F$ signal calculated across all pixels involved in at least one of the three seizures shown. Red lines demarcate seizure boundaries. Images below show still frames of mouse body video and simultaneous $\Delta F/F$ frames of the dorsal cortex, with black lines marking the Allen Mouse CCF area borders. Time relative to seizure start is indicated above each pair of frames

a treadwheel made from a Styrofoam cylinder mounted on a custom axle. During imaging, wheel position was recorded with an optical shaft encoder (US Digital S6-2500-236-IE-S-B) and a PCIe DAQ system (National Instruments, Part #s 781045-01, 782536-01, 192061-01).

Video of the mouse's body should also be collected during imaging. This will allow for the correlation of the behavioral manifestations of seizure with the neural signal. Our body video was collected under infrared illumination with an Allied Vision GC75 camera and synchronized with the video of GCaMP fluorescence.

4 Data Analysis

One of the most challenging aspects of mesoscale Ca^{++} imaging is how to analyze and interpret the large and complex data sets collected with this modality. Because we are interested in two distinct regimes of brain activity, both ictal and interictal, and these regimes are different in terms of their characteristics, we will discuss data analysis in four parts; (1) preprocessing, (2) detection and analysis of epileptiform activity, (3) detection of non-epileptiform activity in both WT and epileptic mice and comparison between genotypes, and (4) relating epileptiform activity to differences in non-epileptic activity between genotypes.

4.1 Basic Image Processing Steps and Analysis

For most applications, raw MeCI data must be preprocessed prior to analysis. Some common preprocessing goals are to suppress noise, correct for variation in indicator expression levels and brain illumination, and compress the data [31, 32]. These can be achieved with a range of operations including the subtraction of a dark frame, $\Delta F/F$ transformation, subtracting an estimate of hemodynamic artifacts, compression and denoising using a dimensionality reduction technique such as singular value decomposition (SVD), detrending, high pass filtering, and bandpass filtering in the heartbeat frequency range. In our work, we use Matlab programs, either running on a local PC in the lab or on our university computing cluster, for preprocessing, although any language capable of executing these straightforward calculations will serve.

Dark frame subtraction is a technique to remove the fixed-pattern noise associated with a camera sensor from each frame MeCI data. This is often done by capturing a short series (>20) of frames either immediately before or after an imaging session in which the illumination source is turned off while all other settings are left untouched. These frames are then stacked, we use a simple average, and subtracted from all frames that were collected with the illumination source turned on. The value and impact of this step depends on the relative magnitude of fixed pattern noise in your frames, which is a function of the camera being used, exposure time, illumination intensity, and temperature among other factors. But, as it is relatively easy to implement and doesn't require much by way of computing or storage resources, it is likely a good idea to incorporate dark frame subtraction into one's data collection and preprocessing pipeline.

The $\Delta F/F$ transformation is widely used with Ca^{++} imaging data of all kinds to correct for differences in fluorescence intensity that result from differences in indicator expression level, illumination intensity, and light scattering. Using transgenic lines or viruses to drive indicator expression throughout the CNS typically does not result in large variations in expression level within a MeCI field of view. However, such variation can occur if a virus was injected directly into neural tissue or transgenic expression is controlled by a Cre line with nonuniform expression across the brain. Additionally, due to the large fields of view produced by tandem lens macroscopes, it is quite difficult to achieve uniform illumination of the tissue under study. Furthermore, because the mouse skull and brain are curved, light that enters the sample at an oblique angle will have to traverse more tissue to reach a given depth relative to the brain's surface than light entering on a perpendicular path [33]. This will impose differing levels of scatter depending on sample geometry, although the magnitude and significance of this effect remains poorly characterized. The $\Delta F/F$ depends on a measure of the fluorescence intensity in the absence of neural activity (baseline fluorescence, F_0) at each pixel or within each region of interest (ROI) being analyzed. Because neural activity is sparse in many brain areas this is often calculated as the average fluorescence taken across all frames in a dataset. Alternatively, it can be calculated as the average over some range of frames in which activity is low, such as the bottom 20th percentile of frames ranked by intensity. This value (F_o) is then subtracted from the value in each frame and the resulting difference is expressed as a fraction of baseline fluorescence, this is then multiplied by 100 to yield a percentage. In this way the $\Delta F/F$ transformation allows for more meaningful comparison of signals from locations that vary in absolute brightness.

Unfortunately, light at both the excitation and emission wavelengths of calcium indicator proteins is absorbed by molecules within the blood, including flavoproteins and hemoglobin. As blood flow and oxygenation change during an imaging session, the degree to which imaging light is absorbed changes, producing signal fluctuations that are independent of indicator fluorescence. Strategies to correct this vary widely across research groups, ranging from doing nothing to subtracting an estimate based on multi-wavelength measurement of blood absorption and modeling [15, 34]. Our approach consists of subtracting an estimate of the hemodynamic signal based on intensity fluctuations measured in images of the cortical surface obliquely illuminated by an 530 nm LED, which does not excite GCaMP but is still absorbed by the blood to a similar degree as imaging light. By alternating between reflectance and epifluorescent illumination, we collect an alternating series of frames with and without GCaMP signal but sharing a similar hemodynamic signal. During preprocessing these frames are deinterleaved and independently used to calculate the fractional change in fluorescence/reflectance at each pixel, in each frame (see $\Delta F/F$ discussion above). We then subtract fractional reflectance, our estimate of the hemodynamic signal, from fractional fluorescence, which contains both GCaMP and hemodynamic signals [30]. This subtraction has been performed in a variety of different ways across studies, but whether its specific form has any meaningful impact on results remains unclear [35–37]. A variation on this strategy takes advantage of the fact that GCaMP fluorescence is Ca^{++} independent when the molecule is excited using violet light (407 nm). Here, instead of collecting reflectance images under oblique illumination at the indicator's emission wavelength, epifluorescent images illuminated with violet provide a measure of all Ca^{++} interdependent signal fluctuations which can then be subtracted from frames illuminated with blue light [38]. Hemodynamic contamination can also be reduced by masking out large blood vessels in images and bandpass filtering signals in the heartbeat frequency range (9–13 Hz).

Signal fluctuations in MeCI data occur on many timescales reflecting both fast and slow changes in neural activity, hemodynamics, and, depending on indicator and imaging conditions, indicator bleach. Because most investigators are interested in relatively fast changes in neural activity, a high pass filter is commonly used to suppress the slow component of these signals. It's worth noting that biological insight can also be gained from the slow component of MeCI signals, and filters should be chosen based on the particular question being investigated [39]. Additionally, detrending can be used to correct for indicator bleaching.

The image stacks produced by MeCI are often quite large and noisy. Both issues can be addressed using a dimensionality reduction technique such as singular value decomposition (SVD). This operation factors the data and allows it to be reconstructed through the multiplication of several much smaller matrices. Noise can be rejected during this reconstruction by using only those portions of these matrices necessary to capture a given fraction of the variance present in the raw data. Many of the preprocessing steps described in this section, such as filtering, can also be performed on the data in factored form.

4.2 Alignment and Segmentation of the Imaging Field

To make meaningful comparisons between MeCI data collected from different imaging sessions and animals, it is often necessary to identify equivalent brain areas across datasets. A common approach is to register images to the Allen Mouse Common Coordinate Framework (CCF), a reference map of mouse brain areas produced by the Allen Institute [40]. We have adopted a simple method for achieving this in our work in which we manually place several control points on prominent anatomical features present in our FOV and then use a nonreflective similarity transformation to bring each session into alignment with the CCF. Based on previously published methods, we selected the anterior edge of the cortex, midway along the mediolateral extent of each olfactory bulb and at the midline, as well as on the posterior edge at the midline for control point placement [38, 41]. The image we use for control point placement is a histogram equalized average of GCaMP6 fluorescence in each session. This process can also be automated and software packages that perform CCF registration have been published [42].

Registration to the CCF provides a segmentation of the brain based on an assumption of stereotypy across animals. While this has proven to be a useful comparative framework, it ignores individual variation in brain parcellation and cannot account for any pathological deviations from normality that may occur in a disease model. There are several analytical tools that can be used to investigate the spatial organization of brain activity within an individual that do not require making these assumptions. One way to examine the relationship between activity at any given location within a MeCI FOV and all other locations can be visualized by calculating seed pixel correlation maps (SPCMs). These are generated by calculating the correlation between signals on a single pixel (the seed pixel) and all other pixels in the FOV. The resulting correlation coefficients are then displayed as an image that indicates regions of tissue where activity is correlated, or anti-correlated, with activity at the seed pixel. In animals with normal cortical arealization, SPCMs often reveal boundaries that correspond closely to area borders in the CCF. Another commonly employed activity-based segmentation technique is independent component analysis (ICA), which identifies maximally independent subcomponents in multivariate signals. In the context of MeCI, these subcomponents are typically interpreted as functional modules within large-scale brain activity [37, 43–45]. However, for comparison across studies, functional units identified using these techniques need to be related to reference brain areas and labels at some point. Finally, ground truth area borders can be determined for some sensory areas such as V1 or barrel cortex using sensory stimulation. While it is not possible to map all areas in this way, it can provide an important test of areal segmentation.

Comparing spatial patterns of activity across sessions and mice in greater detail than area labels benefits from bringing all data into alignment, such that a given pixel in one image covers the same part of the brain in another. We have achieved this in our work using a three-step alignment procedure, beginning with a within-animal affine registration based on histogram equalized average GCaMP6 images from each session. Once all sessions from a given mouse were registered to each other, we aligned sessions across mice. Because details of skull anatomy, vasculature, and the cranial window differ substantially across mice, we performed this step not on average GCaMP images but rather SPCMs of primary visual cortex, a prominent and well-defined cortical area present in all animals. To do this, we generated SPCMs using a 50-by-50 grid of seed pixels covering the whole FOV. We then took the average across maps with seed pixels in primary visual cortex (VISp) in each session and the grand average across all sessions from a given mouse. Using these maps, we registered image sets from all mice in our study to each other with a similarity transformation. Finally, we refined the cross-animal registration by recalculating a similarity transformation to align each mouse's grand average VISp SPCM to the average taken across mice following the second step.

4.3 Detecting and Analyzing Spontaneous Seizures

Because seizures are sudden uncontrolled bursts of brain activity with behavioral manifestations, they can be identified based on several features of the data collected in MeCI experiments. In the Ca^{++} imaging data itself we can leverage the fact that neural activity rapidly increases to abnormally high levels during seizure onset to use the rate of change of $\Delta F/F$ signals as well as their absolute magnitude as criteria in seizure identification. However, these direct measurements of neural activity need to be considered in light of the animal's behavior which should be recorded with a separate camera in all MeCI sessions. In our work, we require behavioral correlates such as tonic arching, tail extension, facial jerking, and forelimb clonus, to classify an event as a seizure.

Our method for seizure identification involves several steps to identify candidate events before visual examination of brain activity and mouse behavior to form a final judgment. Because seizures are rare, we set the bar low for events to be considered as potential seizures, reasoning that its more important to catch all seizures in our data than save time watching videos. We begin by calculating the average pixel-wise $\Delta F/F$ difference between successive frames in a session, as well as the fraction of pixels in each frame above a hand-tuned threshold of 20% $\Delta F/F$. After smoothing these vectors with moving average filters, we calculate the envelope of the average difference vector and identify peaks in it and the fraction of pixels above threshold vector. Peaks that are coincident in both vectors identify periods of time when the $\Delta F/F$ signal was changing rapidly and a large number of pixels reported high intensity activity. We found that defining coincidence as occurring within 15 s of one another works well. This is a window wide enough to accommodate smoothing and a reasonable temporal offset in peak values of these two measurements. Having found a set of coincident peaks, we then need to identify the span of frames containing the potential seizure. We do this based on the peak present in the smoothed average frame difference envelope, by finding the first and last frames on either side of the peak above this signal's average. After identifying these potential seizures, we screened videos of the mouse's behavior for confirmation of a behavioral seizure by watching simultaneous cortical $\Delta F/F_0$ and mouse body videos. In this system, events that meet both $\Delta F/F$ and behavioral criteria are then classified as seizures (*see* **Note 3**).

After seizures are identified, many analyses, such as mapping propagation, depend on being able to identify precisely when the seizure began at any given location in the FOV (*see* **Note 4**). We find the precise borders of seizure activity at each pixel by using the non-seizure data we have from the mouse to define a $\Delta F/F_0$ seizure threshold set to the 99th percentile of all non-seizure, non-IED event peaks from all sessions. This threshold is calculated and applied on an animal-to-animal basis. For each seizure, we identify the first frame in which at least 1% of pixels were above threshold

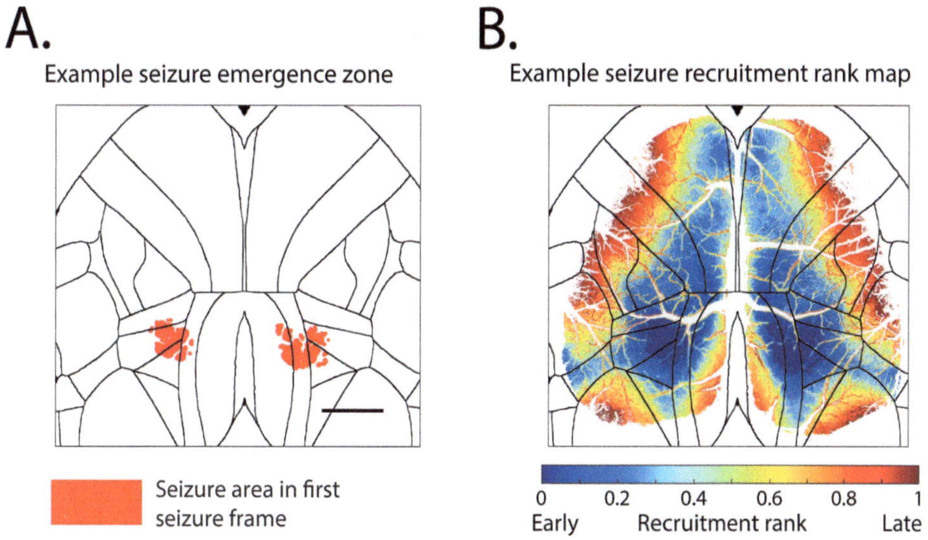

Fig. 2 Analysis of seizure propagation patterns. (**a**) An example seizure emergence zone map. Red pixels are those above seizure threshold in the first frame containing seizure activity. Black lines mark the borders of the Allen Mouse CCF areas. Scale bar is 1 mm. (**b**) An example seizure recruitment rank heatmap (same seizure as panel **a**). The value of each pixel is its recruitment time rank

(seizure start, Fig. 2a), the first frame in which the number of pixels above threshold was maximal (end of seizure growth phase), and the first trailing frame in which less than 1% of pixels was above threshold (seizure termination). Because IEDs occurring close to seizures were sometimes picked up while identifying seizure boundaries, any "seizures" that met our IED criteria were then eliminated. The order in which pixels or areas are recruited into the seizure can be represented as a recruitment map (Fig. 2b).

4.4 Detecting and Analyzing Interictal Activity

Although recurrent spontaneous seizures are the defining feature of epilepsy, it is increasingly recognized that seizures do not occur on top of otherwise normal brain activity [46]. Although the complex relationship between cause and effect is not well understood, in the case of genetic epilepsies it is almost certainly the case that the underlying genetic change modifies brain activity outside of seizures in ways that predispose the brain to seizure activity. Studying animal models of chronic epilepsy provides the unique opportunity to compare brain activity outside of seizure between individuals that share the same genetic background and life history and differ only in disease status. To accomplish this with MeCI requires a method for detecting interictal brain activity, categorizing it in a way that permits equivalent forms of activity to be identified across individuals, measuring features of this activity that permit quantitative analysis, and statistical testing to evaluate any group differences that are measured.

Brain activity in MeCI data appears as a complex and dynamic pattern of shifting fluorescence. Because there are no clear quanta of activity, such as spikes or discrete Ca^{++} events, delineating, categorizing, and compiling statistics on this activity is uniquely challenging. There is by no means a clear consensus on how best to handle this data and extract biologically meaningful measurements. Although this is an active area of research, new and increasingly sophisticated methods are being published with some regularity at the time of this writing. In a recent study, we investigated interictal activity in a mouse model of KCNT1-related epilepsy using MeCI [13]. Our goal in doing so was to determine whether cortical activity differed systematically between epileptic and healthy mice, and if so, how any such differences related to the epileptic activity the mice suffered. We observed numerous spontaneous seizures and many IEDs among the epileptic mice, and from this learned that some cortical areas and individual mice were far more vulnerable to epileptic activity than others. To find out whether this vulnerability correlated with disrupted interictal activity, we took a simple approach to detecting, categorizing, and comparing cortical events in this data.

We started by identifying moments during which cortical activity levels were elevated somewhere within our FOV (Fig. 3). Because the number of pixels in our data was high, we first spatially down-sampled our data by averaging pixel values in a grid of rectangular ROIs laid over each frame from a session (Fig. 3b). We then used the built-in Matlab function findpeaks() to identify peaks in the $\Delta F/F_0$ trace associated with each ROI (Fig. 3c). As mentioned above, episodes of activity in MeCI data are spatially extensive and evolve through time. For this reason, episodes of activity often produced peaks in multiple ROIs and frames. We categorized any peaks that occurred within 250 ms of each other as belonging to the same "event." Once we had detected events in ROIs, we repeated event detection with findpeaks() at full resolution for all pixels in the FOV over a time window that spanned the event. This produced a list of pixels involved in event, and for each a peak time, magnitude in $\Delta F/F_0$, and duration at half peak. We then quantified the spatial profile of each event as the fraction of activity that occurred in each CCF cortical area within our FOV, producing a vector for each event. To group events with similar profiles, we pooled all these vectors, from all sessions and mice, both wild type and KCNT1 mutant, and clustered them using hierarchical clustering, deciding where to cut the cluster tree based on silhouette analysis. We considered events in the same cluster as being of the same type, allowing us to compare equivalent events in different animals and across genotypes (Fig. 4). However, the significance of these differences needed to be tested, a task made difficult by the fact that the data did not conform to well-described distributions and contained many layers of dependencies.

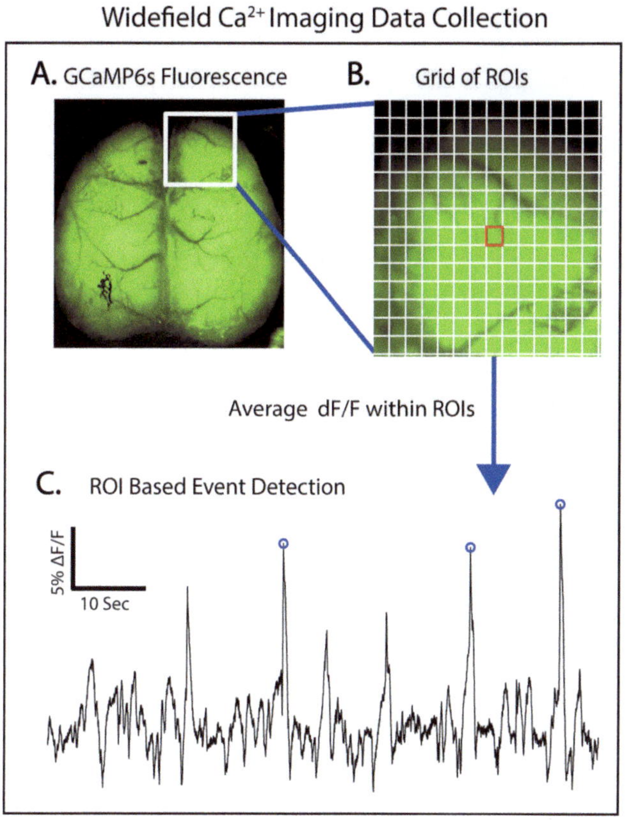

Fig. 3 Identification of neural events in MeCI data. (**a**) GCaMP6s fluorescence in the brain of a Kcnt1$^{m/m}$ mouse expressing GCaMP6s under control of the *Snap-25* promoter. (**b**) Illustration of the grid of regions of interest (ROIs) used in event detection superimposed on a portion of the GCaMP6s fluorescence image. (**c**) $\Delta F/F$ peaks with a prominence greater than 15% were identified in each ROI

To contend with these issues, we devised a series of permutation tests in which shuffling and resampling were used to break the association of interest before recalculating the test statistic. By comparing the true test statistic to distributions attained through repeated shuffling, we were able to evaluate the significance of our measurements (Fig. 5). This work showed that interictal activity in vulnerable cortical areas occurred much more frequently in epileptic mice and that the peak intensity of cortical events was elevated in mice with high disease burden, regardless of where they occurred. This demonstrates that interictal activity features can be used to predict the localization and severity of epileptic activity in the chronically epileptic brain. It also suggests that seizures and IEDs are simply the most obvious and dramatic examples of how this gene variant reshapes brain activity in the disease state, opening the door to future investigations and ultimately a more complete understanding of how molecular dysfunction alters the large-scale

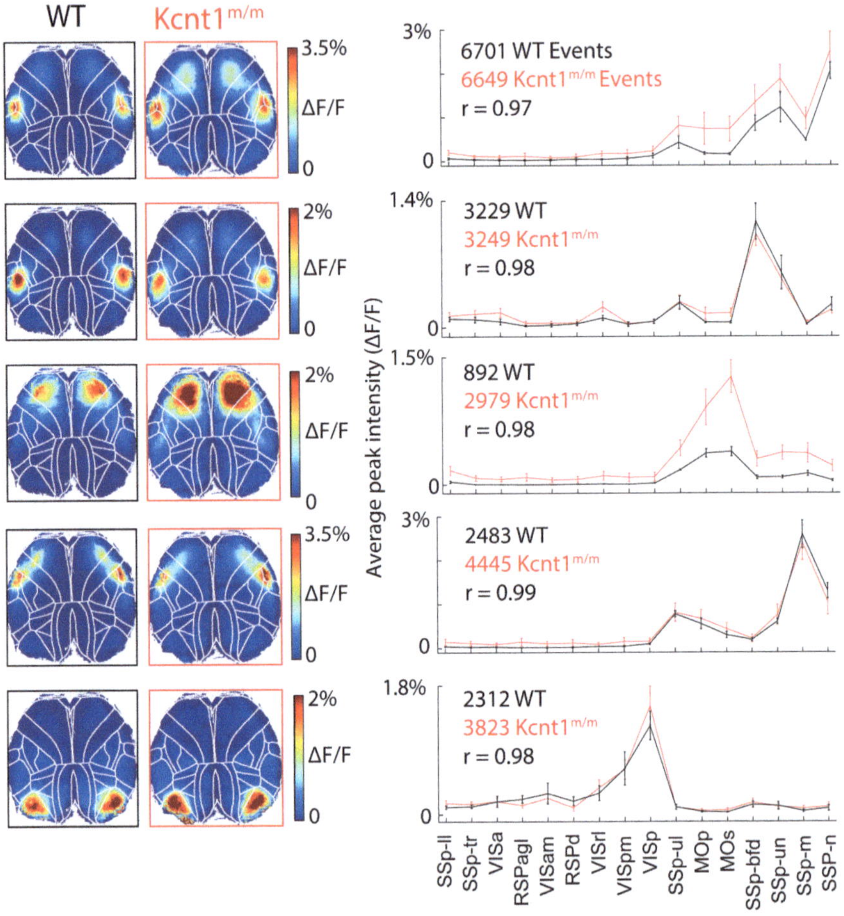

Fig. 4 Spatially clustering events in WT and genetic mouse model of epilepsy to compare interictal activity features. The figure shows five representative event clusters from both WT and the KCNT1 Y796H mouse model. Images on the left show the average event peak intensity at all pixels for each genotype and cluster. These were calculated using event frames in which each pixel was assigned its Δ*F*/*F* peak intensity value, and pixels not involved in the event were set to zero. The line plots show the average peak intensity of neural activity within each Allen Mouse CCF area imaged. These values were constructed by taking the average Δ*F*/*F* peak value across all pixels involved in an event in each area. Pixels not involved in the event were set to zero. The lines represent the averages for each group, taken across session averages for each mouse. Error bars show SEM across mouse averages. Text above each plot gives the total number of events in the cluster from each group and the $Kcnt1^{m/m}$ to WT Pearsons correlation coefficient. The high correlation between the lines indicates that the spatial footprint of the clusters is highly similar, even though elevated event intensity and event rate are present in some event clusters, especially the middle cluster involving the secondary motor cortex

brain activity that underlies patient symptoms. In the present context, this work offers an example of how epileptic and interictal activity can be measured in a chronically epileptic brain, analyzed, and integrated to yield a more holistic view of the disease state than studying either in isolation allows.

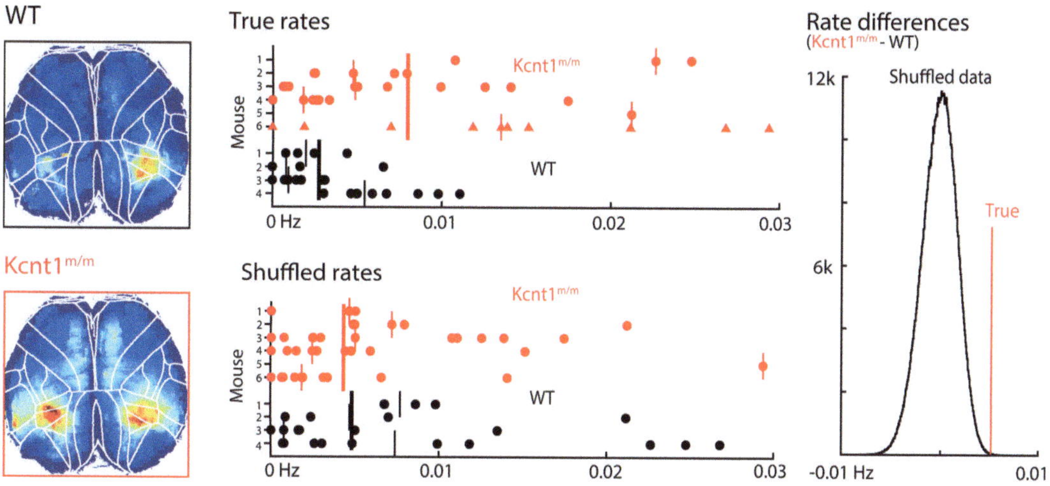

Fig. 5 Example of the permutation test approach to determining whether the rate of neural event types differs between genotypes. Event peak intensity images from WT and Kcnt1 $^{m/m}$ mice are shown on the left. The upper scatter plot shows event rates of all sessions in each mouse. Each row on the *y*-axis represents all sessions from one mouse and each dot represents the mean event rate from one imaging session. The lower plot shows the same data after a random shuffle in which the event rate, calculated in each session from each mouse, was pooled, shuffled, and then redrawn such that each mouse had the same number of sessions before and after the shuffle. Short lines are animal medians and long lines are group means weighted by session number. Histogram to the right shows the distribution of weighted group average differences obtained from one million shuffles; the red line indicates the true group difference (>99.96% of shuffled group differences)

However, this approach is also limited in a number of important ways. By pooling event peaks that occurred at different locations with short latency, we discarded the details of how activity unfolds through time across the cortex. A promising approach used by other groups to study this evolution relies on calculating optic flow in MeCI data [47]. Furthermore, the clustering that we performed yielded a relatively small number of clusters (10) that undoubtedly represent only the coarsest level of the true taxonomy, if any such exists.

5 Conclusion

Only a handful of studies have used mesoscale Ca^{++} imaging of neural activity to study epilepsy in mouse models, and even fewer have used chronic epilepsy models. This approach, however, holds great promise for addressing key questions about seizure initiation, spread, and termination. When applied to chronic models, and especially the large and ever-growing number of mouse models with orthologous human genetic variants that cause epilepsy, it can be used to study not only seizures but also patterns of interictal activity. This opens up the possibility of comparing interictal brain activity to ictal activity within epileptic animals, and to systematically compare interictal activity in epilepsy models to healthy controls. This information can identify cortical areas that are particularly susceptible, or resistant, to pathological activity, and how severely brain activity deviates from normal across large portions of the brain. As mesoscale Ca^{++} imaging is relatively inexpensive to implement, it could be adopted as part of a broader phenotyping strategy for epilepsy models. At its most basic level, the results can be used to target mechanistic studies to specific cortical regions in these models, such as patch-clamp recordings, cellular level Ca^{++} imaging, gene expression, and morphological measurements. Additionally, mesoscale Ca^{++} imaging could also be used to determine where and how brain activity needs to be modulated to holistically restore normal function, reveal off-target treatment effects, and facilitating therapy optimization to deliver maximal benefit with minimal side-effect potential.

6 Notes

1. Ca^{++} indicators with red-shifted excitation-emission spectra are now available [48, 49]. Although they are not as bright as their GFP-based counterparts, in our experience they are serviceable for MeCI. They also open up the possibility of simultaneous MeCI of different neuronal populations or biosignals in two different colors.
2. The main advantage offered by MeCI is its large FOV, which makes it easy to simultaneously monitor large swatch of brain tissue at low cost. To maximize the amount of information each experiment yields most investigators desire widespread indicator expression. However, this is by no means necessary, the large field of view can just as profitably be exploited to image a set of separate brain regions, each expressing indicator protein introduced with a focal virus injection. Furthermore, it's also possible to isolate signals arising from the projections of a defined neuronal population of neurons by targeting it with focal injection of indicator virus, this includes deep brain regions that project superficially. Depending on the specifics of the model under study, many questions can likely be answered, perhaps most appropriately, without widespread or pan-neuronal indicator expression.
3. On numerous occasions we have observed that the intense seizures, both spontaneous and pharmacologically induced, occasionally terminate with intense, slow-moving waves of fluorescence that likely correspond to cortical spreading depression. It is interesting to note that, unlike seizures, these waves appear to propagate uniformly away from their point of origin, unmodified by the underlying synaptic connectivity of

the tissue they pass through. Yet these waves are also often bilaterally symmetrical, raising an interesting question of how this symmetry is generated if propagation is independent of synaptic connectivity. With regard to the issues here at hand, we found that these waves have a strong effect on the singular values we derived from sessions in which they occurred and distorted the SVD reconstructions we obtained. Our solution was simply to remove them from the data before applying SVD, although more elegant solutions likely exist.

4. We have summarized the spatial organization of seizures using two measurements: emergence zones and recruitment maps. We define emergence zones simply as the group of pixels above seizure threshold in the first frame containing seizure activity. These regions may be the site of seizure initiation or simply the first location a seizure appears within a FOV. Comparing onset in the emergence zone to that of behavioral seizure symptoms can provide some evidence as to which is more likely. To construct recruitment maps, we generate an image in which each pixel value is assigned as the time that pixel was recruited into the seizure (Fig. 2b). Transforming recruitment times into ranks highlights the spatial propagation pattern independent of propagation rate, which typically varies significantly between periods when the extent of seizing tissue remains relatively constant and periods when new territory is drawn into the seizure.

References

1. Scharfman HE (2007) The neurobiology of epilepsy. Curr Neurol Neurosci Rep 7(4): 348–354. https://doi.org/10.1007/s11910-007-0053-z
2. Holmes GL (2016) Effect of seizures on the developing brain and cognition. Semin Pediatr Neurol 23(2):120–126. https://doi.org/10.1016/j.spen.2016.05.001
3. Noebels J (2017) Precision physiology and rescue of brain ion channel disorders. J Gen Physiol 149(5):533–546. https://doi.org/10.1085/jgp.201711759
4. Fisher RS, Scharfman HE, deCurtis M (2014) How can we identify ictal and interictal abnormal activity? Adv Exp Med Biol 813:3–23. https://doi.org/10.1007/978-94-017-8914-1_1
5. Thio BJ, Grill WM (2023) Relative contributions of different neural sources to the EEG. NeuroImage 275:120179. https://doi.org/10.1016/j.neuroimage.2023.120179
6. Ekstrom A (2010) How and when the fMRI BOLD signal relates to underlying neural activity: the danger in dissociation. Brain Res Rev 62(2):233–244. https://doi.org/10.1016/j.brainresrev.2009.12.004
7. Rossi LF, Kullmann DM, Wykes RC (2018) The enlightened brain: novel imaging methods focus on epileptic networks at multiple scales. Front Cell Neurosci 12:82. https://doi.org/10.3389/fncel.2018.00082
8. Luo P, Yang F, Li J, Niemeyer JE, Zhan F, Estin J, Zhao M, Li D, Lin W, Liou JY, Ma H, Schwartz TH (2023) Excitatory-inhibitory mismatch shapes node recruitment in an epileptic network. Epilepsia 64:1939. https://doi.org/10.1111/epi.17638
9. Rossi LF, Wykes RC, Kullmann DM, Carandini M (2017) Focal cortical seizures start as standing waves and propagate respecting homotopic connectivity. Nat Commun 8(1):217. https://doi.org/10.1038/s41467-017-00159-6
10. Liou JY, Ma H, Wenzel M, Zhao M, Baird-Daniel E, Smith EH, Daniel A, Emerson R, Yuste R, Schwartz TH, Schevon CA (2018) Role of inhibitory control in modulating focal

seizure spread. Brain 141(7):2083–2097. https://doi.org/10.1093/brain/awy116

11. Kandratavicius L, Balista PA, Lopes-Aguiar C, Ruggiero RN, Umeoka EH, Garcia-Cairasco-N, Bueno-Junior LS, Leite JP (2014) Animal models of epilepsy: use and limitations. Neuropsychiatr Dis Treat 10:1693–1705. https://doi.org/10.2147/NDT.S50371
12. Loscher W (2011) Critical review of current animal models of seizures and epilepsy used in the discovery and development of new antiepileptic drugs. Seizure 20(5):359–368. https://doi.org/10.1016/j.seizure.2011.01.003
13. Tobin WF, Weston MC (2023) Distinct features of interictal activity predict seizure localization and burden in a mouse model of childhood epilepsy. J Neurosci 43:5076. https://doi.org/10.1523/JNEUROSCI.2205-22.2023
14. Ren C, Komiyama T (2021) Characterizing cortex-wide dynamics with wide-field calcium imaging. J Neurosci 41(19):4160–4168. https://doi.org/10.1523/JNEUROSCI.3003-20.2021
15. Vanni MP, Murphy TH (2014) Mesoscale transcranial spontaneous activity mapping in GCaMP3 transgenic mice reveals extensive reciprocal connections between areas of somatomotor cortex. J Neurosci 34(48):15931–15946. https://doi.org/10.1523/JNEUROSCI.1818-14.2014
16. Connor SA, Ammendrup-Johnsen I, Chan AW, Kishimoto Y, Murayama C, Kurihara N, Tada A, Ge Y, Lu H, Yan R, LeDue JM, Matsumoto H, Kiyonari H, Kirino Y, Matsuzaki F, Suzuki T, Murphy TH, Wang YT, Yamamoto T, Craig AM (2016) Altered cortical dynamics and cognitive function upon haploinsufficiency of the autism-linked excitatory synaptic suppressor MDGA2. Neuron 91(5):1052–1068. https://doi.org/10.1016/j.neuron.2016.08.016
17. Ratzlaff EH, Grinvald A (1991) A tandem-lens epifluorescence macroscope: hundred-fold brightness advantage for wide-field imaging. J Neurosci Methods 36(2–3):127–137. https://doi.org/10.1016/0165-0270(91)90038-2
18. Inoue M (2021) Genetically encoded calcium indicators to probe complex brain circuit dynamics in vivo. Neurosci Res 169:2–8. https://doi.org/10.1016/j.neures.2020.05.013
19. Burrows DRW, Samarut E, Liu J, Baraban SC, Richardson MP, Meyer MP, Rosch RE (2020) Imaging epilepsy in larval zebrafish. Eur J Paediatr Neurol 24:70–80. https://doi.org/10.1016/j.ejpn.2020.01.006
20. Wang W, Frankel WN (2021) Overlaps, gaps, and complexities of mouse models of Developmental and Epileptic Encephalopathy. Neurobiol Dis 148:105220. https://doi.org/10.1016/j.nbd.2020.105220
21. Kadam SD, D'Ambrosio R, Duveau V, Roucard C, Garcia-Cairasco N, Ikeda A, de Curtis M, Galanopoulou AS, Kelly KM (2017) Methodological standards and interpretation of video-electroencephalography in adult control rodents. A TASK1-WG1 report of the AES/ILAE Translational Task Force of the ILAE. Epilepsia 58(Suppl 4):10–27. https://doi.org/10.1111/epi.13903
22. Maheshwari A (2020) Rodent EEG: expanding the spectrum of analysis. Epilepsy Curr 20(3):149–153. https://doi.org/10.1177/1535759720921377
23. Shore AN, Colombo S, Tobin WF, Petri S, Cullen ER, Dominguez S, Bostick CD, Beaumont MA, Williams D, Khodagholy D, Yang M, Lutz CM, Peng Y, Gelinas JN, Goldstein DB, Boland MJ, Frankel WN, Weston MC (2020) Reduced GABAergic neuron excitability, altered synaptic connectivity, and seizures in a KCNT1 gain-of-function mouse model of childhood epilepsy. Cell Rep 33:108303
24. Oakley JC, Kalume F, Yu FH, Scheuer T, Catterall WA (2009) Temperature- and age-dependent seizures in a mouse model of severe myoclonic epilepsy in infancy. Proc Natl Acad Sci USA 106:3994. https://doi.org/10.1073/pnas.0813330106
25. Steinmetz NA, Buetfering C, Lecoq J, Lee CR, Peters AJ, Jacobs EAK, Coen P, Ollerenshaw DR, Valley MT, de Vries SEJ, Garrett M, Zhuang J, Groblewski PA, Manavi S, Miles J, White C, Lee E, Griffin F, Larkin JD, Roll K, Cross S, Nguyen TV, Larsen R, Pendergraft J, Daigle T, Tasic B, Thompson CL, Waters J, Olsen S, Margolis DJ, Zeng H, Hausser M, Carandini M, Harris KD (2017) Aberrant cortical activity in multiple GCaMP6-expressing transgenic mouse lines. eNeuro 4(5):ENEURO.0207. https://doi.org/10.1523/ENEURO.0207-17.2017
26. Chan KY, Jang MJ, Yoo BB, Greenbaum A, Ravi N, Wu WL, Sanchez-Guardado L, Lois C, Mazmanian SK, Deverman BE, Gradinaru V (2017) Engineered AAVs for efficient noninvasive gene delivery to the central and peripheral nervous systems. Nat Neurosci 20(8):1172–1179. https://doi.org/10.1038/nn.4593
27. Hamodi AS, Martinez Sabino A, Fitzgerald ND, Moschou D, Crair MC (2020) Transverse sinus injections drive robust whole-brain

expression of transgenes. eLife 9:e53639. https://doi.org/10.7554/eLife.53639

28. Goldey GJ, Roumis DK, Glickfeld LL, Kerlin AM, Reid RC, Bonin V, Schafer DP, Andermann ML (2014) Removable cranial windows for long-term imaging in awake mice. Nat Protoc 9(11):2515–2538. https://doi.org/10.1038/nprot.2014.165
29. Ren C, Komiyama T (2021) Wide-field calcium imaging of cortex-wide activity in awake, head-fixed mice. STAR Protoc 2(4):100973. https://doi.org/10.1016/j.xpro.2021.100973
30. Wekselblatt JB, Flister ED, Piscopo DM, Niell CM (2016) Large-scale imaging of cortical dynamics during sensory perception and behavior. J Neurophysiol 115(6):2852–2866. https://doi.org/10.1152/jn.01056.2015
31. Nietz AK, Popa LS, Streng ML, Carter RE, Kodandaramaiah SB, Ebner TJ (2022) Wide-field calcium imaging of neuronal network dynamics in vivo. Biology (Basel) 11(11):1601. https://doi.org/10.3390/biology11111601
32. Cai Y, Wu J, Dai Q (2022) Review on data analysis methods for mesoscale neural imaging in vivo. Neurophotonics 9(4):041407. https://doi.org/10.1117/1.NPh.9.4.041407
33. Waters J (2020) Sources of widefield fluorescence from the brain. eLife 9:e59841. https://doi.org/10.7554/eLife.59841
34. Valley MT, Moore MG, Zhuang J, Mesa N, Castelli D, Sullivan D, Reimers M, Waters J (2020) Separation of hemodynamic signals from GCaMP fluorescence measured with wide-field imaging. J Neurophysiol 123(1):356–366. https://doi.org/10.1152/jn.00304.2019
35. Ma Y, Shaik MA, Kim SH, Kozberg MG, Thibodeaux DN, Zhao HT, Yu H, Hillman EM (2016) Wide-field optical mapping of neural activity and brain haemodynamics: considerations and novel approaches. Philos Trans R Soc Lond Ser B Biol Sci 371(1705):20150360. https://doi.org/10.1098/rstb.2015.0360
36. Jacobs EAK, Steinmetz NA, Peters AJ, Carandini M, Harris KD (2020) Cortical state fluctuations during sensory decision making. Curr Biol 30(24):4944–4955.e4947. https://doi.org/10.1016/j.cub.2020.09.067
37. West SL, Aronson JD, Popa LS, Feller KD, Carter RE, Chiesl WM, Gerhart ML, Shekhar AC, Ghanbari L, Kodandaramaiah SB, Ebner TJ (2022) Wide-field calcium imaging of dynamic cortical networks during locomotion. Cereb Cortex 32(12):2668–2687. https://doi.org/10.1093/cercor/bhab373
38. Musall S, Kaufman MT, Juavinett AL, Gluf S, Churchland AK (2019) Single-trial neural dynamics are dominated by richly varied movements. Nat Neurosci 22(10):1677–1686. https://doi.org/10.1038/s41593-019-0502-4
39. Mitra A, Kraft A, Wright P, Acland B, Snyder AZ, Rosenthal Z, Czerniewski L, Bauer A, Snyder L, Culver J, Lee JM, Raichle ME (2018) Spontaneous infra-slow brain activity has unique spatiotemporal dynamics and laminar structure. Neuron 98(2):297–305.e296. https://doi.org/10.1016/j.neuron.2018.03.015
40. Wang Q, Ding SL, Li Y, Royall J, Feng D, Lesnar P, Graddis N, Naeemi M, Facer B, Ho A, Dolbeare T, Blanchard B, Dee N, Wakeman W, Hirokawa KE, Szafer A, Sunkin SM, Oh SW, Bernard A, Phillips JW, Hawrylycz M, Koch C, Zeng H, Harris JA, Ng L (2020) The Allen mouse brain common coordinate framework: a 3D reference atlas. Cell 181(4):936–953.e920. https://doi.org/10.1016/j.cell.2020.04.007
41. Allen WE, Kauvar IV, Chen MZ, Richman EB, Yang SJ, Chan K, Gradinaru V, Deverman BE, Luo L, Deisseroth K (2017) Global representations of goal-directed behavior in distinct cell types of mouse neocortex. Neuron 94(4):891–907.e896. https://doi.org/10.1016/j.neuron.2017.04.017
42. Xiao D, Forys BJ, Vanni MP, Murphy TH (2021) MesoNet allows automated scaling and segmentation of mouse mesoscale cortical maps using machine learning. Nat Commun 12(1):5992. https://doi.org/10.1038/s41467-021-26255-2
43. Weiser SC, Mullen BR, Ascencio D, Ackman JB (2023) Data-driven segmentation of cortical calcium dynamics. PLoS Comput Biol 19(5):e1011085. https://doi.org/10.1371/journal.pcbi.1011085
44. Makino H, Ren C, Liu H, Kim AN, Kondapaneni N, Liu X, Kuzum D, Komiyama T (2017) Transformation of cortex-wide emergent properties during motor learning. Neuron 94(4):880–890.e888. https://doi.org/10.1016/j.neuron.2017.04.015
45. Nietz AK, Streng ML, Popa LS, Carter RE, Flaherty EB, Aronson JD, Ebner TJ (2023) To be and not to be: wide-field Ca2+ imaging reveals neocortical functional segmentation combines stability and flexibility. Cereb Cortex 33(11):6543–6558. https://doi.org/10.1093/cercor/bhac523
46. Stafstrom CE, Carmant L (2015) Seizures and epilepsy: an overview for neuroscientists. Cold Spring Harb Perspect Med 5(6):a022426.

https://doi.org/10.1101/cshperspect.a022426

47. Linden NJ, Tabuena DR, Steinmetz NA, Moody WJ, Brunton SL, Brunton BW (2021) Go with the FLOW: visualizing spatiotemporal dynamics in optical widefield calcium imaging. J R Soc Interface 18(181):20210523. https://doi.org/10.1098/rsif.2021.0523
48. Dana H, Mohar B, Sun Y, Narayan S, Gordus A, Hasseman JP, Tsegaye G, Holt GT, Hu A, Walpita D, Patel R, Macklin JJ, Bargmann CI, Ahrens MB, Schreiter ER, Jayaraman V, Looger LL, Svoboda K, Kim DS (2016) Sensitive red protein calcium indicators for imaging neural activity. eLife 5:e12727. https://doi.org/10.7554/eLife.12727
49. Dana H, Novak O, Guardado-Montesino M, Fransen JW, Hu A, Borghuis BG, Guo C, Kim DS, Svoboda K (2018) Thy1 transgenic mice expressing the red fluorescent calcium indicator jRGECO1a for neuronal population imaging in vivo. PLoS One 13(10):e0205444. https://doi.org/10.1371/journal.pone.0205444

Chapter 11

Mesoscale Imaging of Neural Dynamics in Epilepsy

James E. Niemeyer, Hongtao Ma, and Theodore H. Schwartz

Abstract

Epilepsy is a disease characterized by recurrent seizures. When focal in onset, these events start in a defined location and then spread through short- and long-range connections to local and widespread interconnected brain networks. Recent advances in neuroimaging of activity-dependent fluorophores have made widefield imaging an ideal tool for tracking the neural dynamics underlying the onset and propagation of seizures. Here we describe the methods and experimental techniques that can be employed to study seizures in awake behaving rodents, with a particular focus on simultaneously applying genetically encoded calcium indicators and hemodynamic imaging. These methods can be easily extended to normal brain mechanisms in healthy brain as well as other neurological diseases.

Key words Calcium imaging, Fluorescence imaging, Brain networks, Ictogenesis, Seizure networks

1 Introduction

Using light to detect neuronal activity has been possible since the mid-1900s, when activity-dependent changes in optical properties of nervous tissue were first observed [1]. Subsequent research uncovered various related signals, such as those associated with metabolic and hemodynamic changes in tissue. These "intrinsic signals" have been used extensively for brain mapping in neuroscience [2–4]. In the 1960s, voltage imaging of the brain became possible after the discovery that dye molecules applied to the brain surface could be used to report voltage changes on cell membranes [5, 6]. These voltage-sensitive dyes create "extrinsic signals," which have also been used extensively for brain mapping [7, 8]. Calcium dyes followed after, providing another dye-related extrinsic signal imaging method [9]. The major advantage of dye imaging is that it quickly reports changes in neuronal membrane potentials or calcium usage, providing a more direct and faster readout of brain activity than intrinsic signals. The disadvantages, however, are that these dyes must be applied to the brain surface in an experimental preparation and that the signals can have a poor signal-to-noise

James E. Niemeyer et al. (eds.), *Awake Behaving Mesoscopic Brain Imaging*, Neuromethods, vol. 214, https://doi.org/10.1007/978-1-0716-4120-0_11, ratio [10]. Further, responses from cell somas, dendrites, and axons can all contribute to the extrinsic signal. Another disadvantage, of both extrinsic and intrinsic signal imaging, is that these methods are not cell type-specific: inhibitory and excitatory neuronal activity cannot be dissociated, and even glial cell activity can contribute [11].

More recently, bioengineers have developed genetically encoded calcium indicators (GECIs) that largely overcome the limitations above [12]. GECIs are activity-dependent fluorophores that can be placed into genetically defined cells, permitting whole brain expression of the calcium indicator in specific cell types. These fluorophores can even be restricted to neuronal somas, helping to prevent neuropil activation [13, 14]. When combined with modern imaging systems, these GECIs can provide robust spatiotemporal descriptions of brain activity in multiple animal species. In particular, widefield imaging, which uses CCD or CMOS cameras in conjunction with widefield illumination, can provide a broad view of neural dynamics over time, even in awake behaving animals undergoing simultaneous electrographic and intrinsic signal recording. This technique has been fruitful in systems neuroscience, and here we focus particularly on the dynamics of a neurological disease: epilepsy.

Epilepsy is defined by spontaneously recurring seizures, which are pathological brain events characterized by aberrant electrical activity. In focal neocortical epilepsy the seizures develop from a particular brain site and then spread to other regions, sometimes generalizing across the entire organ. This propagation can occur through anatomical connections, contiguous expansion, or pathologic pathways that develop over the course of the disease. By performing widefield illumination and imaging over a large region of cortex, researchers can examine how seizures initiate and then recruit other interconnected brain sites and networks [15–21]. Employing GECIs during this imaging provides a cell type-specific description of these networks throughout the course of seizures. Further, this mesoscale imaging can be performed repeatedly in the same animals, providing longitudinal access to the disease and its mechanisms.

To accomplish these types of studies several methods must be combined. First, commercially available mice are bred to express activity-dependent fluorophores. At the desired age point these animals must undergo surgeries that permit widefield imaging: this usually involves a craniectomy and implant of a clear imaging window or coverslip, along with a headstage. These headstages serve as connectors to either head-fixing systems [20] or directly to small mobile camera systems [22]. The cranial window provides optical access in the awake animal, obviating the need for confound-inducing anesthetics that are known to affect the dynamics of seizure events [20, 23]. Experimental imaging involves carefully timed presentation of light to excite the activity-dependent fluorophore followed by camera capture of the emitted fluorescence. This exciting light is often combined with multiple other wavelengths for hemodynamic-correction or for simultaneous intrinsic signal imaging, which can add complication to the post-processing analyses. These experiments are often performed with simultaneous electrophysiological recordings. The culmination of these procedures results in a rich data set that can be used to track neuronal activity across the brain during and across multiple seizures.

In this chapter, we outline various applications of mesoscale imaging of seizures, with an emphasis on calcium imaging of bulk neural activity across the neocortex. This work initially began with imaging of intrinsic optical signals under anesthesia. A seminal finding [15] showed that epileptiform activity can be faithfully mapped with optical techniques, which provide a much better spatial picture of the pathological events. That work also revealed that a region of inhibition can be observed around seizures during focal neocortical events. The "inhibitory surround" had been detected in electrical recordings [24] but had never been observed in such spatiotemporal detail as a ring of inhibition. This phenomenon likely plays a critical role in preventing the spread of seizures and is thus of great clinical interest in epileptology.

These findings highlight that important mechanisms of epilepsy can be readily observed and measured with mesoscale imaging. However, it was also subsequently considered that this imaging can be paired with other techniques, such as the voltage and calcium indicators mentioned above, to provide even more clarity about the processes underlying seizure generation and propagation. The following sections will highlight the materials and methods used in several recent applications of these techniques, in which the goal is to elucidate these mechanisms that can be leveraged to produce better therapies for human epilepsy patients.

2 Materials

2.1 Experimental Animals and Genetically Encoded Calcium Indicators

For mesoscale imaging of seizures, we use commercially available C57-background mice, though alternative strains are available and the options are continually growing in scope. Mating pairs can be obtained and bred in a short time: gestation requires ~21 days, litter sizes can range from 2 to 12 pups, and mating pairs will often re-conceive prior to weaning of the previous litter [25]. These features result in steady production of research animals.

Various strains of mice are currently available. In our experiments, we use Thy1-GCaMP mice, which express the calcium indicator GCaMP in Thy1-positive brain tissue. In neural tissue, Thy1 is expressed predominately in pyramidal excitatory projection neurons [26], making these animals useful for tracking excitatory neural activity across the brain. To image inhibitory neuronal activity, we use a breeding strategy involving Cre-recombinase and conditional expression of GCaMP to Cre-positive tissue. For Cre-positive tissue animals, we breed mice expressing Cre in parvalbumin-positive inhibitory interneurons (PV cells). This requires a single cross of two mouse strains: one strain must express Cre recombinase in parvalbumin-positive tissue, and one strain must pass on genes that turn on expression of GCaMP in Cre-positive tissue. The result is that some of the offspring in these mice will express GCaMP in PV interneurons (*see* **Note 1**).

GCaMP is an activity-dependent fluorophore commonly used in both single cell and widefield neuroimaging [27]. The acronym stands for GFP (G) Calmodulin (CaM) Peptide (P). When this molecule binds calcium, an ion reliably recruited into neurons during action potential generation, GCaMP fluoresces with an emission wavelength around 515 nm after excitation by ~470 nm light [27]. This fluorophore can faithfully detect changes in neural excitation [28]. Since its original development in the early 2000s various other forms with improved kinetics, brightness, and sensitivity have been invented. Different colored reporters are also available, opening the doors to simultaneous multicolor imaging of different cell types [29]. As mentioned above, these GCaMP molecules can be expressed in particular genetically defined cells [30]. Regardless of the fluorophore used in an experiment, their application in mesoscale imaging is generally the same: light is applied to the brain and then the emitted activity-dependent fluorescence is captured by camera. In the experiments discussed here we use GCaMP6f, a relatively fast GCaMP variant that provides sufficient fluorescence to track neural activity across the brain (*see* **Note 2**).

2.2 Cranial Windowing

Following animal generation, a surgery is performed to implant a cranial window, providing optical access to the brain, and a headstage used to fix the window relative to the camera during experiments. We outline these surgical procedures below. These steps were approved by the Weill Cornell Medical College Institutional Animal Care and Use Committee (IACUC).

Adult C57/BL6-background mice of 5–8 weeks age are used in these procedures. Animals are anesthetized with 3% isoflurane and maintained between 1.0% and 1.5% throughout the surgery. Body temperature is maintained at 37 degrees Celsius with a temperature-regulated heating pad (Harvard Apparatus, Holliston, MA). The mouse's head is fixed in a stereotaxic frame (Kopf) and held level. Hair is removed from the top of the head by scissor trimming, and multiple betadine wipe-and-wash steps are used to clean the scalp. Next, an injection of 0.05 mL of 0.25% bupivacaine is administered under the skin and left for 2 min prior to incision.

Then sterile scissors are used to remove a region of skin at the center of the head. Due to our cranial window size, we typically use a circular skin cut of about 8 mm diameter. The skull is then wiped by Q-tip and lightly scraped by forceps to allow for identification of bregma. A photo image can be taken before the craniectomy to allow for stereotactic injections after surgery. Next, a handheld drill (Osada, EXL-M40) with fine bit (Meisinger, HM1 size 005) is used to remove a circular region of skull (8 mm).

Once the skull is removed, the clear cranial window can be implanted. Our preparation uses a polymer window (see Subheading 2.3), but the following steps can be used regardless of window material. The window should be sized to fit tightly within the craniectomy and is attached to the skull perimeter by Vetbond (3 M). Next, the headstage is placed on top of the animal and attached by Metabond adhesive cement (C&B, Parkell). A second photo can be taken at this time, allowing for co-registration with the pre-craniectomy photo so that precise Bregma coordinates can be used during experiments. Finally, analgesics (Meloxicam, 5 mg/kg; optionally Buprenorphine, 0.5 mg/kg) are provided to the animal following IACUC-approved protocols.

2.3 Cranial Window Construction Procedure

A secure cranial window is critical for chronic imaging in awake rodents. In some experiments a thinned-skull preparation may be sufficient, but if electrode or drug access is needed to the brain, then a soft, penetrable film must be used for simultaneous optical access. Our lab uses polydimethylsiloxane (PDMS), a silicone-based material that can be mixed in the lab (Sylgard 184 silicone elastomer, Dow Corning) and then used for chronic optical imaging of the brain [20, 31]. A base and catalyzing component are mixed in a petri dish. To remove bubbles from the film the dish is placed in a vacuum container (*see* **Note 3**). Once the PDMS is hardened (~1–2 days) it can be removed and cut into appropriate-sized pieces. For our circular windows we use an 8 mm diameter leather stamping tool to create perfectly circular 8 mm windows. These can be stored in Cidex disinfectant solution or alcohol but must be sterile-rinsed before implantation during surgery. Care must be taken to implant PDMS windows without allowing air under the brain; even with care, however, bubbles can form over the brain and impede imaging. These can sometimes be removed (*see* **Note 4**).

2.4 Widefield Imaging Apparatus (Camera and Illumination)

Our lab uses scientific imaging cameras (e.g., Ximea, Germany) and fiber LEDs (Thorlabs, USA) to perform all brain imaging. The animals are head-fixed on an air-supported platform (Neurotar, Finland). The camera is placed directly overhead and exciting LEDs provide light from the sides. Custom MATLAB (Mathworks, USA) scripts serve as our control system, coordinating the timing of the experiment [20]. An Arduino board, also with a custom script, coordinates the precise timing of camera frames with appropriate LED strobes: for GCaMP frames, a 470 nm light is presented to excite the fluorophore; for oxygenated hemoglobin (HbO), a 530 nm light is used; for deoxygenated hemoglobin (HbR), a 610 nm light is used.

Electrophysiology recordings are often included in our epilepsy studies. These recordings are performed with glass pipette electrodes with internal silver wire attached to a microinjector (Nanoject II, Drummond, PA, USA), which is placed through the soft cranial window and into the brain. Stereotactic coordinates, referenced from an atlas [32] and the images taken during surgery, can be used to target specific brain regions.

The glass pipette is backfilled with a chemoconvulsant. We have published extensively using 4-aminopyridine and bicuculine, which are reliable drugs for inducing focal seizures and interictal-like activity in the awake rodent, respectively [19–21]. Pharmacologic injection of chemoconvulsants permits the creation of interictal and ictal foci in discrete predetermined areas of mouse neocortex. The electrophysiology recordings travel from brain to a headstage preamplifier, then a second stage amplifier (A-M Systems, WA, USA) before being digitized at 1 kHz (CED Systems, Cambridge, UK) and sent to a computer program (Spike2, CED, Cambridge, UK) for monitoring and PC storage. The frame times, recorded by the control system and initiated by the Arduino code, are also sent to the electrophysiology program. This system ensures that all data can be compared on a millisecond-timescale in subsequent processing steps.

3 Methods

Seizures can be modeled with various methods. In our studies we usually employ chemoconvulsant drugs placed in discrete, predetermined areas of the neocortex to provide reliable, periodic seizures or interitcal discharges in awake mice. These drugs are applied to stereotactic coordinates by an injecting electrode during imaging. We outline here the typical experimental procedures and post-processing steps required to examine mesoscale epileptiform events. We then highlight several published applications of these methods in the context of other mechanistic experiments aimed at understanding the development and spread of neocortical seizures.

3.1 Animal and Electrode Preparation

We use a Neurotar (Helsinki, Finland) head-fixing system which includes a small air-supported arena to provide the mouse the ability to walk while head-fixed. The animals are habituated to this environment for 20-min sessions for several days prior to the experiment. This habituation period decreases the animal's stress on the day of the experiment, minimizing potential confounds associated with stress responses during the study.

Before beginning experiments the injecting electrode should be prepared. These electrodes contain an internal pump and tubing that must be completely filled with mineral oil. Small air bubbles can prevent accurate injection, so care must be taken to properly fill and test them before beginning the experiments. At one end of the Nanoinject II device is a small wire electrode. We place this electrode into a single-barrel glass microelectrode (50–100 micrometer tip opening) that has been backfilled with chemoconvulsant.

This electrode, referenced to the animal's skin, is attached to an amplifier (amplified 1000× and band-pass filtered 1–500 Hz) and digitized before being sent to a computer running Spike2. When inserted into the brain (300–500 microns below cortex surface), this electrophysiology channel provides the local field potential (LFP). The signals detected from this channel are critical to determining seizure onsets and duration, which (1) provide corresponding times when widefield imaging data can be analyzed and (2) comparison of the electrographic signature of the epileptiform events with cell type-specific mesoscale imaging data. Analyses and integration of these signals are discussed below.

3.2 4-Aminopyiridine Seizures

While spontaneous focal human seizures can arise from a myriad of causes and occur unpredictably, sometimes even arising from more than one location, it is advantageous in the laboratory to have a model where seizures arise from a single location in a reliable and replicable fashion. 4-Aminopyridine (4-AP) is a potassium channel blocker that, in neuronal tissue, increases action potential duration and neurotransmitter release (Buckle et al. 1982), and which induces in vivo focal seizures that arise spontaneously, i.e., not triggered by stimulation, and which either remain localized or spread throughout the brain [19, 20, 33, 34].

4-AP is purchased in a powder form (Sigma-Aldrich, MO, USA) and then mixed with sterile water. In our experiments we use concentrations up to 15 mM [16], though lower concentrations (down to 1 mM) can reduce the probability of status epilepticus in awake, non-anesthetized, rodents [20]. A small injection of ~200 nL of 4-AP around 300–500 micrometers below the cortical surface will typically induce seizures within 20 min. Seizure durations generally range from 10 to 90 s, and seizures will often repeat several times over the course of 1–2 h. During this time, widefield imaging of GCaMP and intrinsic signals are simultaneously performed.

3.3 Activity Analyses

Custom analysis scripts are written in MATLAB. Based on events detected in the electrographic data, we extract the corresponding imaging frames from specific time periods of interest.

The readout of activity-dependent fluorescence from GCaMP is slightly biased by hemodynamic changes in the brain [35]. To counteract these changes in the GCaMP signal we apply a formula for removing the bias at all timepoints,

$$\frac{Ftrue(t)}{Ftrue(tb)} = \left(\frac{F(t)}{F(tb)}\right) / \left(\frac{I(t)}{I(tb)}\right)$$

where *Ftrue* is the GCaMP signal after removing the hemodynamic bias at frame or timepoint t, F is the observed fluorescence, I is the intrinsic optical signal observed at 530 nm following 2 Hz lowpass filtering, and tb is the baseline time. This baseline should come from a period immediately prior to the start of an epileptiform event. We note that other methods of signal separation are used in our field (*see* **Note 5**).

Following hemodynamic correction, the next step is to calculate the amount of change in activity observed across the seizure period. In widefield imaging this must be performed on a pixel-by-pixel basis. We measure this activity as the change in fluorescence over time in units of baseline activity:

$$\frac{dF}{F} = \frac{Ft - Fb}{Fb}$$

where Ft is the hemodynamic-corrected fluorescence at frame t and Fb is the baseline fluorescence. The resulting dF/F activity can be visualized across space and time during experiments.

Seizure propagation can be derived from the imaging data using a variety of different analysis techniques. One option is to search for all pixels with dF/F crossing a set threshold. While the chosen threshold is arbitrary, one that we have found useful is defined as a fraction of the maximal recorded activity [36]. In one recent experiment we extracted all pixels during seizures that showed at least 30% of the maximal activity measured at a 4-AP seizure focus [20]. This simple quantification provides one measure of how far the seizure activity spreads across a wide area of brain (Fig. 1), which can be used to compare seizures between different conditions, such as under anesthesia vs. awake states [20].

Another option for measuring areal spread and propagation can be derived from a seed correlation technique. This method involves extracting the dF/F at some location, typically the seizure focus, and calculating the correlation coefficient (e.g., Pearson's r) of every other pixel in the widefield imaging data. The resulting coefficient values can be visualized as a spatial heatmap showing the degree to which all pixels correlate with the seizure focus (Fig. 1c). In the time domain, the dF/F signal can also be correlated with the electrical signal derived from the LFP. For example, in the 4-AP model, seizures can initiate in one of two patterns easily distinguished in the LFP: low-voltage fast onsets and sentinel spike

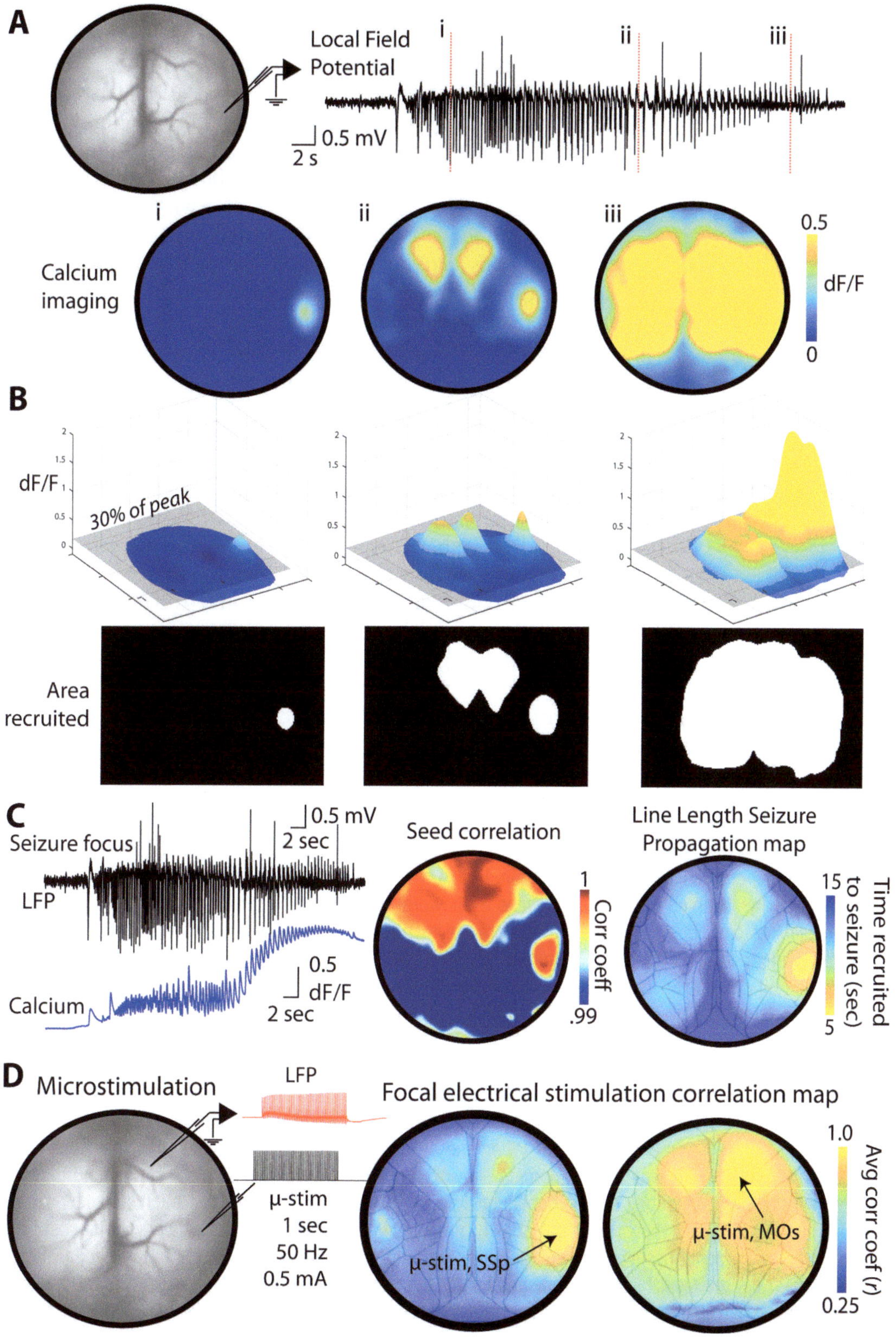

Fig. 1 Mapping seizure spread by change in fluorescence and correlation. (**a**) Electrographic and GCaMP fluorescence data are shown for a seizure that initiated in the right somatosensory cortex following injection of

onsets. Fortuitously, both occur in human epilepsy and indicate that the 4-AP model recapitulates this aspect of human epilepsy [37]. The region of brain active during these two types of events can be highlighted by performing a linear correlation between the GCaMP signal and LFP [20]. Another method to map individual seizure events, adapted from electroencephalography studies, is to use "line length" of signals over time [38]. The line length refers to the sum of change of signal in a specified time window, which can be calculated in a moving boxcar over a time series. In the case of imaging data, this can be performed on each individual pixel throughout the seizure. After a threshold is set (e.g., 2 standard deviations above mean during baseline, non-seizing periods), then a "seizure onset" time can be determined for each pixel in the image (Fig. 1c) to create a two-dimensional map of where seizure invasion occurs across the cortex.

One feature of mesoscale imaging is the accessibility that it provides into the brain. When the cranial window is penetrable (as in our described PDMS preparation), the imaging can be easily combined with electrophysiology and neurostimulation methods. In recent studies, we have begun using our seizure mapping methods above to evaluate how these events develop in networks of the neocortex. One component of the evaluation is to characterize how neurostimulation spreads in these networks, which can yield interesting results that highlight how the identical stimulation in the same animal can recruit cortical activity quite differently depending on the location of stimulation (Fig. 1d).

Fig. 1 (continued) 4-Aminopyridine. (**b**) Area of seizure recruitment can be measured by the change in fluorescence that crosses a threshold over time. In this case, neural activity that reaches above 30% of the maximum activity observed at the seizure focus is shown at three different time points (i–iii), along with an image mask that can be used to quantify the area of seizure invasion. As can be seen, discrete nodes in secondary motor cortex are recruited bilaterally. (**c**) Seed correlation between the seizure focus or local field potential and all other pixels can also provide a map of recruitment strength throughout the seizure. In this figure, pixel correlation with the seed is calculated for the entire seizure, but segments can also be analyzed, as shown later in Fig. 5. Similar seizure spread maps are obtained with both methods. However, calcium signals can also be treated like electro-encephalogram (EEG) data to detect the onset of seizure invasion across the cortex. By using a moving calculation of line length (the amount of change in calcium signal in 500 ms windows), a 2D map can be created to show where and when these line length changes cross a threshold to indicate seizure invasion. (**d**) When the seizure in panel *C* is considered in the context of neuroanatomy (neocortical brain region borders shown), it becomes clear that seizures can spread through distant connections just as they spread contiguously outward. We have begun using microstimulation with calcium imaging to probe how electrical activity propagates through particular networks. This network consists of primary somatosensory cortex (parietal SSp) and prefrontal cortex (prefrontal secondary motor cortex, MOs). Note the clear bilateral representation of the network in both the 4-AP seizure (C) and during microstimulation, but also the clear difference in bilateral propagation when the same stimulation is applied in parietal SSp versus prefrontal MOs

These basic tools can be used to measure network activity propagation, as well as the spatiotemporal onset and termination of a variety of different compartments i.e., excitatory and inhibitory neurons, glia, hemodynamic signals, membrane voltage, throughout large areas of the neocortex distributed through both hemispheres. In the next section we will provide a variety of examples of how we have used mesoscopic imaging to gain a better understanding of the way that focal neocortical seizures spread through the brain.

3.4 Voltage-Sensitive Dye Imaging During Anesthetized Seizure Activity

Early work demonstrated that epileptic events can be measured and studied with intrinsic optical signal imaging [15], which highlighted spatially separate regions of inhibition and excitation during focal seizures. However, these hemodynamic signals do not directly report the activity of neurons. To address this, our lab employed voltage-sensitive dye (VSD) imaging of epileptic activity in anesthetized rodent cortex. This technique uses dyes (e.g., RH-795, 0.6 mg/mL in 0.9% NaCl saline) that can be applied to the dura following a craniotomy. After 45 min of exposure, the dye can be washed away before imaging [39, 40] and simultaneous electrographic recording.

In one study, we used this technique to analyze voltage changes across the neocortex during interictal epileptiform events induced by focal injection of bicuculine methiodide (5 mM in 165 mM NaCl). Mesoscale imaging was then performed with filtered light (530 +/− 10 nm for RH-795 dye) and an Imager 3001 camera operating at a frame rate of 340 Hz. Fluorescence emission was captured after passing through a longpass (610 nm) filter. This fast imaging was necessary for the rapid voltage changes observed in the brain, and the resultant data provides a clear description of the spatiotemporal propagation of interictal spikes, including rapid excitatory activity in the focus followed by a delayed ring of inhibition (Fig. 2).

Fast imaging of cortical voltage changes can also be applied to seizures. In one study [41], a voltage sensitive dye (RH-1691 and 1692) was applied to the exposed brain of anesthetized rodents and then seizures were induced with focal injection of 4-AP(15 mM, 0.5 uL) through an injecting microelectrode. (Fig. 2). VSD imaging revealed broad spatial recruitment of brain tissue immediately following an initial spike, in contrast to the more focal activation of low-voltage fast activity. During seizure evolution, VSD imaging showed variable directionality of individual spike-wave patterns, which was notably different from the classical views of slow progressive "Jacksonian march" propagation of seizures [42](Fig. 2). This data also revealed that a transient ring of inhibition occurred immediately before each excitatory burst, and that the precise onset of each spike-wave discharge varied from spike to spike. As opposed to the traditional model of seizure initiation from a single focus,

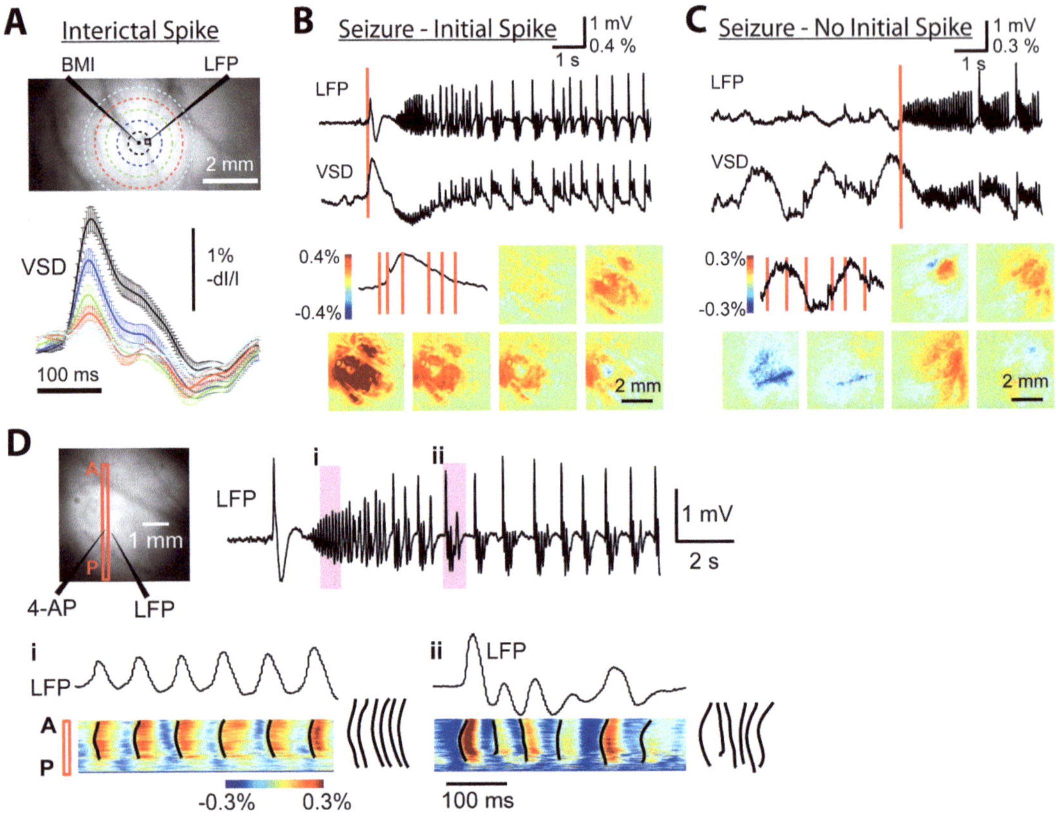

Fig. 2 Voltage-sensitive dye (VSD) imaging of epileptiform activity. (**a**) Interictal spike induced by BMI injection, showing the clear spread of VSD signal over the cortex followed by a delayed ring of inhibition. (**b**) 4-AP often induces seizures that begin with a large initial spike. VSD shows the clear spatial spread of these events. (**c**) In a seizure that does not begin with an initial spike, VSD captures the small electrographic fluctuations detected in the local field potential, but no widespread spike occurs as in *B*. (**d**) VSD imaging during the discharges of a seizure show obvious changes in voltage across the cortex, but also show how these events can be somewhat consistent in propagation pattern during early seizure phases (i) and, later in the seizure, can vary fairly dramatically (ii). During the later spike-and-wave periods, even the initiation of individual waves can vary (see black lines that indicate the VSD wavefront along the anteroposterior (A-P) axis). (Adapted from Ref. [39, 41])

this data presents a more nuanced picture of how ictal events evolve.

3.5 Intrinsic Optical Image Signals and Voltage Imaging during Anesthetized Epileptiform Activity

Hemodynamic signals can also be measured concurrently with voltage changes during epileptiform activity. Their similarities and differences provide insight into neurovascular coupling, a crucial component of brain function in health and disease. In one study [39], bicuculine methiodide (5 mM) was injected into the neocortex that had been prepared for voltage-sensitive dye and intrinsic optical signal imaging. Total hemoglobin and deoxygenated hemoglobin were derived from the intrinsic signals and compared to changes in voltage across the cortex during interictal spikes. The hemodynamic signals were typically 1.5–2 times the spatial size of

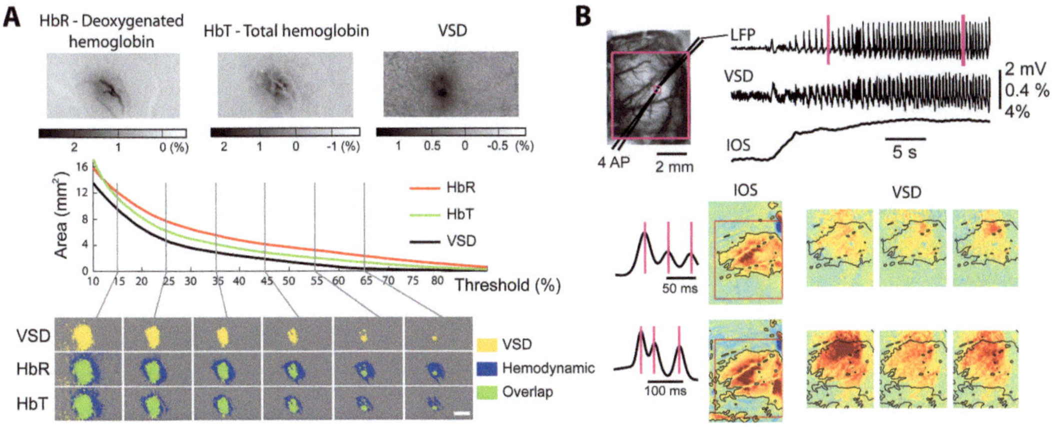

Fig. 3 Voltage sensitive dye and concurrent hemodynamic imaging. (**a**) Images of VSD and hemodynamic signals showing more widespread HbR and HbT during interictal spikes compared to voltage. These differences were strong and constant even as threshold was varied for the interictal detection. (**b**) During seizures, VSD signals correspond well to the individual discharges observed in the electrographic recording, while the IOS signal slowly increases in strength before reaching a plateau. Panels below show hemodynamic signal (IOS) and voltage over the cortex during discharges within a single seizure. The IOS and VSD areas are not identical, and, like the interictal spiking, the VSD area is smaller. However, both IOS and VSD signals increase with time throughout the seizure. (Adapted from Ref. [39, 41])

the VSD signal, which is much larger compared to the signal changes observed by other groups in experiments that used somatosensory stimulation to examine neurovascular coupling [43]. This difference was evident even over a wide range of thresholds used for detecting interictal spikes (Fig. 3). However, despite this hemodynamic overshoot, the signals still correlated well with the extent of excitatory interictal activity across multiple time points during the spikes, which indicates the potential for using hemodynamic surrogates to map interictal activity in human patients.

Full seizures can also be examined with simultaneous voltage and intrinsic optical signal imaging. In one study [41], 4-AP seizures were initiated in the neocortex (15 mM) of anesthetized rodents following preparation for VSD imaging (RH-1691 and 1692). During the first spike of a seizure, the VSD wave propagated far beyond the hemodynamic signal, but then, similar to the interictal spiking data, the area of blood volume signal was larger than the spread of individual spike-wave discharges during the seizure (Fig. 3). Later, during seizure evolution, the signals overlapped but the VSD signal showed a clear reflection of the individual spike-wave discharges while the hemodynamic signal increased and plateaued. This data showed that, despite obvious spatial correlations, the hemodynamic and neural signals are separate processes that each possess a unique evolution during the course of a seizure. Finally, during seizure termination, VSD signals showed an abrupt termination, but hemodynamic signals continued to slowly decline over ~8 s before returning to baseline.

3.6 Widefield Experiments to Compare Seizures Under Awake and Anesthetized Conditions

Widefield imaging paradigms can also be used for elucidating the impact of anesthesia on hemodynamic and neuronal activities. Here we will describe techniques published in a recent study [20] in which we examined seizure propagation in GCaMP-expressing mice in anesthetized and awake states.

We used the C57BL/6J-Tg(Thy1-GCaMP6f)GP5.5Dkim/J mouse line (Jackson Labs #024276), which expresses GCaMP in Thy1+ neurons. These cells are predominantly excitatory pyramidal cells and are accurate reporters of excitatory neuronal activity [26]. Following the surgical steps and preparations described above, we implanted large, ~5 × 7 mm cranial windows and head-bars in these animals. After surgical recovery (2 weeks) and habituation (30-min sessions 3 days in a row) to the head-fixation apparatus, we performed seizure experiments by focally injecting 4-AP through the cranial window. The animals were separated into two groups: in one group the experiments were performed in awake conditions in which the mouse was head-fixed and unanesthetized; in the other group the animals were anesthetized under isoflurane throughout the experiment.

To ensure that seizure severity was comparable, we screened the lowest dose of 4-AP that could elicit repeated seizures that lasted over 10 s in duration with 1-min intervals between events. In the awake behaving state, this optimal dose was 2 mM 4-AP at volumes of 200 nL. In the anesthetized state, the optimal dose was 5 mM 4-AP at the same 200 nL volume. In both groups, the local field potential was recorded by the injecting electrode as described above. Imaging was also performed to capture simultaneous calcium and hemodynamic activity in the awake and anesthetized brain state during seizures.

Seizures were electrographically similar between awake and anesthetized animals. Events began either with or without an initial sentinel spike and low-voltage fast activity (LVFA), with no significant difference in durations (which were typically ~45 s), though the frequency of seizures was slightly higher in the awake state. For both types of seizure onsets there was an observable difference between the anesthetized and awake states. The calcium signal in the awake state demonstrated temporal fluctuations that more closely approximated the electrographic signature of the seizure as recorded in the LFP, as compared with the anesthetized state. This suggests an anesthetic-induced blunting of the calcium signal under anesthesia. Further differences were also observed by analyzing the areal extent of activation at seizure onset [36]. Taking all pixels at 30% of maximum activity (at the focus), it was found that the cortical area recruited at seizure onset in the anesthetized state was significantly larger than the area recruited in the awake state.

Multiwavelength imaging allowed us to examine both calcium (neuronal) and hemodynamic activity across the bilateral dorsal cortex (Fig. 4). In awake animals, the calcium signal reached a maximum area about 10 s after onset and then slowly decreased in area over the course of the seizure. Hemodynamic signals, however, showed different effects. In awake animals, increases in blood perfusion provided sufficient oxygenation such that the seizure focus was only occasionally and weakly hypoxic, while the surround regions were hyperperfused with oxygen. In anesthetized animals, on the other hand, total hemoglobin influx was blunted leading to an increase in deoxygenated blood, which created a more dramatic "epileptic dip" than in awake conditions. While previous findings had suggested that this "epileptic dip" in the oxygenation of brain tissue might serve as a tool for detecting the seizure focus [44], our mesoscale imaging data indicates that this dip is much weaker in the awake state. Thus, hemodynamic seizure localization in awake animals will have to rely on other signals rather than the early dip. As will be apparent, CBV or total hemoglobin provides a much more accurate mapping signal in awake animals.

We also investigated the propagation of ictal events to the contralateral hemisphere. Interestingly, in congruence with the observed hemodynamic signal blunting, no seizure events propagated contralaterally in the anesthetized mice. Meanwhile, about half of the awake-state seizures propagated contralaterally. Two patterns of spread were observed: in most of these spreading seizures (6/8), activity first propagated across the ipsilateral cortex before appearing at a homotopic (mirror) location in the opposite hemisphere. In a small minority of events (2/8), the seizures spread throughout ipsilateral cortex before propagating contiguously into the contralateral cortex. We also observed that the total hemoglobin signal provided a more accurate picture of the spatial spread of neural activity than the oxygenated or deoxygenated blood signals. As these latter signals are more comparable to the BOLD signals measured by fMRI, this mesoscale brain imaging data suggests that caution should be exercised when interpreting fMRI BOLD in the context of seizure mapping.

Finally, we tested the theory that anesthesia-induced blunting of vasodilation might be the cause of the observed increased deoxygenated blood found under anesthesia. To test this, we injected dextran-conjugated fluorescein isothiocyanate (FITC-dx) in mice and then focally photoactivated this compound over the brain surface. This photoactivation of FITC-dx, performed with 470 nm light, disrupts endothelial cells and restricts vasodilation, but does not impact neuronal activity [45].

We first recorded 3–5 seizure events in awake mice and then anesthetized the mice and injected 50 uL of 10% FITC-dx (in sterile saline) into the retro-orbital sinus. Following endothelial disruption by FITC-dx, the oxygenated blood signal was significantly

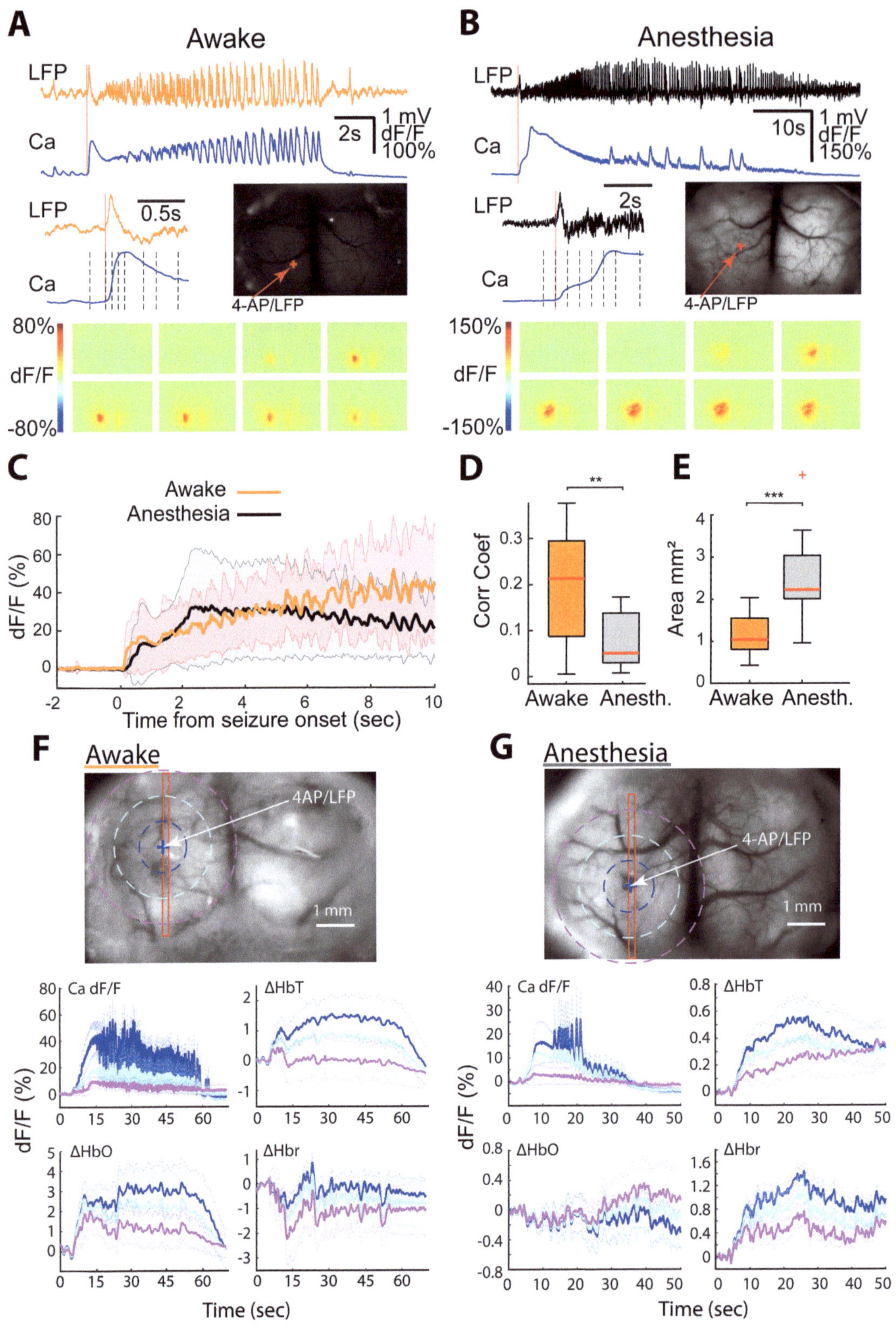

Fig. 4 Mesoscale imaging of neuronal activity and hemodynamics in awake and anesthetized brain states. (**a**) A 4-AP seizure in an awake head fixed Thy1GCaMP mouse undergoing widefield imaging. Electrophysiology

diminished while the deoxygenated blood signal significantly increased. Our overall findings suggested that the disrupted vasoconstriction leads to a decrease in total hemoglobin, which is accompanied by a hypoperfusion of the brain during a seizure, generally mimicking the effect that was observed in anesthetized animals. Thus it appears that, during seizures, isoflurane anesthesia can inhibit the brain's hyperemic response, causing significant differences in neurovascular coupling between anesthetized and awake brain states.

3.7 Analyses of Epileptiform Activity in Mesoscale Brain Networks

One of the fundamental mechanisms whereby seizures initiate and propagate emerges from the breakdown of excitatory and inhibitory (E/I) balance [46, 47]. In one recent study from our group [19], mesoscale imaging was performed using activity-dependent dye (Oregon Green 488 BAPTA-1 AM, Life Technologies) injected into neocortex. Next, seizures were initiated by 4-AP injected focally through a micropipette electrode as described above. These ictal events were spatially constricted around the site of 4-AP injection, which can be observed (Fig. 5) where the correlation of all visible pixels is calculated relative to a seed trace at the injection site.

In order to understand how breakdown of E/I balance might underlie seizure spread, we injected an inhibitory receptor blocker, bicuculline methiodide (BMI, 5 mM), into a distal (~5 mm) cortical site. Thus, the experiment consisted of a focal 4-AP seizure and a distant region of compromised inhibition. The electrographic and calcium imaging data showed that two separate seizure foci were created that initiated nearly simultaneously in the two regions

Fig. 4 (continued) and calcium signals show the seizure development. (**b**) Same data as in A except the animal is placed under isoflurane anesthesia during the experiment. Note the broader region of brain (heatmap panels) that is active during the seizure. (**c**) Population activity of all animals and events (8 awake mice with 17 total seizures, 7 anesthetized mice with 14 seizures). Seizures in the awake brain state show a continuous increase in calcium activity throughout the events, while anesthetized seizures peaked in activity around 2–3 s and then slowly decreased. (**d**) Correlation between the calcium activity and the electrographic LFP was significantly higher in the awake vs. anesthetized condition. (**e**) The area of brain recruited to seizures was significantly higher in the anesthetized condition compared to awake conditions. (Two-tailed unpaired t-test; ** $p < 0.01$; *** $p < 0.001$). (**f, g**) An ROI-based analysis was used to examine neural and hemodynamic signals at varying spatial extents across the cortex in both awake and anesthetized conditions. Concentric rings show three different circular regions with corresponding imaging data in the lower panels. These plots show the calcium (neuronal), total hemoglobin (HbT), oxygenated hemoglobin (HbO), and deoxygenated hemoglobin (HbR) observed during the seizures. In both brain states, the seizure induces a clear increase in calcium and HbT. However, in the awake condition (**f**), note the increased HbO signal and decreased HbR signal, indicating hyperoxygenation in the seizure surround. In the anesthetized condition (**g**), we observe relatively higher HbR signal compared to HbO, indicating an "epileptic dip" in tissue oxygenation. (Adapted from Yang et al. [20])

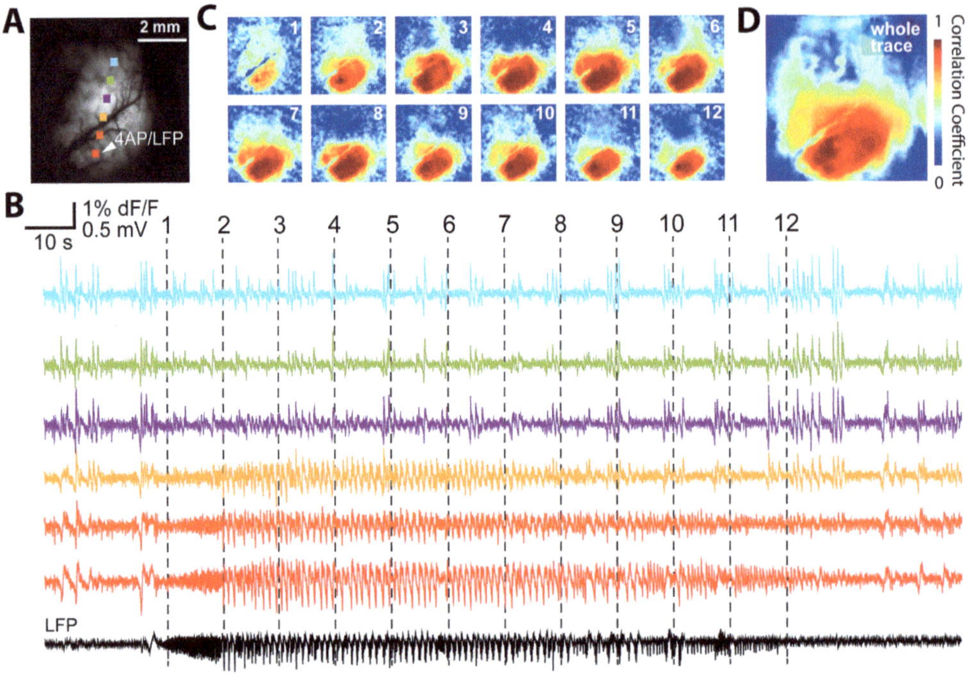

Fig. 5 Widefield seizure imaging. (**a**) This image shows the field of view from an experiment in which a 4-Aminopyridine seizure is initiated in neocortex (at site of red square) in an anesthetized rat by an injecting electrode. Colored squares represent pixel sites where activity was sampled for the traces in B. (**b**) Calcium activity is measured as change in fluorescence relative to pre-seizure baseline and plotted here to show the evolution of an individual seizure. Local field potential, measured at the red square in A, is shown at bottom. (**c**) The activity traces for all pixels are compared with the trace measured at the injection site (red square) at 12 different time points during the seizure. The 12 corresponding correlation maps are shown as heatmaps with values representing Pearson's correlation coefficient. (**d**) The correlation of all pixels throughout the whole seizure is also shown relative to a seed at the seizure focus. Note that in this preparation the correlation of pixel activity, panels in C-D, compared with the seizure focus, depends primarily on the distance from that focus. (Adapted from Liou et al. [19])

(Fig. 6). Seed correlation maps demonstrated that the two foci correlated with localized brain tissue ~2 mm from each focus.

Widefield imaging also presented the opportunity to test whether a secondary focus can develop in contralateral homotopic cortex (a mirror-opposite site in the brain). To examine this, our group used the same paradigm of 4-AP seizure initiation but then placed the BMI pipette at the same anteroposterior position but in the opposite hemisphere. Interestingly, in this configuration, no secondary focus developed, which can be seen in the LFP and calcium traces shown in Fig. 6e; the seeded correlation maps also demonstrate this difference. This imaging data, which we confirmed with multiunit recordings, indicates that cross-callosal connections are not as effective as ipsilateral connections for recruiting a secondary ictal focus in this seizure model.

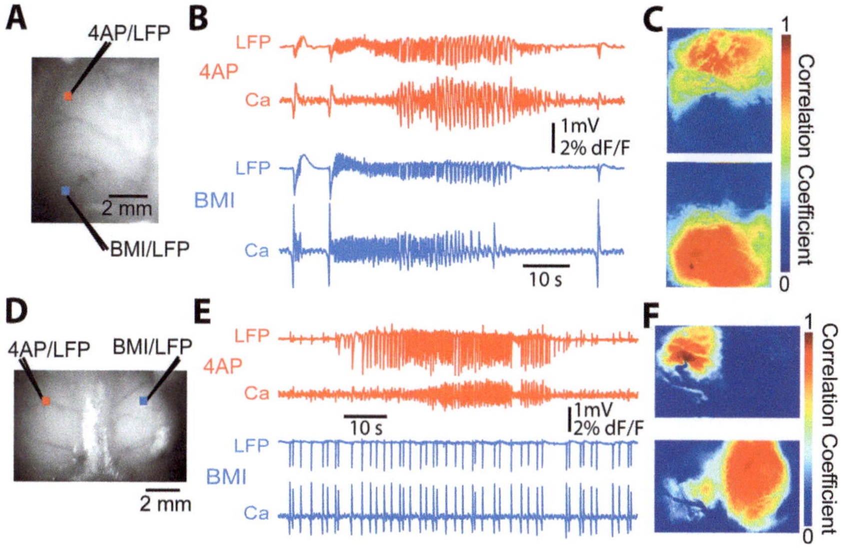

Fig. 6 Widefield imaging of a dual-focus seizure and failure of contralateral propagation. (**a**) Image of brain and schematic locations of the 4-AP and BMI injection sites. Electrodes were used to inject as well as record electrographic data. (**b**) Local field potential and calcium data from the sites shown in A, which demonstrate seizure development at both regions, indicating the creation of a second ictal focus at the BMI site. (**c**) Correlation coefficient maps were created by calculating correlation relative to the seed locations for 4-AP and BMI injections shown in panel A as red and blue squares, respectively. Note that a second seizure focus developed at the BMI site, though the activity was not strongly correlated between the two foci. (**d**–**f**) Same layout of data as A-C except that the BMI injecting electrode was placed at mirror-opposite position in the right hemisphere of the brain. Panel E shows that, unlike ipsilateral BMI injection, the contralateral BMI injection only shows interictal spikes and no contralateral propagation of the 4-AP seizure. The difference between the two hemispheres can also be appreciated by the seeded correlation maps in F, where both hemispheres show only ipsilateral correlation. (Adapted from Liou et al. [19])

This data highlights how seizure propagation patterns can be informed by brain connectivity. More recently, we have begun to examine similar questions in a seizure network paradigm in which we explore the impact of long-range projections on seizure spread. Following the protocol listed in our Methods above, we bred mice to express GCaMP in either excitatory (Thy1+) or inhibitory (PV+) cell types, and then implanted soft PDMS cranial windows at age 8 weeks. These windows were 8 mm in diameter, permitting a wide field of view over both hemispheres. We next injected bicuculline methiodide (BMI, 5 mM in 0.9% NaCl, 50–150 nL) into a subregion of primary somatosensory cortex (S1) that is monosynaptically connected with ipsilateral secondary motor cortex (M2), as well as contralateral S1.

Electrographic data showed that, on average, individual interictal spikes propagated from the S1 focus first to ipsilateral M2 (iM2), then to contralateral M2 (cM2), and finally to contralateral S1 (cS1). Similarly, we observed higher amplitude interictal spikes in iM2, followed by cM2, and finally cS1 (Fig. 7) This was

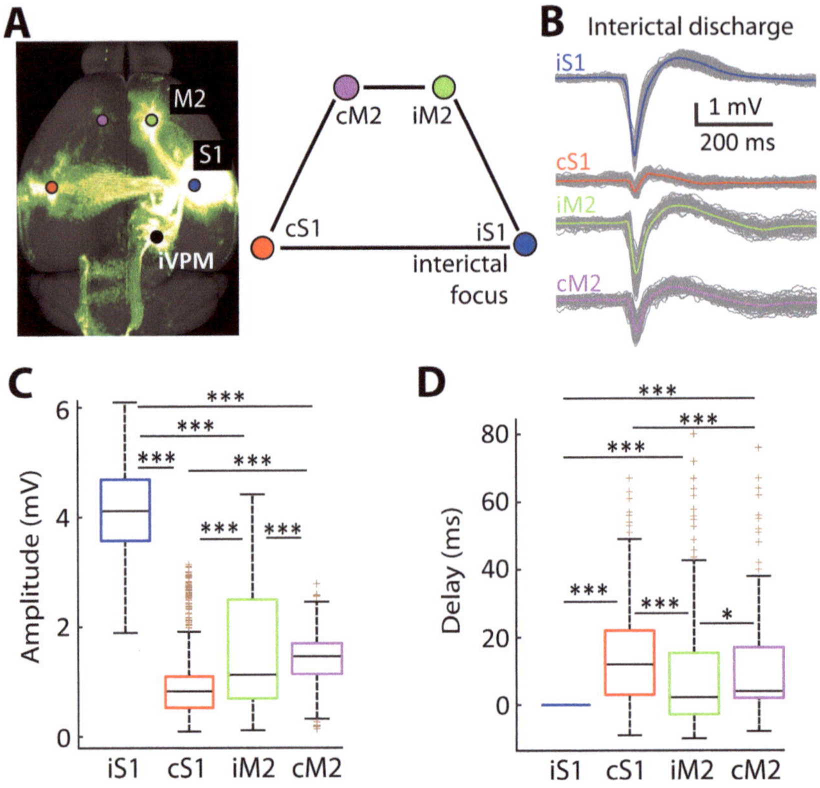

Fig. 7 Epileptiform activity mapping in an anatomical network. (**a**) An anatomical network was selected after browsing the viral tracing experiments in the Allen Institute's brain atlas. The chosen S1-M2 network consists of a subregion of primary somatosensory cortex (SSp-bfd, shortened to S1), its contralateral counterpart (cS1), and ipsilateral secondary motor cortex (MOs, labelled iM2 here). Ipsilateral ventral posteromedial nucleus (VPM) is also strongly connected to S1 but cannot be imaged in the widefield paradigm. This network also highlights a di-synaptic site, contralateral M2 (cM2). We consider it di-synaptic because the tracing experiments do not show direct connection with iS1, but it is well connected with cS1 and iM2. The amplitude of spikes is strongest in the focus, followed by iM2, then cM2, then cS1. (**b**) Interictal spiking was induced in iS1 by focal application of bicuculline methiodide (BMI, 5 mM) through an injecting electrode. Electrodes were placed in the other three neocortical nodes and interictal spike propagation was measured across the brain. (**c**, **d**) Interictal spike amplitude (maximal voltage recorded per spike) and propagation delay time were measured. The spikes were detected on all electrodes and their onset was determined to be the time of maximum slope (peak of first derivative of the voltage trace). This onset was compared with the onset detected at the interictal focus in iS1 to measure the delay of the propagating interictal spike. We observed that, in this S1-M2 network, interictal spikes propagate bilaterally, initiating in S1 and then propagating, on average, to iM2, then cM2, and finally cS1. Interestingly, in both the amplitude and delay data, it seems that cM2, despite being a di-synaptically connected brain site, is more involved in interictal spikes than cS1, a brain site connected monosynaptically to the interictal focus. This electrographic characterization led us to probe the network with cell-type-specific imaging. Statistical comparisons are shown using one-way ANOVA with a post hoc Tukey-Kramer test: *$p < 0.05$, *** $p < 0.001$. Outlier values are shown with "+" symbols. (Adapted from Luo et al. [21])

somewhat surprising given the direct monosynaptic connections between ipsilateral and contralateral S1. We next performed wide-field imaging and found that excitatory Thy1+ cell activity recapitulated this amplitude pattern: GCaMP signals were strongest in iS1, followed by iM2, then cM2, and finally cS1 (Fig. 8). However, when we performed imaging of PV+ cell activity, we found that the amplitude of cS1 was, on average, higher than the amplitude of cM2. In other words, the excitatory cell activity was relatively stronger in cM2, while inhibitory cell activity was relatively stronger in cS1. In agreement with this, we observed a significantly higher participation rate of PV+ cells in cS1 versus cM2 compared to Thy1+ cells. This finding suggests that cross-callosal projections in sensory cortex may preferentially innervate inhibitory cells in contralateral cortex, whereas cross callosal connections in motor cortex may be excitatory a principle of seizure networks that was previously proposed by in vitro work [48]. If this principle exists in human seizure networks then it could potentially be leveraged to prevent bilateral spread of seizures.

3.8 Analyses of Glial, Hemodynamic, and Neuronal Activity

In addition to neuronal activity, mesoscale imaging can be used to probe the relationship between hemodynamic responses and glial cell activity. Both glia [49–51] and neurovascular coupling changes [15, 17, 39, 52] have been implicated in seizure initiation. Various studies have also reported that glial activity is critical for neurovascular coupling and may also be important in seizure spread [53–55]. To explore this three-way interaction, our lab recently studied how focal seizures affect hemodynamic, glial, and neuronal activity during widefield imaging [33].

We applied the calcium indicator Oregon Green 488 BAPTA-1AM (OGB-1, Life Technologies) by convection enhanced delivery to bulk load the neocortex [18] of anesthetized adult Sprague-Dawley rats. The OGB-1 solution was created by diluting 50 ug of OGB-1 in 5 uL of artificial cerebral spinal fluid (ACSF). This solution was then injected by a glass electrode (50–100 um diameter opening) inserted 1 mm below the brain surface using a WPI micropump at 100 nL/min. A volume of 8 microliters of OGB-1 solution was sufficient to permit calcium imaging across a 5 × 8 mm craniotomy.

In separate animals, glial astrocyte activity was imaged following topical application of the calcium indicator Rhod-2 AM, which selectively labels astrocytes [56, 57]. The dye solution was created by diluting 50 ug of Rhod-2 in 5 uL of DMSO-F127 and then in 50 uL of ACSF. Next, a well was created by excreting a small rim of silicone glue around the margins of the skull craniotomy. The dye was then topically applied and allowed to permeate the brain surface for 90 min, after which a solution of ACSF was used to wash away any unbound dye. Note that genetically encoded calcium indicators are now available to target these same cell types (*see* **Note 6**).

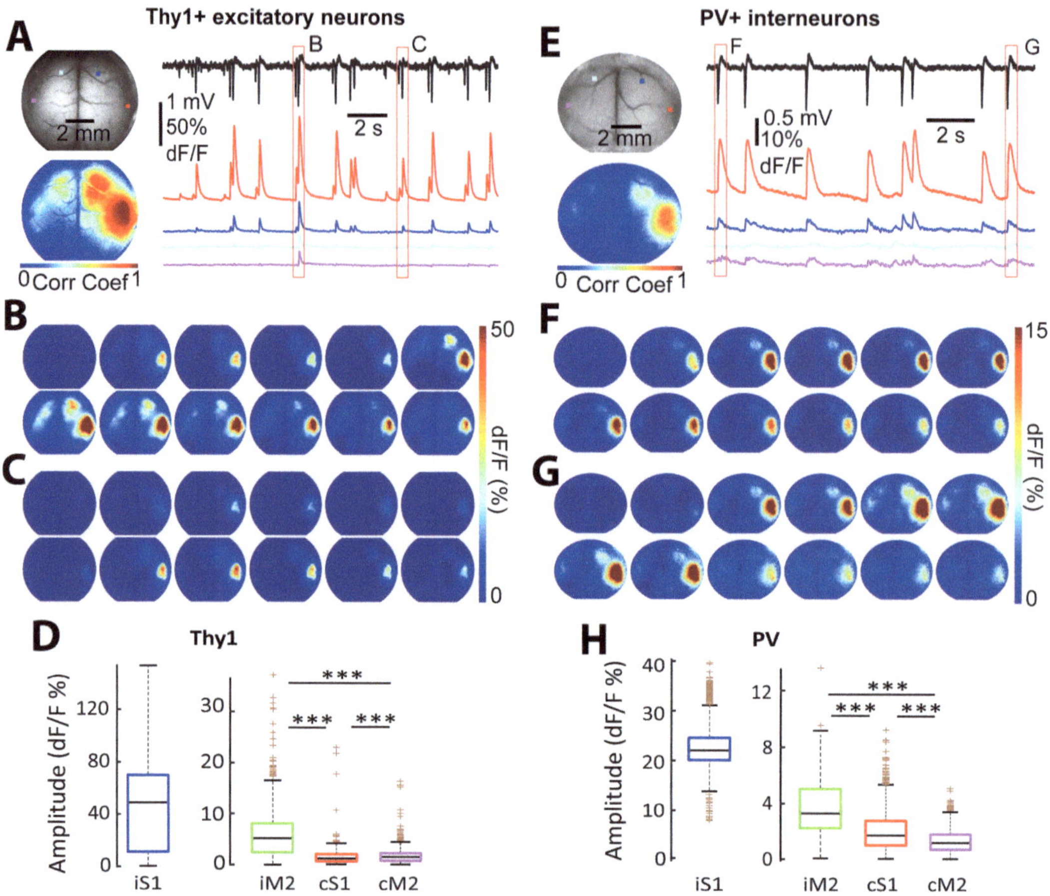

Fig. 8 Mesoscale network imaging of excitatory and inhibitory activity during interictal spikes. Following upon electrographic results, we applied widefield calcium imaging to Thy1-positive excitatory (**a**–**d**) and parvalbumin-positive inhibitory (E-H) cell types. A and E show electrographic (black trace) data recorded from the interictal focus where BMI was injected. Brain images are shown with squares demonstrating pixel regions where calcium activity traces are shown for iS1 (red), iM2 (blue), cM2 (cyan), and cS1 (pink). Heatmap shows the correlation coefficient (Pearson's *r*) calculated for all pixels relative to the interictal focus at iS1. Note the clear correlation of iM2, and weaker but still apparent correlation in cM2 and cS1. (**b**, **c**) Widefield imaging frames (taken every 33 ms) showing the change in fluorescence occurring during spikes highlighted in A (red boxes). Note that the larger spike clearly propagates to all network nodes, while the smaller spike (**c**) remains mostly focal. Panels E-G show the data in the same format but for inhibitory PV cell activity. Calcium signals were taken from each detected interictal spike and shown in panels (**d**, **h**) for the two cell types. iS1 showed the strongest activity, as expected, followed by iM2. Interestingly, in Thy1 excitatory cell activity, we observed significantly stronger cM2 activity vs. cS1; meanwhile, in PV excitatory cell activity, we observed significantly stronger cS1 activity vs. cM2. This difference suggests that cross-callosal neurons projecting from iS1 to cS1 may synapse preferentially on interneurons, while cross callosal neurons from iM2 to cM2 may synapse preferentially on excitatory neurons. This hypothesis is supported by recently published histology and electrophysiology data [74] and other evidence suggesting varying amounts of feedforward inhibition along the anteroposterior axis in contralateral neurons during bilateral seizure propagation [48]. (Adapted from Luo et al. [21])

OGB-1 was imaged with a 470 nm illumination and a 510 long-pass filter; Rhod-2 was imaged with a 530 nm illumination and 560 nm long-pass filter. Cerebral blood volume was measured by application of higher wavelength lights in both groups. Seizures were then induced by focal application of 4-AP through an injecting electrode placed 300–500 μm below the somatosensory cortical surface using a Nanoinject II injector. The electrode measured neural activity, which was bandpass filtered (1–500 Hz) using a Grass amplifier and then digitized by a CED Power 1401 before being recorded on a PC running Spike2 software. This configuration allows one to measure the electrographic events with concurrent imaging of astrocytic or neuronal activity.

The main goal of this study was to determine the differential roles of neurons, glial, and the pattern of neurovascular coupling during focal-onset seizures (Fig. 9). Using custom programs (written in MATLAB), we processed the imaging and electrophysiology data to examine and compare the time courses of these signals. First, to increase signal-to-noise ratios, we performed spatial smoothing (with a two-dimensional Gaussian function with sigma = 6 pixels). The neural imaging data was then high-pass filtered with a 1-Hz cutoff. The fluorescence activity was computed relative to baseline as described above (dF/F), using a baseline period when no epileptiform activity was seen in the electrode recording.

The onset and termination times of the seizures were defined separately for the LFP, neuronal, and glial activity traces as the moment when the signal crossed over and then returned past 3 standard deviations above the baseline (*see* **Note 7**). Our data showed that glial activation during seizure onsets was notably delayed, lagging both the hemodynamic and neuronal activity changes at seizure initiation. To quantify spatial spread, a seed trace correlation method was applied. The seed trace was taken from a region of interest in which the activity waveform closely resembled the electrographic signal (the 4-AP injection site where seizures were generated). Correlation coefficients were calculated between this seed trace and all the other pixels in the field of view, allowing creation of a heat map of these values. To quantify the spatial spread of the seizure, the maximal correlation was determined and a value of 50% of this peak number was used as a threshold. Any pixels with a correlation above this 50% cutoff were considered to have been invaded by the seizure. Similar to the onset timing, the glial wave was poorly correlated with the spatial and temporal development of the seizure and ended prior to the termination of neuronal or hemodynamic activity.

One advantage of widefield imaging is that the experimenter has broad access to the cortex during experiments, allowing for chemical manipulations. For example, to assess the importance of ictal glial waves glial activity during seizures, we applied the glial-

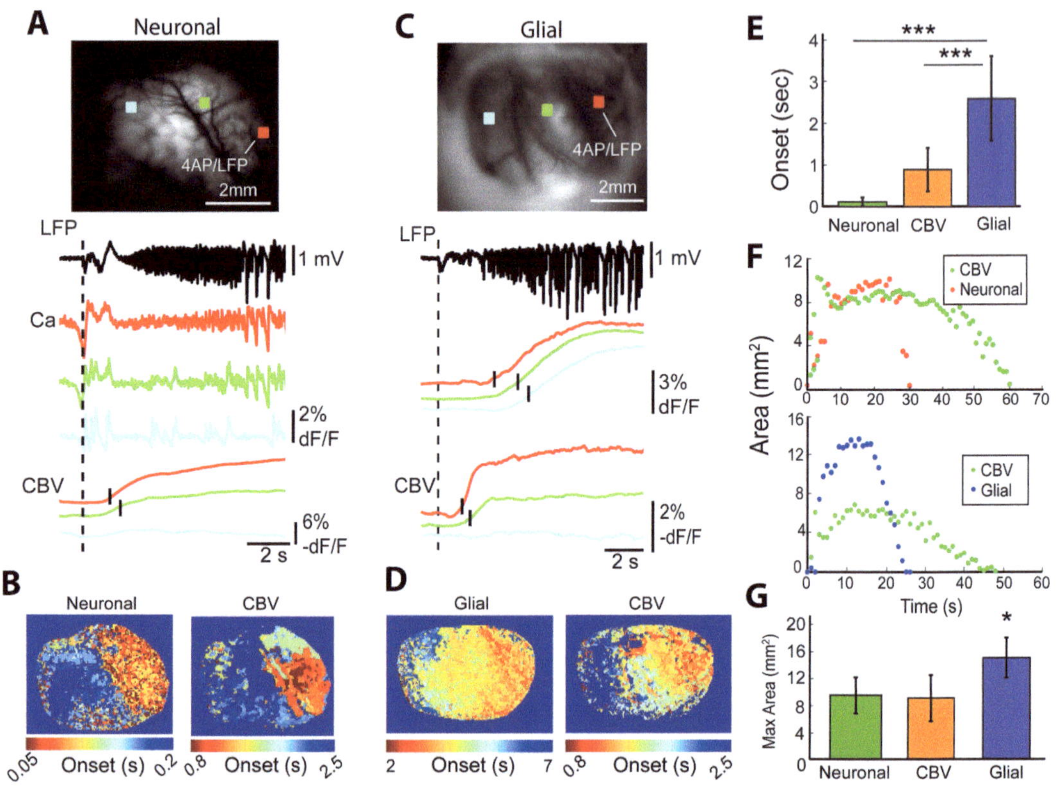

Fig. 9 Neurovascular coupling of glial and neuronal activity during focal seizures. (**a**) Mesoscale imaging of neuronal calcium activity and cerebral blood volume (CBV) measured at three sites over the dorsal cortex. Colored squares indicate pixel regions for calculating dF/F of imaging signals below. Black trace shows the electrographic seizure at the 4-AP seizure focus. (**b**) Time of onset detected in neuronal and hemodynamic signals relative to the electrographic seizure onset, calculated at each pixel to create the heatmap. Note the large delay in CBV versus neuronal signal in both the traces and the heatmaps. (**c**, **d**) Data presented in same format as A-B but for glial cell imaging following Rhod-2 application. Note the large delay in glial activity relative to CBV and much larger delay in the traces and heatmaps. (**e**) Onset time quantified over multiple seizures and animals. Glial activation was significantly delayed relative to both hemodynamic and neuronal onsets ($p < 0.001$). Hemodynamic signals were significantly delayed relative to neuronal activation ($p < 0.001$). (**f**) Areal recruitment across two seizures. Top panel shows a seizure during neuronal calcium imaging and hemodynamics. The bottom panel shows areal recruitment during glial cell imaging with hemodynamics. Note the spatial match between neuronal and hemodynamic recruitment while glial activation showed a very broad and steep increase followed by decrease. (**g**) Summary of area recruited during seizures across all animals and events. As in panel F, glial activation was more spatially broad while the neuronal and hemodynamic signals were more well correlated. Statistical significance comes from a one-way ANOVA followed by a Tukey-Kramer post hoc analysis (* $p < 0.01$, *** $p < 0.001$). (Adapted from Baird-Daniel et al. [33])

specific inhibitor fluoroacetate (FA) to prevent glial activity. This toxin is an established astrocyte-specific inhibitor [58, 59], so this blockade did not directly impact either neuronal or hemodynamic activity. FA was applied after calcium dye staining but before 4-AP injection. After blocking glial activity, seizure activity appeared

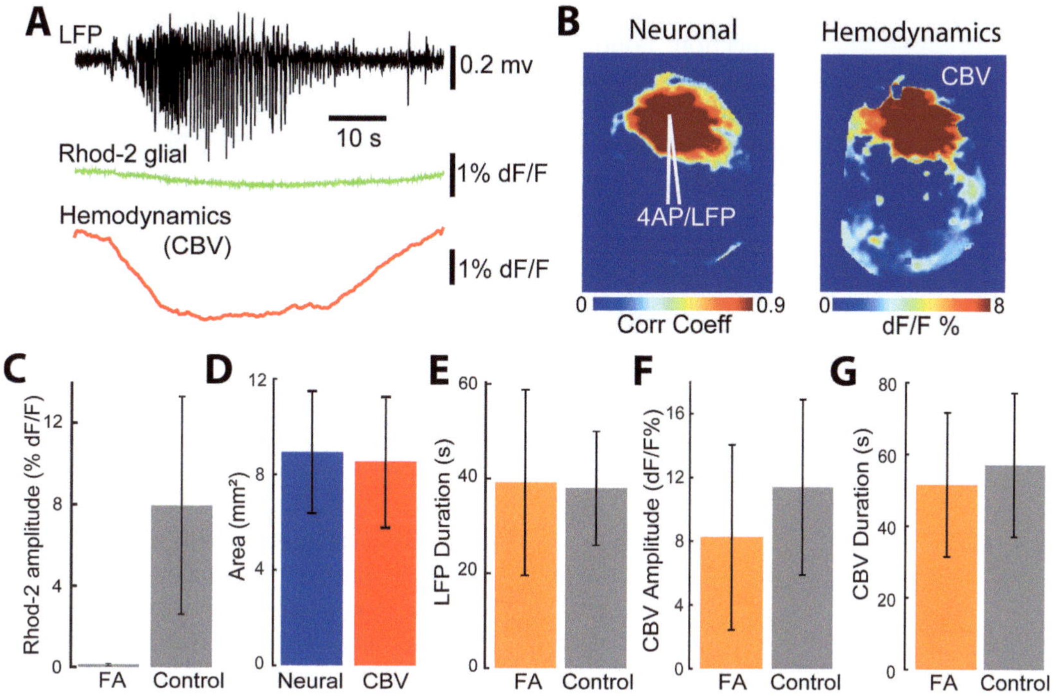

Fig. 10 Pharmacological tests during mesoscale imaging. The results from our imaging of glial, neuronal, and hemodynamic activity suggested that glial responses are significantly delayed and more widespread. To test whether glial cell activity is important for seizure initiation and propagation, our group used the glial cell inhibitor fluoroacetate (FA) to block this response. (**a**) Electrographic, glial (Rhod-d), and hemodynamic activity measured during widefield imaging following FA application demonstrates an abolished glial response but intact electrophysiological and hemodynamic seizure response. (**b**) Neural and hemodynamic heatmaps show the seizure is largely similar to conditions without FA application (Fig. 9). (**c**) The glial cell activity amplitude (dF/F) was reduced to nearly zero following FA application, while control (no FA) conditions showed clear seizure-related responses. (**d**) Neuronal and hemodynamic area of seizures remained similar following FA. (**e**) Electrographic seizure duration was roughly similar between FA and control seizures, as was hemodynamic amplitude (**f**) and duration of significant hemodynamic change (**g**). These analyses show that glial cell activity is not critical for seizure onset in the 4-AP model. Overall, these data demonstrate the utility of widefield imaging for examining how different components of brain activity can be manipulated to determine their relative importance in the development of seizures. (Adapted from Baird-Daniel et al. [33])

unaffected as recorded with the LFP and OGB-1. Likewise, the simultaneously measured intrinsic optical signals also showed an intact hemodynamic response (Fig. 10). Importantly, the amplitudes of these signals and the seizure durations were also unchanged. This data indicates that glial activity was not essential either for seizure activity or ictal hemodynamic coupling.

Taking further advantage of our widefield paradigm, our group then asked the question whether gap junctions, which are required for glial waves, influenced seizure activity. Gap junctions are small intercellular channels that form a syncytium of connected cells. In this experiment, the gap junction blocker, carbenoxolone (CBX,

100 uM), was applied to the cortical surface prior to the 4-AP injection. Following CBX application, seizures persisted, as in the prior set of experiments, but the amplitude of the glial wave was significantly reduced. Thus, functional gap junctions did not appear to be necessary for seizure onset and propagation but were necessary for the glial wave associated with seizures.

In a final test of glial-related activity during seizures, we blocked neuronal activity by applying the sodium channel blocker tetrodotoxin (TTX, 2uM) before 4-AP injection. The OGB-1 calcium and electrographic activity were both abolished, as expected. More interestingly, the activity of glial cells was also nearly abolished, and significantly lower than in control conditions. Thus, the onset and spread of the glial wave during seizures appears to be triggered by neuronal activity and nonessential for ictal activity and for hemodynamic coupling.

3.9 Conclusion

Understanding seizure generation and spread is critical to developing new treatments for human epilepsy. Widefield imaging is a powerful tool that allows researchers to examine neuronal, hemodynamic, and electrographic data simultaneously during seizures in a controlled laboratory environment. Some of the major findings using these techniques are described above, such as the significant role that anesthesia can play in affecting neural responses and neurovascular coupling across the cortex during seizures, and the different activity patterns of glial and both excitatory and inhibitory neurons at distant sites in the brain. These studies also highlight some the various ways in which the method of widefield imaging can be leveraged to not just track neuronal and nonneuronal activity during seizures, but also how amenable the preparation is to mechanistic experiments such as application of pharmacologic agonists and antagonists, anesthetics, and drugs that can disrupt vascular reactivity. As bioengineers continue to develop faster and better imaging sensors, particularly genetically encoded fluorophores, novel experiments will provide even more opportunities for mesoscale imaging. For example, researchers have begun using multicolor fluorophores to report neuronal vs. neurotransmitter activity across the cortex [29, 60], a paradigm that in the future may provide better understanding of how different cell types and chemicals may permit or inhibit seizures in awake animals. These techniques may be particularly useful in evaluating network-oriented treatments of epilepsy that involve neurostimulators [61–64], surgery [65, 66], or focal gene therapy [67–70], in which mesoscale effects should be considered and analyzed.

4 Notes

1. Many strains of mice are available, worldwide, from commercial breeders and individual investigators. Genotyping, either in-house or from commercial services like Transnetyx, is important to ensure the desired animal type (mutant) is being used in experiments.
2. While GCaMP variants may be sufficient for most widefield imaging, care must be taken to verify that adequate fluorescence is attainable when the fluorophore is genetically encoded. For example, if the fluorophore is only expressed in a very small percentage of cells then a stronger excitation light may be required.
3. Many bubbles are introduced to the fluid during elastomer mixing. The vacuum chamber is critical because the bubbles will not rise out in normal atmospheric pressure conditions. The petri dish holding the PDMS must be kept level, which can be accomplished with a two-dimensional leveler placed inside the chamber. Leveling the chamber can be performed by placing shims at appropriate locations under the container.
4. The best way to prevent bubbles forming between the film and brain is to implant the PDMS without allowing any air to remain under its surface by gently pressing on the film as vetbond is applied to the margin. In some cases, however, the brain may separate from the PDMS (possibly due to the brain recovering from inflammation during the surgery) and create a space between the PDMS and dura. We have solved these situations by injecting a small amount of sterile saline (under anesthesia) through the PDMS film. In some cases we have used two micropipette tips inserted into the PDMS: one tip allows air to escape while the other pipette tip is used to inject sterile saline. If a small bubble of air remains it can often be pushed away from the field of view by gently pressing the PDMS with a Q-tip.
5. GCaMP imaging depends on exciting fluorophores in tissue and capturing their emission. Both stages of this process can involve absorption by the tissue, creating a situation in which the collected emitted light contains a mixture of signals. In the brain, it is known that oxygenated and deoxygenated blood are two major sources of this mixture, and our lab uses the Beer-Lambert law to correct for these. Some other groups will simply use a violet wavelength (405 nm) interleaved with the 470 GCaMP exciting light; this 405 wavelength is isosbestic for GCaMP, which means that the emitted fluorescence is not activity dependent, thus providing a calcium-independent baseline that can be subtracted from the observed GCaMP

imaging frames [71, 72]. Our group prefers the hemodynamic-oriented correction method because it is based in known signal sources.

6. Astrocytic calcium imaging can be performed in mouse expressing genetically encoded fluorophores as well. See, for example, Jackson Laboratory mouse strain #029655: Aldh1l1-Cre/ERT2 BAC transgenic mouse (donated by Baljit Khakh, UCLA), which provide a Cre-inducible platform for expressing fluorophores specifically in astrocytes.
7. Using standard deviation cutoffs is a common method for determining when a signal reaches a value above a baseline. However, there may be times when weak signal changes are still of interest. In those cases, taking the cross correlation of two activity traces can more accurately capture small events [73].

References

1. Hill DK, Keynes RD (1949) Opacity changes in stimulated nerve. J Physiol 108(3):278–281
2. Frostig RD et al (1990) Cortical functional architecture and local coupling between neuronal activity and the microcirculation revealed by in vivo high-resolution optical imaging of intrinsic signals. Proc Natl Acad Sci USA 87(16):6082–6086
3. Grinvald A et al (1986) Functional architecture of cortex revealed by optical imaging of intrinsic signals. Nature 324(6095):361–364
4. Bonhoeffer T, Grinvald A (1991) Iso-orientation domains in cat visual cortex are arranged in pinwheel-like patterns. Nature 353(6343):429–431
5. Tasaki I et al (1968) Changes in fluorescence turbidity and birefringence associated with nerve excitation. Proc Natl Acad Sci USA 61(3):883
6. Salzberg BM, Davila HV, Cohen LB (1973) Optical recording of impulses in individual neurons of an invertebrate central nervous-system. Nature 246(5434):508–509
7. Grinvald A, Petersen CC (2015) Imaging the dynamics of neocortical population activity in behaving and freely moving mammals. Adv Exp Med Biol 859:273–296
8. Sharon D, Grinvald A (2002) Dynamics and constancy in cortical spatiotemporal patterns of orientation processing. Science 295(5554): 512–515
9. Tsien RY (1989) Fluorescent-probes of cell signaling. Annu Rev Neurosci 12:227–253
10. Braubach O, Cohen LB, Choi Y (2015) Historical overview and general methods of membrane potential imaging. Adv Exp Med Biol 859:3–26
11. Chemla S, Chavane F (2010) Voltage-sensitive dye imaging: technique review and models. J Physiol Paris 104(1–2):40–50
12. Storace D et al (2015) Genetically encoded protein sensors of membrane potential, vol 859. Membrane potential imaging in the nervous system and heart, pp 493–509
13. Chen Y et al (2020) Soma-targeted imaging of neural circuits by ribosome tethering. Neuron 107(3):454–469 e6
14. Shemesh OA et al (2020) Precision calcium imaging of dense neural populations via a cell-body-targeted calcium indicator. Neuron 107(3):470–486.e11
15. Schwartz TH, Bonhoeffer T (2001) In vivo optical mapping of epileptic foci and surround inhibition in ferret cerebral cortex. Nat Med 7(9):1063–1067
16. Zhao M et al (2009) Spatiotemporal dynamics of perfusion and oximetry during ictal discharges in the rat neocortex. J Neurosci 29(9):2814–2823
17. Ma HT et al (2009) The importance of latency in the focality of perfusion and oxygenation changes associated with triggered afterdischarges in human cortex. J Cereb Blood Flow Metab 29(5):1003–1014
18. Ma HT et al (2014) Wide-field in vivo neocortical calcium dye imaging using a convection-enhanced loading technique combined with simultaneous multiwavelength imaging of voltage-sensitive dyes and hemodynamic signals. Neurophotonics 1(1):015003

19. Liou JY et al (2018) Role of inhibitory control in modulating focal seizure spread. Brain 141: 2083–2097
20. Yang F et al (2021) Mesoscopic mapping of ictal neurovascular coupling in awake behaving mice using optical spectroscopy and genetically encoded calcium indicators. Front Neurosci 15:704834
21. Luo P et al (2023) Excitatory-inhibitory mismatch shapes node recruitment in an epileptic network. Epilepsia 64:1939
22. Rynes ML et al (2021) Miniaturized head-mounted microscope for whole-cortex mesoscale imaging in freely behaving mice. Nat Methods 18(4):417–425
23. Li J et al (2024) Mesoscopic mapping of hemodynamic responses and neuronal activity during pharmacologically induced interictal spikes in awake and anesthetized mice. J Cereb Blood Flow Metab 44:911
24. Prince DA, Wilder BJ (1967) Control mechanisms in cortical epileptogenic foci – surround inhibition. Arch Neurol 16(2):194–202
25. Jackson RB, Laboratory M, Green EL (1966) Biology of the laboratory mouse, 2nd edn. Blakiston Division, New York, xii 706 p
26. Dana H et al (2014) Thy1-GCaMP6 transgenic mice for neuronal population imaging in vivo. PLoS One 9(9):e108697
27. Nakai J, Ohkura M, Imoto K (2001) A high signal-to-noise Ca(2+) probe composed of a single green fluorescent protein. Nat Biotechnol 19(2):137–141
28. Huang L et al (2021) Relationship between simultaneously recorded spiking activity and fluorescence signal in GCaMP6 transgenic mice. elife 10:10
29. Lohani S et al (2022) Spatiotemporally heterogeneous coordination of cholinergic and neocortical activity. Nat Neurosci 25(12): 1706–1713
30. Ji GJ et al (2004) Ca2+-sensing transgenic mice – postsynaptic signaling in smooth muscle. J Biol Chem 279(20):21461–21468
31. Heo C et al (2016) A soft, transparent, freely accessible cranial window for chronic imaging and electrophysiology. Sci Rep 6:27818
32. Franklin KBJ, Paxinos G (2013) Paxinos and Franklin's The mouse brain in stereotaxic coordinates, 4th edn. Academic, Amsterdam, an imprint of Elsevier. 1 volume (unpaged)
33. Baird-Daniel E et al (2017) Glial calcium waves are triggered by seizure activity and not essential for initiating ictal onset or neurovascular coupling. Cereb Cortex 27(6):3318–3330
34. Schafer EW Jr, Brunton RB, Cunningham DJ (1973) A summary of the acute toxicity of 4-aminopyridine to birds and mammals. Toxicol Appl Pharmacol 26(4):532–538
35. Ma Y et al (2016) Wide-field optical mapping of neural activity and brain haemodynamics: considerations and novel approaches. Philos Trans R Soc B Biol Sci 371(1705):20150360
36. Chen-Bee CH et al (1996) Areal extent quantification of functional representations using intrinsic signal optical imaging. J Neurosci Methods 68(1):27–37
37. Saggio ML et al (2020) A taxonomy of seizure dynamotypes. elife 9:e55632
38. Esteller R et al (2001) Line length: An efficient feature for seizure onset detection. In: Proceedings of the 23rd annual international conference of the IEEE engineering in medicine and biology society, Vols 1–4, vol 23, pp 1707–1710
39. Ma HT et al (2009) Hemodynamic surrogates for excitatory membrane potential change during interictal epileptiform events in rat neocortex. J Neurophysiol 101(5):2550–2562
40. Ma HT, Wu CH, Wu JY (2004) Initiation of spontaneous epileptiform events in the rat neocortex in vivo. J Neurophysiol 91(2):934–945
41. Ma H, Zhao M, Schwartz TH (2013) Dynamic neurovascular coupling and uncoupling during ictal onset, propagation, and termination revealed by simultaneous in vivo optical imaging of neural activity and local blood volume. Cereb Cortex 23(4):885–899
42. Jackson JH (1958) Selected writings. Basic Books, New York
43. Takashima I, Kajiwara R, Iijima T (2001) Voltage-sensitive dye versus intrinsic signal optical imaging: comparison of optically determined functional maps from rat barrel cortex. Neuroreport 12(13):2889–2894
44. Bahar S et al (2006) Intrinsic optical signal imaging of neocortical seizures: the 'epileptic dip'. Neuroreport 17(5):499–503
45. Chen BR et al (2014) A critical role for the vascular endothelium in functional neurovascular coupling in the brain. J Am Heart Assoc 3(3):e000787
46. Trevelyan AJ, Schevon CA (2013) How inhibition influences seizure propagation. Neuropharmacology 69:45–54
47. Connors BW (1984) Initiation of synchronized neuronal bursting in neocortex. Nature 310(5979):685–687
48. Walker J et al (2012) Propagation of epileptiform events across the corpus callosum in a cingulate cortical slice preparation. PLoS One 7(2):e31415

49. Devinsky O et al (2013) Glia and epilepsy: excitability and inflammation. Trends Neurosci 36(3):174–184
50. Robel S, Sontheimer H (2016) Glia as drivers of abnormal neuronal activity. Nat Neurosci 19(1):28–33
51. Gomez-Gonzalo M et al (2010) An excitatory loop with astrocytes contributes to drive neurons to seizure threshold. PLoS Biol 8(4): e1000352
52. Zhao MR et al (2007) Focal increases in perfusion and decreases in hemoglobin oxygenation precede seizure onset in spontaneous human epilepsy. Epilepsia 48(11):2059–2067
53. Hillman EM (2014) Coupling mechanism and significance of the BOLD signal: a status report. Annu Rev Neurosci 37:161–181
54. McCaslin AF et al (2011) In vivo 3D morphology of astrocyte-vasculature interactions in the somatosensory cortex: implications for neurovascular coupling. J Cereb Blood Flow Metab 31(3):795–806
55. Iadecola C (2004) Neurovascular regulation in the normal brain and in Alzheimer's disease. Nat Rev Neurosci 5(5):347–360
56. Wang XH et al (2006) Astrocytic Ca2+ signaling evoked by sensory stimulation in vivo. Nat Neurosci 9(6):816–823
57. Ghosh A, Wyss MT, Weber B (2013) Somatotopic astrocytic activity in the somatosensory cortex. Glia 61(4):601–610
58. Fonnum F, Johnsen A, Hassel B (1997) Use of fluorocitrate and fluoroacetate in the study of brain metabolism. Glia. 21:106–113
59. Peña-Ortega F, Rivera-Angulo AJ, Lorea-Hernández JJ (2016) Pharmacological Tools to Study the Role of Astrocytes in Neural Network Functions. In: Glial Cells in Health and Disease of the CNS. Springer 47–66
60. Sun FM et al (2020) Next-generation GRAB sensors for monitoring dopaminergic activity in vivo. Nat Methods 17(11):1156
61. Doss DJ, Johnson GW, Englot DJ (2024) Imaging and stereotactic electroencephalography functional networks to guide epilepsy surgery. Neurosurg Clin N Am 35(1):61–72
62. Piper RJ et al (2022) Towards network-guided neuromodulation for epilepsy. Brain 145(10): 3347–3362
63. Fan JM et al (2022) Network connectivity predicts effectiveness of responsive neurostimulation in focal epilepsy. Brain Commun 4(3): fcac104
64. Khambhati AN et al (2021) Long-term brain network reorganization predicts responsive neurostimulation outcomes for focal epilepsy. Sci Transl Med 13(608):eabf6588
65. Amorim-Leite R et al (2020) History of the network approach in epilepsy surgery. Neurosurg Clin N Am 31(3):301–308
66. Sinha N et al (2017) Predicting neurosurgical outcomes in focal epilepsy patients using computational modelling. Brain 140(2): 319–332
67. Sullivan KA et al (2023) Drug-inducible gene therapy effectively reduces spontaneous seizures in kindled rats but creates off-target side effects in inhibitory neurons. Int J Mol Sci 24(14):11347
68. Wykes RC et al (2012) Optogenetic and potassium channel gene therapy in a rodent model of focal neocortical epilepsy. Sci Transl Med 4(161):161ra152
69. Shaimardanova AA et al (2022) Gene and Cell therapy for epilepsy: a mini review. Front Mol Neurosci 15:868531
70. Knowles JK et al (2022) Precision medicine for genetic epilepsy on the horizon: recent advances, present challenges, and suggestions for continued progress. Epilepsia 63(10): 2461–2475
71. Musall S et al (2019) Single-trial neural dynamics are dominated by richly varied movements. Nat Neurosci 22(10):1677
72. Couto J et al (2021) Chronic, cortex-wide imaging of specific cell populations during behavior. Nat Protoc 16(7):3241
73. Niemeyer JE et al (2022) Seizures initiate in zones of relative hyperexcitation in a zebrafish epilepsy model. Brain 145(7):2347–2360
74. Naskar S et al (2021) Cell-type-specific recruitment of GABAergic interneurons in the primary somatosensory cortex by long-range inputs. Cell Rep 34(8):108774

Chapter 12

Mesoscale Imaging of Stroke

Adam Santorelli, Colin T. Sullender, Christopher Smith, and Andrew K. Dunn

Abstract

Preclinical research allows neuroscientists and engineers to investigate both physiological effects of stroke and the subsequent recovery, as well as design, test, and optimize novel imaging methods and devices. Anesthesia is widely used to help sedate animals; however, the use of anesthesia has been shown have systemic effects on neuronal and vascular function. Thus, awake imaging in rodents has gained popularity. Specifically, awake imaging for stroke enables a better understanding of the process of ischemic stroke formation, the mechanisms of neuronal death, and the subsequent recovery period. In this protocol, we provide a guide on the development of a laser speckle contrast imaging system that allows for the implementation of a novel dual-modality system that allows for awake imaging and a targeted photothrombosis method. Laser speckle contrast imaging (LSCI) is a label-free optical imaging technique that can provide continuous full-field images of the blood flow dynamics of the cortical surface. We harness this technique to provide continuous monitoring of the vasculature of the cortex to allow for user-defined targeted regions for photothrombosis in awake mice. Furthermore, we show how the system can be used for chronic awake imaging of stroke mice to assess the revascularization of the infarct region.

Key words Cerebral blood flow, Optical imaging, Laser speckle contrast imaging, Multi-exposure speckle imaging, Neurosurgery, Awake stroke imaging

1 Introduction

The transition to fully awake imaging eliminates the systemic effects of general anesthesia. Anesthesia is widely used in preclinical neuroscience research to sedate animals while imaging despite systemic effects on neuronal and vascular function [1]. Isoflurane has been shown to reduce excitatory synaptic transmission [2], impair oxygen autoregulation [3], suppress the magnitude and speed of neurovascular coupling [4], and cause abnormal increases in cerebral blood flow (CBF) [5, 6]. Isoflurane has also been shown to convey neuroprotective effects that may reduce the severity of ischemic lesions [7, 8] and suppress the occurrence and frequency of spreading depolarizations [9]. These effects can mask the benefits of

James E. Niemeyer et al. (eds.), *Awake Behaving Mesoscopic Brain Imaging*, Neuromethods, vol. 214,
https://doi.org/10.1007/978-1-0716-4120-0_12,

neuroprotective agents and potentially impact the outcomes of long-term studies [10, 11]. Additionally, the use of different general anesthetics (urethane vs. isoflurane) can lead to conflicting vascular measurements [12–14].

Laser speckle contrast imaging (LSCI) is a full-field, label-free, optical imaging technique that can provide continuous maps of blood flow; thus, it can be used in an extensive number of applications across neuroscience, dermatology, dentistry, and ophthalmology [15, 16]. Studies have covered research topics such as monitoring CBF during ischemic stroke induction [17], chronic monitoring of the vasculature remodeling after stroke induction [9, 18, 19], investigating the effects of isoflurane-based vasodilation [20], and studying the impact of obesity on CBF [21]. There have also been advancements in research to integrate speckle contrast measurements within a surgical microscope for intraoperative use [22, 23].

Multi-exposure speckle imaging (MESI) was developed as an extension of LSCI to provide a more robust estimate of the correlation time constant (τ_c), ultimately allowing for accurate chronic monitoring of vascular blood flow. MESI requires collecting LSCI images over a wide range of camera exposure times to properly sample the underlying flow dynamics. While the complexity of the MESI hardware has been challenging to integrate into a clinical setting [23], MESI has been used for a wide array of in vivo mouse studies. MESI can create flow maps, commonly using the Inverse Correlation Time (ICT, $1/\tau_c$) as a quantifiable metric to represent flow, thus creating ICT maps, over the entire region of interest (ROI) that is being imaged. MESI is noninvasive (once a cranial window has been implanted), and does not require any dye injection, thus we are able to perform continuous imaging both during the stroke induction and any chronic imaging during the subsequent recovery period, the repeated measurements do not induce any unneeded stress on the mouse.

Animal models of ischemic stroke are extensively used to study the mechanisms of neuronal death and recovery and to perform preliminary testing on neuroprotective interventions. While there are numerous techniques for inducing focal ischemia, the majority rely upon occlusion of the middle cerebral artery (MCA) and its branches. The MCA is the largest cerebral artery in the brain and the most common vessel involved with human ischemic events [24]. The models that can most reliably reproduce the lesions and pathophysiology of human stroke (e.g. ischemic core and penumbra) offer the best experimental platforms for preclinical research.

Intraluminal MCA occlusion (MCAo) is the most widely used technique and is performed by introducing a monofilament suture into the internal carotid artery to block blood flow to the MCA [25]. This model is capable of inducing both permanent and transient focal ischemia similar to that of human stroke and does not require craniotomy. The procedure results in large-scale infarct volumes (21–45% of ipsilateral hemisphere) that most closely resemble malignant infarction in humans [26]. However, the majority of human strokes are much smaller in size (4.5–14%) [26, 27], making traditional MCAo a poor model for studying recovery at a similar scale.

The photothrombosis model of stroke induction uses intravascular photooxidation to generate well-defined cortical lesions [28]. Photosensitive dyes such as rose bengal are injected intravenously and irradiated with light to produce singlet oxygen, which causes localized endothelial damage initiating platelet aggregation and thrombus formation [29]. Rose bengal has been extensively utilized as a photothrombotic agent [30, 31] and has well-characterized pharmacokinetics with fast clearance from the body [32]. A significant advantage of the photothrombotic model is the ability to stereotactically control the position and size of the infarct to target specific functional regions. However, the technique results in rapid vasogenic edema, which is thought to restrict the development of the ischemic penumbra and local reperfusion [26].

A digital micromirror device (DMD) is an optical semiconductor device that consists of a two-dimensional array of thousands of individually addressable mirrors that can be tilted to spatially modulate light. The DMD offers a new method for targeted photothrombosis that allows for increased control over the stroke induction process compared to conventional techniques that only illuminate a single focal volume. Entire vessels, arbitrarily shaped regions, or even multiple locations can be simultaneously occluded by using the DMD to pattern the irradiating light. By specifically targeting vessels, collateral photooxidative damage to the surrounding tissue can be minimized.

The elimination of anesthesia from neuroimaging experiments has grown increasingly popular in recent years with two primary strategies taking the forefront. The first is the use of head-mounted miniaturized components [33–35]. While this technique allows for freely moving tethered imaging, it requires extensive optical engineering, increasing the complexity and associated costs, additionally, it introduces significant motion artifacts caused by normal animal behavior [34]. The second strategy, which is implemented in this chapter, permits the use of existing imaging platforms and involves restraining the animal's head while positioned on a treadmill [4, 36–39] or confined in a small chamber [40]. This technique allows for walking or running in place while minimizing motion of the head and imaging region.

The procedure in this protocol can be summarized as follows; first, a chronic cranial window, with an attached headbar, must be implanted in the mouse subject; secondly, the imaging and targeted-stroke induction system must be modified to allow for awake imaging, this includes the integration of a low-profile continuous belt treadmill and the optics required for the MESI system and DMD-targeted photothrombosis. Finally, we outline the process to simultaneously collect blood flow images from awake mice during the stroke induction, and the subsequent recovery period, including details for chronic imaging to assess recovery.

2 Materials

2.1 Cranial Window Implantation

1. Use medical air vaporized isoflurane (2.0%) via nose-cone inhalation to anesthetize the mouse.
2. Use a feedback heating pad (such as DC Temperature Controller, FHC) to maintain body temperature at 37 °C.
3. Monitor vital signs via pulse oximetry (MouseOx, Starr Life Sciences).
4. Place the mouse in a head-fixed stereotaxic frame (Narishige Scientific Instrument Lab) and administer carprofen (5 mg/kg, subcutaneous) for anti-inflammation and dexamethasone (2 mg/kg, intramuscular) to reduce the severity of cerebral edema following removal of the skull.
5. Sterilize all tools and the artificial cerebrospinal fluid (ACSF, buffered pH 7.4) in an autoclave.
6. Shave and resect the scalp to expose skull between the bregma and lambda cranial coordinates. Apply a thin layer of cyanoacrylate (Vetbond Tissue Adhesive, 3 M) to the exposed skull to facilitate the adhesion of dental cement.
7. Using a dental drill, (0.8 mm burr, Ideal Microdrill, Fine Science Tools) remove a 2–3 mm diameter portion of the skull over the frontoparietal cortex while leaving the dura intact. Ensure regular ACSF perfusion to prevent overheating.
8. Place a 5 mm round cover glass (#1.5, World Precision Instruments) over the exposed brain.
9. Deposit a dental cement mixture along the perimeter while applying gentle pressure to the cover glass. This process bonds the cover glass to the surrounding skull to create a sterile, air-tight seal around the craniotomy and allows for restoration of intracranial pressure.
10. Align the circular cutout in the headbar with the cranial window and rotate it laterally until parallel with the cover glass. This ensures that the cranial window will be perpendicular to the imaging system's optical axis when the animal is restrained in the awake imaging setup.
11. Apply dental cement around the headbar to permanently attach it to the animal's skull.

12. Apply a second layer of cyanoacrylate over the dental cement to further seal the cranial window (*see* **Note 1** for additional details).

2.2 Awake Imaging Treadmill System

The design of the awake imaging treadmill system is based on [39, 41], and allows for one-dimensional self-propelled movement. This low-profile system allows for ease of use with most optical systems designed in labs. Additionally, this system minimizes motion artifacts due to vertical variations that were encountered with the foam wheel treadmill design. The mouse was restrained by fixing the head plate into the holder.

Materials

1. Four 25 mm square optical construction rails (Thorlabs XE25L09).
2. Two ½″ diameter posts (Thorlabs TR12).
3. Two right-angle mounts for ½″ diameter rods (Thorlabs RA90).
4. Two ½″ diameter brass or steel rods to act as headplate bars (custom machined for holding headplate; schematic available at http://github.com/blinklab).
5. Two 2–56 screws for attaching headplate to brass rods.
6. One 1.5″ wide black polyester ribbon flat belt (Creative Ideas GRO1102-030).
7. Two sets of LEGO tires and axles (front and rear roller) that are 0.25″ and 0.375″ respectively.
8. Vero Black, for 3-D printing the main body (Janelia Labs, J005549 NRB Main Body schematic).

Construction

1. Form the base of the treadmill using four optical construction rails (Fig. 1), which allows for integration with the optical table and optical components.
2. Print the main body of the treadmill using a high-quality 3-D printer.
3. Screw the main body of the treadmill into the optical rail housing.
4. Integrate the LEGO tires and wheel with the main body of the treadmill.
5. Wrap the ribbon belt across the wheels to form the flat belt that will allow for self-propelled motion.

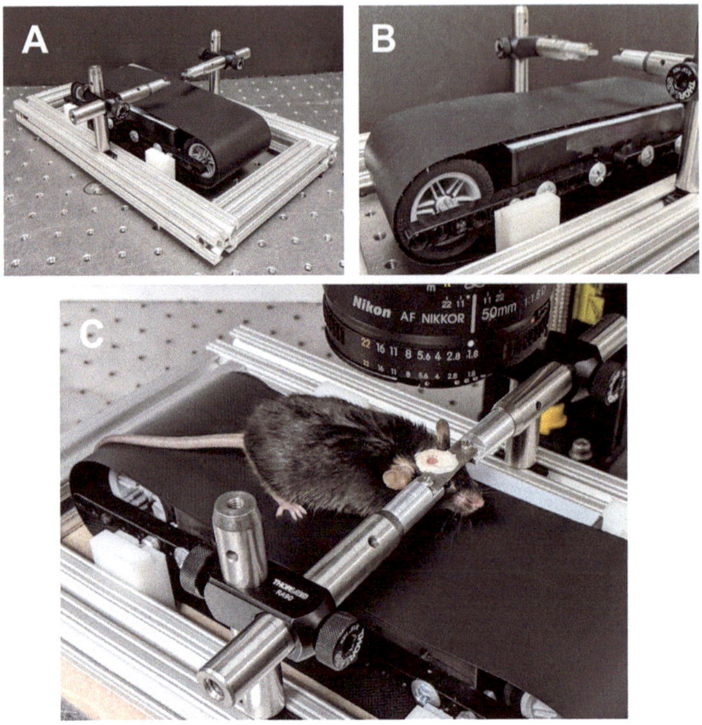

Fig. 1 Photographs of the awake imaging system showing the (**a**) low-profile, self-propelling belt treadmill, with (**b**) a close-up of the pulley system, and (**c**) a mouse placed under the optical imaging system with its head-restrained on the belt treadmill using the permanently attached headbar

6. Screw the two ½″ diameter Thorlabs posts into the optical rails, approximately centered over the midpoint of the treadmill.
7. Place one right-angle mount on each post, about 1½″ above the surface of the treadmill, so that the headplate bars can be passed through the mounts and meet above the treadmill. Adjust the right-angle mounts until the headplate bars can hold a headplate without straining the metal. Rotate the headplate bars so that they are at a known angle relative to the surface of the vibration isolation table. Also make sure that this angle will hold the mouse's head in a reasonable position (*see* **Note 2** for treadmill maintenance).

2.3 Optical Imaging Setup Requirements

A detailed explanation of the design, requirements, and implementation of the MESI system has been reported previously [42]. Here we include the modifications necessary for the targeted stroke induction (Fig. 2).

1. Stroke induction laser: Use a continuous wave green laser (typically 532 nm such as the 200 mW AixiZ LLC) to induce the targeted photothrombosis (when coupled with rose bengal

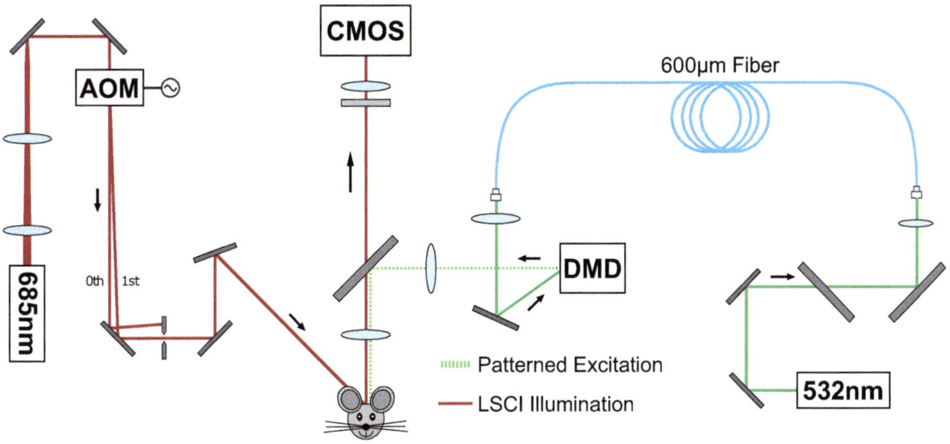

Fig. 2 A detailed schematic of the dual modality system that includes the MESI system and the DMD targeted stroke systems

dye). The packaged diode laser has a 2 mm collimated output that operates at a fixed current with convection cooling.

2. Use a neutral density filter (OD 1.0, NE10A-A, Thorlabs, Inc.) to attenuate the laser power. Usually only about 20 mW is needed.
3. Couple the green light into a fiber optic patch cord.
4. To relay the laser light, prior to creating the pattern, use a fiber optic patch cord with a 600 µm core size. This light will be used to illuminate the DMD.
5. A DLP® LightCrafter™ Evaluation Module (Texas Instruments Inc.) was modified to expose the bare DMD (DLP3000, 608 × 684 pixels, 7.6 µm pitch, Texas Instruments Inc.), *see* **Note 3**, for illumination. The spatially patterned modulated light was then relayed to the sample with 0.5× magnification.
6. Imaging light source: A wavelength-stabilized laser diode (ideally in the 600–850 nm wavelength range), capable of generating a decent amount of power, to ensure sufficient light at the shortest exposure time (ideally on the order of 50 µs), is the ideal choice of illumination for MESI (for example, 50 mW, HL6750MG, Thorlabs, Inc.).
7. Temperature-controlled housing: Mount the chosen laser diode in a laser diode mount with integrated temperature control (for example, TCLDM9, Thorlabs, Inc.).
8. Temperature controller: For stability and repeatability, use a temperature controller (TEC, such as TED200C, Thorlabs, Inc.) to set and maintain the laser diode operating temperature.

9. Laser diode controller: For the stability and repeatability of the illumination light, drive the laser diode with a constant current, controlled by a laser diode controller (such as the LDC202, Thorlabs, Inc.).
10. Light modulation: Use a free space AOM (for example, AOMO 3100-125, Gooch and Housego) and an RF driver (97-03307-34, Gooch and Housego) to modulate the intensity of the collimated laser light, ensuring the successful implementation of MESI. Use an iris to isolate the first order diffracted light from the free space AOM.
11. Field of view control: An aspheric lens (C240TME-B, Thorlabs, Inc.) can be used to control the beam diameter (here it was reduced to 1 mm).
12. Collection optics: The scattered light was relayed through a pair of dichroic beamsplitters and a bandpass filter (685 ± 40 nm, S685/40m, Chroma Technology Corp.). Use a pair of camera lenses (for example, 50 mm Nikon DSLR lenses) to direct the collected light to the camera sensor.
13. Camera: Use a high-speed CMOS camera that can be triggered from external source (such as the acA1920-155um Basler AG). The camera can be monochromatic, as only pixel intensity is required.
14. Control electronics: Use a multifunction I/O device (USB-6363, National Instruments Corp.), referred to as the data acquisition hardware (DAQ), to produce the camera exposure trigger signals and AOM modulation voltages. The drivers and associated libraries (NI-DAQmx library) for the DAQ will need to be installed (*see* **Note 4** for further details about MESI).
15. Processing software: Compute the speckle contrast, at each exposure time, from the collected raw signals. Solve the MESI equation, often using via nonlinear least squares curve fitting, to compute the ICT maps. These steps can easily be done in common post-processing software such as MATLAB.

3 Methods

3.1 Awake Imaging During Targeted Stroke Induction

This section will cover the steps necessary for awake imaging during targeted stroke induction. Note that the stroke induction procedure itself required the brief use of anesthesia for the injection of the photothrombotic agent. While it is unclear what minimum dosage of isoflurane is needed to convey its neuroprotective effects [10], the photothrombosis was performed under the lingering influence of isoflurane. This obfuscates the true changes in blood flow and oxygenation caused by the photothrombotic occlusion itself. Transitioning to a tail vein or intraperitoneal injection prior to mounting on the treadmill would allow for the complete elimination of isoflurane from the stroke induction process. The resulting infarct was less severe than the anesthetized demonstration and was likely mitigated by the availability of collateral blood supply.

1. Initialize the dual modality system and ensure all optical components are functional and at a stable operating point.
2. The subject mouse is fixed to the headplate holder of the system (*see* **Note 5** for details about potential motion artifacts due to locomotion) as in Fig. 1. Allow several minutes for habituation.
3. Launch the speckle imaging software to ensure system functionality and that the cortical surface is in focus.
4. The subject was briefly anesthetized on the treadmill via nose-cone inhalation of isoflurane (3.0%).
5. Rose bengal was administered intravenously via retro-orbital injection (50 μL, 15 mg/mL).
6. Stop anesthesia and remove the nose-cone. The mouse will typically regain consciousness after about 3 min.
7. Check the live speckle image view to ensure that the cortical surface remains in focus after the dye injection.
8. Using the speckle software choose the specific region of interest (ROI) that will be illuminated by the patterned DMD light. This is the area that will undergo clot formation (*see* **Note 6** for additional details about DMD pattern control).
9. Expose the subject to the DMD-patterned green light for approximately 5–15 min to induce a photothrombotic occlusion. Descending arterioles were the primary targets because they serve as bottlenecks in the cortical oxygen supply. Target vessels were identified based on vascular orientation and a posteriori knowledge. After selecting a target vessel, adjust the subject such that the target vessel is approximately in the center of the FOV. This target area should also be the region of best focus if the surface is curved and consistent focus across the FOV is impossible.
10. Use the live speckle view in the software to monitor clot formation within the targeted area. An example of this process is shown in Fig. 3; details about this series of figures is described below. The series of speckle contrast images depict the progression of photothrombosis with the targeted vessel fully occluding after less than 3 min of exposure.
    (i) Images in Fig. 3a, b are the speckle contrast values over the entire FOV. Areas in black represent regions of low speckle contrast (high flow), whereas regions in white are indicative of no flow.

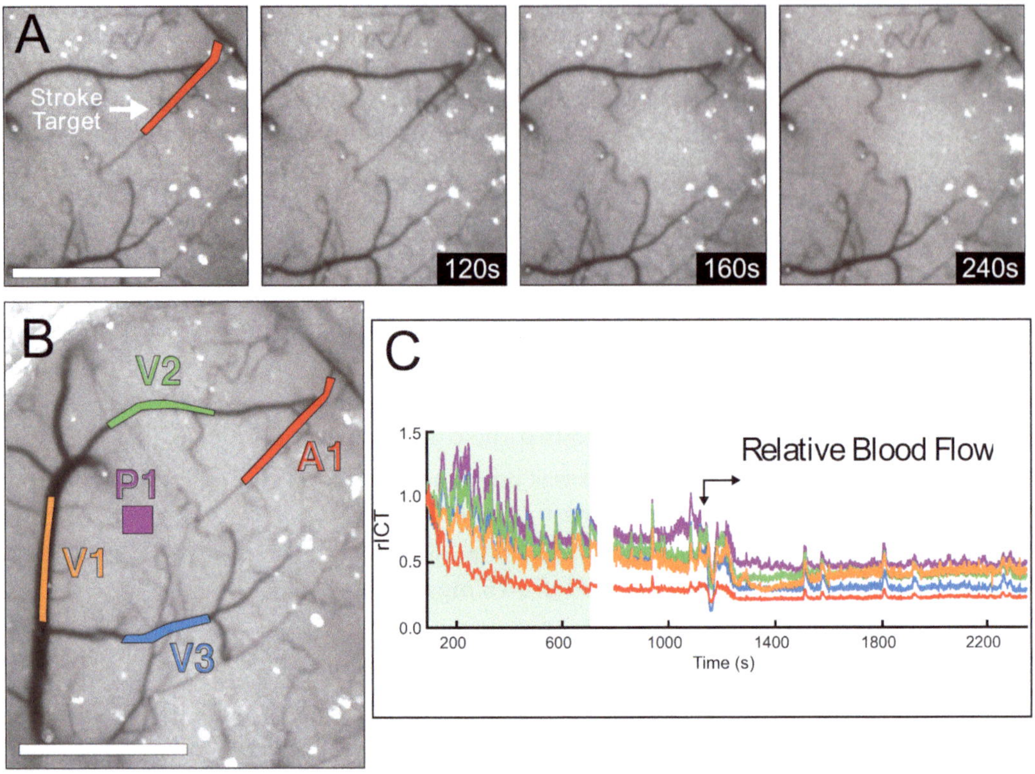

Fig. 3 (**a**) Single-exposure speckle contrast images depicting the occlusion of a descending arteriole using DMD-targeted photothrombosis. The red overlay indicates the 0.06 mm^2 region simultaneously illuminated for occlusion. (**b**) One arteriole (A1), three veins (V1, V2, V3), and one parenchyma region (P1) were targeted for flow measurements after stroke induction. (**c**) Relative blood flow within the targeted regions during and after photothrombosis. The green-shaded section indicates irradiation of the targeted arteriole. The gap in data occurred because of technical difficulties. The arrows indicate the propagation of an ischemia-induced depolarization event (White scale bars = 1 mm)

(ii) Figure 3a presents a series of speckle contrast images for the first 4 min post-photothrombosis (with the target vessel highlighted in red at $t = 0$ s). This time series of speckle images confirms the occlusion of the targeted vessel.

(iii) We highlight the five regions in Fig. 3b (one arteriole, three veins, and one parenchyma) we monitored for dynamic relative blood flow. The arteriole region (A1) is the same vessel targeted for photothrombotic occlusion. The resulting timecourses of relative blood flow (using the relative ICT values) can be seen in Fig. 3c. Compared to anesthetized photothrombosis measurements, these time-courses are significantly more variable because of walking and other animal motions. However, the subject had no visible reaction to the photothrombosis. By $t = 200$ s, flow within the targeted arteriole had decreased to <50% of baseline. However, the animal was likely still experiencing the effects of anesthesia, making it impossible to attribute the changes exclusively to the photothrombotic occlusion.

(iv) The propagation of an ischemia-induced depolarization event can be seen beginning at $t = 1150$ s, with sharp reductions in relative blood flow across all ROIs. The magnitude of these changes are smaller than those seen in the anesthetized data [20], which is consistent with the results of a calcium imaging study performed during awake ischemic stroke [43]. As the depolarization subsided, flow within the targeted arteriole further decreased to <20% of baseline. Flow in all other regions remained depressed following the depolarization and were relatively steady until the end of the 40-min imaging session. There was also a reduction in animal motion compared to pre-depolarization behavior as seen by the lack of regular spikes in relative blood flow.

11. If stroke induction is not effected after 15 min, repeat the steps for injection and stroke injection to help the stroke take. This should not be repeated more than once at this dosage.

3.2 Chronic Awake Post-Stroke Hemodynamics

This section covers the monitoring of the chronic progression of the ischemic lesion using awake imaging for 8 days following photothrombosis, with the same mouse subject as in Subheading 3.1 (*see* **Note 7** for details on the upkeep of the mouse for chronic imaging). Anesthesia was not utilized in any of the post-photothrombosis imaging sessions. Data was only acquired when the animal was completely stationary because motion could interfere with the measurements and the resulting hemodynamic response would not be indicative of the resting state. Each imaging session followed the same protocol, namely:

1. Initialize the MESI imaging system (the DMD system is not needed for chronic awake imaging) and ensure all optical components are functional and at a stable operating point.
2. The subject mouse is fixed to the headplate holder of the system (*see* **Note 5** for details about potential motion artifacts due to locomotion) as in Fig. 1. Allow several minutes for habituation.
3. Launch the speckle imaging software to ensure system functionality and that the cortical surface is in focus.
4. Collect MESI sequences, typically, 1 min of data is sufficient.
5. Repeat the process for desired number of days/sessions in the study.

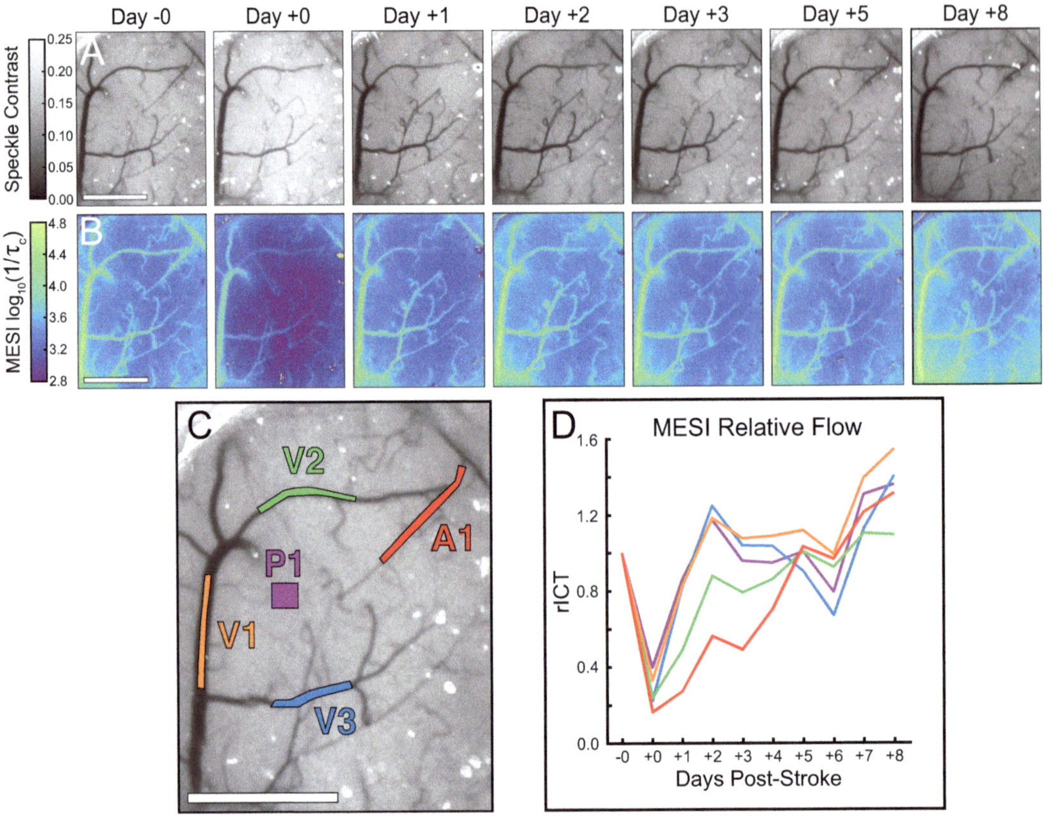

Fig. 4 Progression of the ischemic lesion over 8 days as imaged with (**a**) single-exposure LSCI displaying the speckle contrast images and (**b**) MESI displaying the ICT images on a log scale. Day −0 measurements were taken immediately prior to photothrombosis induction and Day +0 measurements were taken immediately after. (**c**) One arteriole (A1), three veins (V1), and one parenchyma region (P1) were targeted for chronic (**d**) relative blood flow. Relative MESI ICT was baselined against Day −0 measurements (White scale bars = 1 mm)

6. Convert the collected MESI data to ICT maps. This is done by solving for τ_c at every pixel in the image. As MESI allows for repeatable τ_c measurements across imaging sessions, we can reliably analyze and compare chronic changes in flow. Thus, we are able to monitor the progression and recovery of the ischemic lesion. In Fig. 4 we demonstrate the ability to use MESI to chronically monitor (over 8 consecutive days) the progression of an ischemic lesion, assess recovery of the vasculature, and the changes in relative blood flow pre- and post-stroke. The figure details are described as follows:

   (i) The perfusion of the occluded arteriole and broader effects on cortical flow were tracked using 5 ms single-exposure LSCI (Fig. 4a) and MESI ICT (Fig. 4b). The gradient between the occluded vessel and the surrounding tissue diminishes over the course of Days +1, +2, and +3 as the occluded arteriole begins to reperfuse. By Day +5, the vessel had fully reperfused leaving little evidence of the infarct in the speckle contrast or ICT imagery.

(ii) The same five regions (one arteriole, three veins, and one parenchyma) used during the acute awake photothrombosis measurements were also targeted for chronic relative blood flow measurements (Fig. 4c, d). The relative blood flow was calculated using the pre-stroke (Day −0) MESI ICT measurements as the baseline. The first post-stroke measurements (Day +0) were taken immediately after the conclusion of the photothrombosis induction and reveal systemic deficits in blood flow, likely caused by the spreading depolarization. Blood flow within the targeted arteriole (A1) and a nearby vein (V2) slowly recovered over the course of 5 days while the remaining ROIs experienced a slight overshoot in relative flow during the same period. Flow continued to increase across all ROIs over the remaining 3 days and was elevated over baseline by the final imaging session on Day +8.

(iii) The speed of the recovery from the photothrombotic infarct is similar to the anesthetized [20, 41] with a full recovery after 5 days. The targeted arteriole remained occluded for 3 days following photothrombosis before fully reperfusing. During this period, another arteriole approaching from midline (bottom of the camera FOV) hyperperfused and likely served as a collateral blood supply to mitigate the severity of the flow deficit. Once flow was restored to the targeted vessel on Day +4, the collateral arteriole returned to near baseline flow.

4 Notes

1. Animals were allowed to recover from anesthesia and monitored for cranial window integrity and normal behavior for at least 2 weeks prior to imaging. Additional carprofen injections (5 mg/kg) were administered subcutaneously 2, 4, and 7 days post-surgery to relieve inflammation from the procedure. Cranial windows were lightly cleaned prior to each imaging session using a cotton swab and 70% ethanol (v/v). If necessary, a topical application of mineral oil was used to improve image quality by index matching. Any discoloration on or around the cranial window was documented and monitored for possible infection. Any cracks in or breaking of the cranial window were also documented and resulted in the immediate euthanasia of the animal.

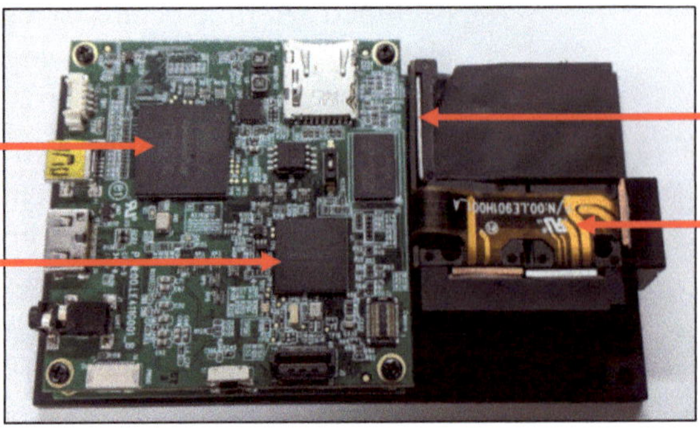

Fig. 5 Overview of TI DLP® LightCrafter™, indicating the four primary components that make up the evaluation board

2. Treadmill general maintenance tips: Thoroughly clean the surface of the wheel with 70% EtOH or a laboratory cleaner of your choice after removing the mouse from the wheel. Clean the area around the foam wheel. Urine and feces are sprayed off of the wheel when the mouse runs. It can be helpful to lay a paper towel or other disposable materials underneath the wheel to collect the majority of the mouse's waste. All parts should be free of oil, grease, or any other contaminants.
3. The Texas Instruments DLP® LightCrafter™ evaluation module contains a DLP3000 DMD, a DM365 embedded processor running Linux, and an RGB LED light engine developed by Young Optics, shown in Fig. 5. In order to utilize a custom light source such as a laser, the light engine must be removed to gain physical access to the DMD.
4. Further details about the MESI optical setup, necessary control electronics, and the required software packages and logic control can be found in a *Methods* paper form our lab [42] and in [20].
5. Despite the efficacy of the head restraint system, correlation time estimates with LSCI remain extremely sensitive to any motion within the imaging area. The animal walking would cause abrupt spikes in ICT that could not definitively be ascribed to neurovascular coupling instead of just a slight motion-related displacement. Excluding data during these periods of heightened activity offers the simplest solution for acquiring reliable instantaneous measurements of blood flow. However, if the hemodynamic response itself is being studied, then further measures would need to be taken in order to more robustly restrict motion of the brain.
6. The Speckle Software was modified to control the DMD via its Ethernet-over-USB command interface. Users can define arbitrarily shaped regions of interest (ROIs) using the LSCI camera as reference and upload the resulting binary masks to the DMD

for the patterning of excitation light. Registration between the camera and the projected pattern can be performed to guarantee alignment with the reference image. Individual patterns or timed pattern sequences can be uploaded and displayed on the DMD, with a TTL pulse emitted on each pattern change.

7. Animals were checked daily to monitor both behavior and the integrity of the cranial window by veterinary staff at the University of Texas at Austin Animal Research Center (ARC). Animals were housed in climate-controlled rooms with timed lighting (12-h light/dark cycles) to maintain a comfortable living environment and given food and water ad libitum. Social housing with multiple animals reduced the risk of overeating commonly seen when solo housing animals. This minimized possible growth in the animal's size and helped maintain the integrity of the cranial window. Any aggression resulted in the removal of the aggressor into a separate cage.

References

1. Janssen BJ, De Celle T, Debets JJ, Brouns AE, Callahan MF, Smith TL (2004) Effects of anesthetics on systemic hemodynamics in mice. Am J Phys Heart Circ Phys 287(4):H1618–H1624
2. Berg-Johnsen J, Langmoen IA (1992) The effect of isoflurane on excitatory synaptic transmission in the rat hippocampus. Acta Anaesthesiol Scand 36(4):350–355
3. Aksenov D, Eassa JE, Lakhoo J, Wyrwicz A, Linsenmeier RA (2012) Effect of isoflurane on brain tissue oxygen tension and cerebral autoregulation in rabbits. Neurosci Lett 524(2): 116–118
4. Pisauro MA, Dhruv NT, Carandini M, Benucci A (2013) Fast hemodynamic responses in the visual cortex of the awake mouse. J Neurosci 33(46):18343–18351
5. Strebel S, Lam A, Matta B, Mayberg TS, Aaslid R, Newell DW (1995) Dynamic and static cerebral autoregulation during isoflurane, desflurane, and propofol anesthesia. Anesthesiology 83(1):66–76
6. Iida H, Ohata H, Iida M, Watanabe Y, Dohi S (1998) Isoflurane and sevoflurane induce vasodilation of cerebral vessels via ATP-sensitive K+ channel activation. Anesthesiology 89(4): 954–960
7. Sakai H, Sheng H, Yates RB, Ishida K, Pearlstein RD, Warner DS (2007) Isoflurane provides long-term protection against focal cerebral ischemia in the rat. Anesthesiology 106(1):92–99
8. Burchell SR, Dixon BJ, Tang J, Zhang JH (2013) Isoflurane provides neuroprotection in neonatal hypoxic ischemic brain injury. J Investig Med 61(7):1078–1083
9. Schrandt CJ, Kazmi SS, Jones TA, Dunn AK (2015) Chronic monitoring of vascular progression after ischemic stroke using multiexposure speckle imaging and two-photon fluorescence microscopy. J Cereb Blood Flow Metab 35(6):933–942
10. Kapinya KJ, Prass K, Dirnagl U (2002) Isoflurane induced prolonged protection against cerebral ischemia in mice: a redox sensitive mechanism? Neuroreport 13(11):1431–1435
11. Seto A, Taylor S, Trudeau D, Swan I, Leung J, Reeson P, Brown CE (2014) Induction of ischemic stroke in awake freely moving mice reveals that isoflurane anesthesia can mask the benefits of a neuroprotection therapy. Front Neuroenerg 6:1
12. Kazmi SS, Salvaggio AJ, Estrada AD, Hemati MA, Shaydyuk NK, Roussakis E et al (2013) Three-dimensional mapping of oxygen tension in cortical arterioles before and after occlusion. Biomed Opt Express 4(7):1061–1073
13. Ponticorvo A, Dunn AK (2010) Simultaneous imaging of oxygen tension and blood flow in animals using a digital micromirror device. Opt Express 18(8):8160–8170
14. Sullender CT, Mark AE, Clark TA, Esipova TV, Vinogradov SA, Jones TA, Dunn AK (2018) Imaging of cortical oxygen tension and blood flow following targeted photothrombotic stroke. Neurophotonics 5(3):035003

15. Boas DA, Dunn AK (2010) Laser speckle contrast imaging in biomedical optics. J Biomed Opt 15(1):011109
16. Dunn AK (2012) Laser speckle contrast imaging of cerebral blood flow. Ann Biomed Eng 40(2):367–377
17. Dunn AK, Bolay H, Moskowitz MA, Boas DA (2001) Dynamic imaging of cerebral blood flow using laser speckle. J Cereb Blood Flow Metab 21(3):195–201
18. He F, Sullender CT, Zhu H, Williamson MR, Li X, Zhao Z, Luan L (2020) Multimodal mapping of neural activity and cerebral blood flow reveals long-lasting neurovascular dissociations after small-scale strokes. Sci Adv 6(21):eaba1933
19. Kazmi SMS, Parthasarthy AB, Song NE, Jones TA, Dunn AK (2013) Chronic imaging of cortical blood flow using multi-exposure speckle imaging. J Cereb Blood Flow Metab 33(6): 798–808
20. Sullender CT, Richards LM, He F, Luan L, Dunn AK (2022) Dynamics of isoflurane-induced vasodilation and blood flow of cerebral vasculature revealed by multi-exposure speckle imaging. J Neurosci Methods 366:109434
21. Soleimanzad H, Gurden H, Pain F (2018) Wide-field speckle imaging and two-photon microscopy for the investigation of cerebral blood flow in vivo in mice models of obesity. In: Biophotonics: photonic solutions for better health care VI, vol 10685. International Society for Optics and Photonics, p 1068508
22. Miller DR, Ashour R, Sullender CT, Dunn A (2021) Laser speckle contrast imaging for visualizing blood flow during cerebral aneurysm surgery: a comparison with indocyanine green angiography. medRxiv. https://doi.org/10.1101/2021.04.29.21254954
23. Richards LM, Kazmi SS, Olin KE, Waldron JS, Fox DJ Jr, Dunn AK (2017) Intraoperative multi-exposure speckle imaging of cerebral blood flow. J Cereb Blood Flow Metab 37(9): 3097–3109
24. Sicard KM, Fisher M (2009) Animal models of focal brain ischemia. Exp Transl Stroke Med 1: 1–6
25. Kozuimi J, Yoshida Y, Nakazawa T, Ooneda G (1986) Experimental studies of ischemic brain edema. 1. A new experimental model of cerebral embolism in rats in which recirculation can be introduced in the ischemic area. Jpn J Stroke 8:1–8
26. Carmichael ST (2005) Rodent models of focal stroke: size, mechanism, and purpose. NeuroRx 2(3):396–409
27. Brott T, Marler JR, Olinger CP, Adams HP Jr, Tomsick T, Barsan WG et al (1989) Measurements of acute cerebral infarction: lesion size by computed tomography. Stroke 20(7): 871–875
28. Watson BD, Dietrich WD, Busto R, Wachtel MS, Ginsberg MD (1985) Induction of reproducible brain infarction by photochemically initiated thrombosis. Ann Neurol 17(5): 497–504
29. Dietrich WD, Watson BD, Busto R, Ginsberg MD, Bethea JR (1987) Photochemically induced cerebral infarction: I. Early microvascular alterations. Acta Neuropathol 72(4): 315–325
30. Grome JJ, Gojowczyk G, Hofmann W, Graham DI (1988) Quantitation of photochemically induced focal cerebral ischemia in the rat. J Cereb Blood Flow Metab 8(1):89–95
31. Parthasarathy AB, Kazmi SS, Dunn AK (2010) Quantitative imaging of ischemic stroke through thinned skull in mice with multi exposure speckle imaging. Biomed Opt Express 1(1):246–259
32. Klaassen CD (1976) Pharmacokinetics of rose bengal in the rat, rabbit, dog, and guinea pig. Toxicol Appl Pharmacol 38(1):85–100
33. Gu W, Jiang W, Brännström T, Wester P (1999) Long-term cortical CBF recording by laser-Doppler flowmetry in awake freely moving rats subjected to reversible photothrombotic stroke. J Neurosci Methods 90(1):23–32
34. Helmchen F, Fee MS, Tank DW, Denk W (2001) A miniature head-mounted two-photon microscope: high-resolution brain imaging in freely moving animals. Neuron 31(6): 903–912
35. Flusberg BA, Jung JC, Cocker ED, Anderson EP, Schnitzer MJ (2005) In vivo brain imaging using a portable 3.9 gram two-photon fluorescence microendoscope. Opt Lett 30(17): 2272–2274
36. Dombeck DA, Khabbaz AN, Collman F, Adelman TL, Tank DW (2007) Imaging large-scale neural activity with cellular resolution in awake, mobile mice. Neuron 56(1):43–57
37. Wienisch M, Blauvelt DG, Sato TF, Murthy VN (2012) Two-photon imaging of neural activity in awake, head-restrained mice. In: Neuronal network analysis: concepts and experimental approaches. Humana Press, pp 45–60
38. Kaifosh P, Lovett-Barron M, Turi GF, Reardon TR, Losonczy A (2013) Septo-hippocampal GABAergic signaling across multiple

modalities in awake mice. Nat Neurosci 16(9): 1182–1184

39. Heiney SA, Ohmae S, Kim OA, Medina JF (2018) Single-unit extracellular recording from the cerebellum during eyeblink conditioning in head-fixed mice. Neuromethods 134:39–71
40. Silasi G, Xiao D, Vanni MP, Chen AC, Murphy TH (2016) Intact skull chronic windows for mesoscopic wide-field imaging in awake mice. J Neurosci Methods 267:141–149
41. Royer S, Zemelman BV, Losonczy A, Kim J, Chance F, Magee JC, Buzsáki G (2012) Control of timing, rate and bursts of hippocampal place cells by dendritic and somatic inhibition. Nat Neurosci 15(5):769–775
42. Santorelli A, Sullender CT, Dunn AK (2023) Multi-exposure speckle imaging for quantitative evaluation of cortical blood flow. In: Karamyan VT, Stowe AM (eds) Neural repair. Methods in molecular biology, vol 2616. Humana Press, New York
43. Balbi M, Vanni MP, Silasi G, Sekino Y, Bolanos L, LeDue JM, Murphy TH (2017) Targeted ischemic stroke induction and mesoscopic imaging assessment of blood flow and ischemic depolarization in awake mice. Neurophotonics 4(3):035001

Index

James E. Niemeyer et al. (eds.), *Awake Behaving Mesoscopic Brain Imaging*, Neuromethods, vol. 214, https://doi.org/10.1007/978-1-0716-4120-0,

MIX
Papier aus verantwortungsvollen Quellen
Paper from responsible sources
FSC® C105338

If you have any concerns about our products,
you can contact us on
ProductSafety@springernature.com

In case Publisher is established outside the EU,
the EU authorized representative is:
Springer Nature Customer Service Center GmbH
Europaplatz 3, 69115 Heidelberg, Germany

Printed by Libri Plureos GmbH
in Hamburg, Germany